Statistics Texts in Statistics

Series Editors
G. Casella
S. Fienberg
I. Olkin

For other titles published in this series, go to
www.springer.com/series/417

Allan Gut

An Intermediate Course
in Probability

Second Edition

Allan Gut
Department of Mathematics
Uppsala University
SE-751 06 Uppsala
Sweden
allan.gut@math.uu.se

Series Editors:

George Casella
Department of Statistics
University of Florida
Gainesville, FL 32611-8545
USA

Stephen Fienberg
Department of Statistics
Carnegie Mellon University
Pittsburgh, PA 15213-3890
USA

Ingram Olkin
Department of Statistics
Stanford University
Stanford, CA 94305
USA

ISSN 1431-875X
ISBN 978-1-4899-8446-3   ISBN 978-1-4419-0162-0 (eBook)
DOI 10.1007/978-1-4419-0162-0
Springer Dordrecht Heidelberg London New York

© Springer Science+Business Media, LLC 2009
Softcover re-print of the Hardcover 2nd edition 2009
All rights reserved. This work may not be translated or copied in whole or in part without the written permission of the publisher (Springer Science+Business Media, LLC, 233 Spring Street, New York, NY 10013, USA), except for brief excerpts in connection with reviews or scholarly analysis. Use in connection with any form of information storage and retrieval, electronic adaptation, computer software, or by similar or dissimilar methodology now known or hereafter developed is forbidden.
The use in this publication of trade names, trademarks, service marks, and similar terms, even if they are not identified as such, is not to be taken as an expression of opinion as to whether or not they are subject to proprietary rights.

Printed on acid-free paper

Springer is part of Springer Science+Business Media (www.springer.com)

Preface to the First Edition

The purpose of this book is to provide the reader with a solid background and understanding of the basic results and methods in probability theory before entering into more advanced courses (in probability and/or statistics). The presentation is fairly thorough and detailed with many solved examples. Several examples are solved with different methods in order to illustrate their different levels of sophistication, their pros, and their cons. The motivation for this style of exposition is that experience has proved that the hard part in courses of this kind usually is the application of the results and methods; to know how, when, and where to apply what; and then, technically, to solve a given problem once one knows how to proceed. Exercises are spread out along the way, and every chapter ends with a large selection of problems.

Chapters 1 through 6 focus on some central areas of what might be called pure probability theory: multivariate random variables, conditioning, transforms, order variables, the multivariate normal distribution, and convergence. A final chapter is devoted to the Poisson process because of its fundamental role in the theory of stochastic processes, but also because it provides an excellent application of the results and methods acquired earlier in the book. As an extra bonus, several facts about this process, which are frequently more or less taken for granted, are thereby properly verified. The book concludes with three appendixes: In the first we provide some suggestions for further reading and in the second we provide a list of abbreviations and useful facts concerning some standard distributions. The third appendix contains answers to the problems given at the end of each chapter.

The level of the book is between the first undergraduate course in probability and the first graduate course. In particular, no knowledge of measure theory is assumed. The prerequisites (beyond a first course in probability) are basic analysis and some linear algebra.

Chapter 5 is, essentially, a revision of a handout by professor Carl-Gustav Esseen. I am most grateful to him for allowing me to include the material in the book.

The readability of a book is not only a function of its content and how (well) the material is presented; very important are layout, fonts, and other aesthetical aspects. My heartfelt thanks to Anders Vretblad for his ideas, views, and suggestions, for his design and creation of the `allan.sty` file, and for his otherwise most generous help.

I am also very grateful to Svante Janson for providing me with various index-making devices and to Lennart Norell for creating Figure 3.6.1.[1] Ola Hössjer and Pontus Andersson have gone through the manuscript with great care at different stages in a search for misprints, slips, and other obscurities; I thank them so much for every one of their discoveries as well as for many other remarks (unfortunately, I am responsible for possible remaining inadvertencies). I also wish to thank my students from a second course in probability theory in Uppsala and Jan Ohlin and his students from a similar course at the Stockholm University for sending me a list of corrections on an earlier version of this book.

Finally, I wish to thank Svante Janson and Dietrich von Rosen for several helpful suggestions and moral support, and Martin Gilchrist of Springer-Verlag for the care and understanding he has shown me and my manuscript.

Uppsala
May 1995

Allan Gut

---

[1] Figure 3.7.1 in this, second, edition.

Preface to the Second Edition

The first edition of this book appeared in 1995. Some misprints and (minor) inadvertencies have been collected over the years, in part by myself, in part by students and colleagues around the world. I was therefore very happy when I received an email from John Kimmel at Springer-Verlag asking whether I would be interested in an updated second edition of the book.

And here it is!

In addition to the cleaning up and some polishing, I have added some remarks and clarifications here and there, and a few sections have moved to new places.

More important, this edition features a new chapter, which provides an introductory outlook into further areas and topics, such as stable distributions and domains of attraction, extreme value theory and records, and, finally, an introduction to a most central tool in probability theory and the theory of stochastic processes, namely the theory of martingales. This chapter is included mainly as an appetizer to the more advanced theory, for which suggested further reading is given in Appendix A. I wish to thank Svante Janson for a careful reading of the chapter and for several remarks and suggestions.

I conclude the preface of this second edition by extending my heartfelt thanks to John Kimmel for his constant support and encouragement—for always being there—over many years.

Uppsala   *Allan Gut*
April 2009

Contents

Notation and Symbols

Ω	sample space
ω	elementary event
$\mathcal{F}$	collection of events
$I\{A\}$	indicator function of (the set) A
$\#\{A\}$	number of elements in (cardinality of) (the set) A
A^c	complement of the set A
$P(A)$	probability of A
$X, Y, Z, \ldots$	random variables
$F(x), F_X(x)$	distribution function (of X)
$X \in F$	X has distribution (function) F
$C(F_X)$	the continuity set of F_X
$p(x), p_X(x)$	probability function (of X)
$f(x), f_X(x)$	density (function) (of X)
$\Phi(x)$	standard normal distribution function
$\phi(x)$	standard normal density (function)
$Be(p)$	Bernoulli distribution
$\beta(r, s)$	beta distribution
$Bin(n, p)$	binomial distribution
$C(m, a)$	Cauchy distribution
$\chi^2(n)$	chi-square distribution
$\delta(a)$	one-point distribution
$Exp(a)$	exponential distribution
$F(m, n)$	(Fisher's) F-distribution
$Fs(p)$	first success distribution
$\Gamma(p, a)$	gamma distribution
$Ge(p)$	geometric distribution
$H(N, n, p)$	hypergeometric distribution
$L(a)$	Laplace distribution

$\mathrm{Ln}(\mu, \sigma^2)$	log-normal distribution
$N(\mu, \sigma^2)$	normal distribution
$N(0, 1)$	standard normal distribution
$\mathrm{NBin}(n, p)$	negative binomial distribution
$\mathrm{Pa}(k, \alpha)$	Pareto distribution
$\mathrm{Po}(m)$	Poisson distribution
$\mathrm{Ra}(\alpha)$	Rayleigh distribution
$t(n)$	(Student's) t-distribution
$\mathrm{Tri}(a, b)$	triangular distribution
$U(a, b)$	uniform or rectangular distribution
$W(a, b)$	Weibull distribution
$X \in \mathrm{Po}(m)$	X has a Poisson distribution with parameter m
$X \in N(\mu, \sigma^2)$	X has a normal distribution with parameters μ and σ^2
$F_{X,Y}(x, y)$	joint distribution function (of X and Y)
$p_{X,Y}(x, y)$	joint probability function (of X and Y)
$f_{X,Y}(x, y)$	joint density (function) (of X and Y)
$F_{Y|X=x}(y)$	conditional distribution function (of Y given that $X = x$)
$p_{Y|X=x}(y)$	conditional probability function (of Y given that $X = x$)
$f_{Y|X=x}(y)$	conditional density (function) (of Y given that $X = x$)
E, $E\,X$	expectation (mean), expected value of X
Var, Var$X$	variance, variance of X
$\mathrm{Cov}(X, Y)$	covariance of X and Y
ρ, $\rho_{X,Y}$	correlation coefficient (between X and Y)
$E(Y \mid X)$	conditional expectation
$\mathrm{Var}(Y \mid X)$	conditional variance
$g(t)$, $g_X(t)$	(probability) generating function (of X)
$\psi(t)$, $\psi_X(t)$	moment generating function (of X)
$\varphi(t)$, $\varphi_X(t)$	characteristic function (of X)
$\mathbf{0}$	zero vector; $(0, 0, \ldots, 0)$
$\mathbf{1}$	one vector; $(1, 1, \ldots, 1)$
$\mathbf{I}$	identity matrix
$\mathbf{A}'$	transpose of the matrix $\mathbf{A}$
$\mathbf{A}^{-1}$	inverse of the matrix $\mathbf{A}$
$\mathbf{A}^{1/2}$	square root of the matrix $\mathbf{A}$
$\mathbf{A}^{-1/2}$	inverse of the square root of the matrix $\mathbf{A}$

$\mathbf{X}, \mathbf{Y}, \mathbf{Z}, \ldots$	random vectors
$E\mathbf{X}$	mean vector of $\mathbf{X}$
$\boldsymbol{\mu}$	mean vector
$\boldsymbol{\Lambda}$	covariance matrix
$\mathbf{J}$	Jacobian
$N(\boldsymbol{\mu}, \boldsymbol{\Lambda})$	multidimensional normal distribution
$X_{(1)}, X_{(2)}, X_{(k)}, \ldots$	order variables
$(X_{(1)}, X_{(2)}, \ldots, X_{(n)})$	order statistic
$X \stackrel{d}{=} Y$	X and Y are equidistributed
$X_n \stackrel{\text{a.s.}}{\longrightarrow} X$	X_n converges almost surely to X
$X_n \stackrel{p}{\longrightarrow} X$	X_n converges in probability to X
$X_n \stackrel{r}{\longrightarrow} X$	X_n converges in r-mean (L^r) to X
$X_n \stackrel{d}{\longrightarrow} X$	X_n converges in distribution to X
a.s.	almost sure(ly)
CLT	central limit theorem
i.a.	inter alia
iff	if and only if
i.i.d.	independent, identically distributed
i.o.	infinitely often
LLN	law of large numbers
w.p.1	with probability 1
$\square$	end of proof or (series of) definitions, exercises, remarks, etc.

Introduction

1 Models

The object of *probability theory* is to describe and investigate mathematical models of random phenomena, primarily from a theoretical point of view. *Statistics* is concerned with creating principles, methods, and criteria to treat data pertaining to such (random) phenomena or to data from experiments and other observations of the real world by using, for example, the theories and knowledge available from the theory of probability.

Modeling is used in many fields, including physics, chemistry, biology, and economics. The models are, in general, *deterministic*. The motion of the planets, for example, may be described exactly; one may, say, compute the exact date and hour of the next solar eclipse.

In probability theory one studies models of *random* phenomena. Such models are intended to describe *random experiments*, that is, experiments that can be repeated (indefinitely) and where future outcomes cannot be exactly predicted even if the experimental situation can be fully controlled; there is some randomness involved in the experiment.

A trivial example is the tossing of a coin. Even if we have complete knowledge about the construction of the coin—for instance, that it is symmetric—we cannot predict the outcome of future tosses. A less trivial example is quality control. Even though the production of some given object (screws, ball bearings, etc.) is aimed at making all of them identical, it is clear that some (random) variation occurs, no matter how thoroughly the production equipment has been designed, constructed, and installed. Another example is genetics. Even though we know the "laws" of heredity, we cannot predict with certitude the sex or the eye color of an unborn baby.

An important distinction we therefore would like to stress is the difference between *deterministic* models and *probabilistic* models. A differential equation, say, may well describe a random phenomenon, although the equation does not capture any of the randomness involved in the real problem; the

differential equation models (only) the *average* behavior of the random phenomenon. One way to make the distinction is to say that deterministic models
describe the *macroscopic* behavior of random phenomena, whereas probabilistic models describe the *microscopic* behavior. The deterministic model gives
a picture of the situation from a distance (in which case one cannot observe
local (random) fluctuations), whereas the probabilistic model provides a picture of the situation close up. As an example we might consider a fluid. From
far away it moves along some main direction (and, indeed, this is what the individual molecules do—on average). At "atomic" distances, on the other, one
may (in addition) observe the erratic (random) movement of the individual
molecules.

The conclusion to be drawn here is that there are various ways to describe a
(random) phenomenon mathematically with different degrees of precision and
complexity. One important task for an applied mathematician is to choose the
model that is most appropriate in his or her particular case. This involves a
compromise between choosing a more accurate and detailed model on the one
hand and choosing a manageable and tractable model on the other. What we
must always remember is that we are modeling some real phenomenon and
that a model is a model and not reality, although a good description of it we
hope.

Keeping this in mind, the purpose of this book is, as the title suggests,
to present some of the theory of probability that goes beyond the first course
taken by all mathematics students, thus making the reader better equipped
to deal with problems of and models of random phenomena. Let us add,
however, that this is not an applied text; it concentrates on "pure" probability
theory. The full discussion of the theory of stochastic processes would require
a separate volume. We have, however, decided to include one chapter (the
last) on the Poisson process because of its special importance in applications.
In addition, we believe that the usual textbook treatment of this process is
rather casual and that a thorough discussion may be of value. Moreover, our
treatment shows the power of the methods and techniques acquired in the
earlier chapters and thus provides a nice application of that theory.

Sections 2 through 9 of this introductory chapter browse through the typical first course in probability, recalling the origin of the theory, as well as
definitions, notations, and a few facts. In the final section we give an outline
of the contents of the book.

2 The Probability Space

The basis of probability theory is the *probability space*. Let us begin by describing how a probability space comes about.

The key idea is the *stabilization of relative frequencies*. In the previous
section we mentioned that a random experiment is an experiment that can

be repeated (indefinitely) and where future outcomes cannot be exactly predicted even if the experimental situation can be fully controlled. Suppose that we perform "independent" repetitions of such an experiment and that we record each time if some "event" A occurs or not (note that we have not yet mathematically defined what we mean by either independence or event). Let $f_n(A)$ denote the number of occurrences of A in the first n trials, and let $r_n(A)$ denote the relative frequency of occurrences of A in the first n trials, that is, $r_n(A) = f_n(A)/n$. Since the dawn of history, one has observed the stabilization of the relative frequencies. This means that, empirically, one has observed that (it seems that)

$$r_n(A) \text{ converges to some real number as } n \to \infty. \tag{2.1}$$

As an example, consider the repeated tossing of a coin. In this case this means that, eventually, the number of heads approximately equals the number of tails, that is, the stabilization of their relative frequencies to $1/2$.

Now, as we recall, the aim of probability theory is to provide a model of random phenomena. It is therefore natural to use relation (2.1) as a starting point for a definition of what is meant by the probability of an event.

The next step is to axiomatize the theory; this was done by the famous Soviet/Russian mathematician A.N. Kolmogorov (1903–1987) in his fundamental monograph *Grundbegriffe der Wahrscheinlichkeitsrechnung*, which appeared in 1933. Here we shall consider some elementary steps only.

The first thing to observe is that a number of rules that hold for relative frequencies should also hold for probabilities. Let us consider some examples in an intuitive language.

(a) Since $0 \le f_n(A) \le n$ for any event A, it follows that $0 \le r_n(A) \le 1$. The probability of an event therefore should be a real number in the interval $[0, 1]$.

(b) If A is the empty set $\emptyset$ ("nothing"), then $f_n(\emptyset) = 0$ and hence $r_n(\emptyset) = 0$. The probability of the empty set should therefore equal 0. Similarly, if A is the whole space Ω ("everything"), then $f_n(\Omega) = n$ and hence $r_n(\Omega) = 1$. The probability of the whole space should therefore equal 1.

(c) Let B be the complement of A within the whole space. Since in each performance either A or B occurs and never both simultaneously, we have $f_n(A) + f_n(B) = n$, and hence $r_n(A) + r_n(B) = 1$. The sum of the probability of an event and the probability of its complement should therefore equal 1.

(d) Suppose that the event A is contained in the event B. This clearly implies that $f_n(A) \le f_n(B)$ and hence that $r_n(A) \le r_n(B)$. It follows that the probability of A should be at most equal to the probability of B.

(e) Suppose that the events A and B are disjoint, and let C be their union. Then $f_n(C) = f_n(A) + f_n(B)$ and hence $r_n(C) = r_n(A) + r_n(B)$, from which we would conclude that the probability of the union of two disjoint events equals the sum of their individual probabilities. This is called *finite additivity*.

(f) A closer inspection of the last property shows that if A and B are not disjoint, then we have $f_n(C) \leq f_n(A) + f_n(B)$ and hence $r_n(C) \leq r_n(A) + r_n(B)$, from which we would conclude that the probability of the union of two events is at most equal to the sum of the individual probabilities.

(g) An even closer inspection shows that, in fact, $f_n(C) = f_n(A) + f_n(B) - f_n(D)$ in this case. Here D equals the intersection of A and B. It follows that $r_n(C) = r_n(A) + r_n(B) - r_n(D)$, from which we would conclude that the probability of the union of two events equals the sum of the individual probabilities minus the probability of their intersection.

It is easy to construct further rules that should hold for probabilities. Further, it is obvious that some of the rules might be derived from others.

The next task is to find the minimal number of rules necessary to develop the theory of probability.

To this end we introduce the *probability space* $(\Omega, \mathcal{F}, P)$. Here Ω, the *sample space*, is some (abstract) space—the set of *elementary events* $\{\omega\}$—and $\mathcal{F}$ is the collection of *events*. In basic terms, $\mathcal{F}$ equals the collection of subsets of Ω. More technically, $\mathcal{F}$ equals the collection (σ-algebra) of *measurable* subsets of Ω. Since we do not require measurability, we adhere to the first definition, keeping in mind, however, that though not completely correct, it will be sufficiently so for our purposes. Finally, P satisfies the following three (Kolmogorov) axioms:

1. For any $A \in \mathcal{F}$, there exists a number $P(A)$, the probability of A, satisfying $P(A) \geq 0$.
2. $P(\Omega) = 1$.
3. Let $\{A_n, n \geq 1\}$ be a collection of pairwise disjoint events, and let A be their union. Then

$$P(A) = \sum_{n=1}^{\infty} P(A_n).$$

One can now show that these axioms imply all other rules, such as those hinted at above. We also remark that Axiom 3 is called *countable additivity* (in contrast to *finite additivity*; cf. (e), which is less restrictive).

3 Independence and Conditional Probabilities

In the previous section we made "independent" repetitions of an experiment. Let us now define this concept properly. Two events A and B are *independent* iff the probability of their intersection equals the product of their individual (the marginal) probabilities, that is, iff

$$P(A \cap B) = P(A) \cdot P(B). \tag{3.1}$$

The definition can be extended to arbitrary finite collections of events; one requires that (3.1) hold for all finite subsets of the collection. If (3.1) holds for all pairs only, the events are called *pairwise independent*.

Another concept introduced in this connection is *conditional probability*. Given two events A and B, with $P(B) > 0$, we define the conditional probability of A given B, $P(A \mid B)$, by the relation

$$P(A \mid B) = \frac{P(A \cap B)}{P(B)}. \tag{3.2}$$

In particular, if $B = \Omega$, then $P(A \mid \Omega) = P(A)$, that is, conditional probabilities reduce to ordinary (unconditional) probabilities. If A and B are independent, then (3.2) reduces to $P(A \mid B) = P(A)$ (of course).

It is an easy exercise to show that $P(\cdot \mid B)$ satisfies the Kolmogorov axioms for a given, fixed B with $P(B) > 0$ (please check!).

We close this section by quoting *the law of total probability* and *Bayes' formula* (Thomas Bayes (1702(?)–1761) was an English dissenting minister).

Let $\{H_k, 1 \leq k \leq n\}$ be a partition of Ω, that is, suppose that $H_k, 1 \leq k \leq n$, are disjoint sets and that their union equals Ω. Let A be an event. The law of total probability states that

$$P(A) = \sum_{k=1}^{n} P(A \mid H_k) \cdot P(H_k), \tag{3.3}$$

and Bayes' formula states that

$$P(H_i \mid A) = \frac{P(A \mid H_i) \cdot P(H_i)}{\sum_{k=1}^{n} P(A \mid H_k) \cdot P(H_k)}. \tag{3.4}$$

4 Random Variables

In general, one is not interested in events of $\mathcal{F}$ per se, but rather in some function of them. For example, suppose one plays some game where the payoff is a function of the number of dots on two dice; suppose one receives 2 euros if the total number of dots equals 2 or 3, that one receives 5 euros if the total number of dots equals 4, 5, 6, or 7, and that one has to pay 10 euros otherwise. As far as payoff is concerned, we have three groups of dots: $\{2, 3\}$, $\{4, 5, 6, 7\}$, and $\{8, 9, 10, 11, 12\}$. In other words, our payoff is a *function* of the total number of dots on the dice. In order to compute the probability that the payoff equals some number (5, say), we compute the probability that the total number of dots falls into the class $(\{4, 5, 6, 7\})$, which corresponds to the relevant payoff (5). This leads to the notion of *random variables*.

A random variable is a (measurable) function from the probability space to the real numbers:

$$X : \Omega \to \mathbf{R}. \tag{4.1}$$

Random variables are denoted by capital letters, such as X, Y, Z, U, V, and W.

We remark, once more, that since we do not presuppose the concept of measurability, we define random variables as functions. More specifically, random variables are defined as *measurable* functions.

For our example, this means that if X is the payoff in the game and if we wish to compute, say, the probability that X equals 5, then we do the following:

$$P(X = 5) = P(\{\omega : X(\omega) = 5\}) = P(\# \text{ dots} = 4, 5, 6, 7) \qquad \left(= \frac{1}{2}\right).$$

Note that the first P pertains to the real-valued object X, whereas the other two pertain to events in $\mathcal{F}$; the former probability is *induced* by the latter.

In order to describe a random variable one would need to know $P(X \in B)$ for all possible B (where we interpret "all possible B" as all subsets of $\mathbf{R}$, with the tacit understanding that in reality all possible (permitted) B constitute a collection, $\mathcal{B}$, of subsets of $\mathbf{R}$, which are the measurable subsets of $\mathbf{R}$; note that $\mathcal{B}$ relates to $\mathbf{R}$ as $\mathcal{F}$ does to Ω). However, it turns out that it suffices to know the value of $P(X \in B)$ for sets B of the form $(-\infty, x]$ for $-\infty < x < \infty$ (since those sets generate $\mathcal{B}$). This brings us to the definition of a *distribution function*.

The distribution function F_X of the random variable X is defined as

$$F_X(x) = P(X \leq x), \quad -\infty < x < \infty. \tag{4.2}$$

A famous theorem by the French mathematician H. Lebesgue (1875–1941) states that there exist three kinds of distributions (and mixtures of them). In this book we are only concerned with two kinds: *discrete* distributions and *(absolutely) continuous* distributions.

For discrete distributions we define the *probability function* p_X as $p_X(x) = P(X = x)$ for all x. It turns out that a probability function is nonzero for at most a countable number of x values (try to prove that!). The connection between the distribution function and the probability function is

$$F_X(x) = \sum_{y \leq x} p_X(y), \quad -\infty < x < \infty. \tag{4.3}$$

For continuous distributions we introduce the *density function* f_X which has the property that

$$F_X(x) = \int_{-\infty}^{x} f_X(y) \, dy, \quad -\infty < x < \infty. \tag{4.4}$$

Moreover, $F_X'(x) = f_X(x)$ for all x that are continuity points of f.

As typical discrete distributions we mention the binomial, geometric, and Poisson distributions. Typical continuous distributions are the uniform (rectangular), exponential, gamma, and normal distributions. Notation and characteristics of these and of other distributions can be found in Appendix B.

5 Expectation, Variance, and Moments

In order for us to give a brief description of the distribution of a random variable, it is obviously not very convenient to present a table of the distribution function. It would be better to present some suitable characteristics. Two important classes of such characteristics are measures of *location* and measures of *dispersion*.

Let X be a random variable with distribution function F. The most common measure of location is the *mean* or *expected value* EX, which is defined as

$$EX = \begin{cases} \sum_{k=1}^{\infty} x_k \cdot p_X(x_k), & \text{if } X \text{ is discrete,} \\ \int_{-\infty}^{\infty} x \cdot f_X(x)\, dx, & \text{if } X \text{ is continuous,} \end{cases} \tag{5.1}$$

provided the sum or integral is absolutely convergent. If we think of the distribution as the (physical) mass of some body, the mean corresponds to the center of gravity. Note also that the proviso indicates that the mean does not necessarily exist. For nonnegative random variables X with a divergent sum or integral, we shall also permit ourselves to say that the mean is infinite ($EX = +\infty$).

Another measure of location is the *median*, which is a number m (not necessarily unique) such that

$$P(X \geq m) \geq \frac{1}{2} \quad \text{and} \quad P(X \leq m) \geq \frac{1}{2}. \tag{5.2}$$

If the distribution is symmetric, then, clearly, the median and the mean coincide (provided that the latter exists). If the distribution is skew, the median might be a better measure of the "average" than the mean. However, this also depends on the problem at hand.

It is clear that two distributions may well have the same mean and yet be very different. One way to distinguish them is via a measure of dispersion—by indicating how spread out the mass is. The most commonly used such measure is the *variance* $\mathrm{Var}X$, which is defined as

$$\mathrm{Var}X = E(X - EX)^2, \tag{5.3}$$

and can be computed as

$$\mathrm{Var}X = \begin{cases} \sum_{k=1}^{\infty} (x_k - EX)^2 \cdot p_X(x_k), & \text{if } X \text{ is discrete,} \\ \int_{-\infty}^{\infty} (x - EX)^2 \cdot f_X(x)\, dx, & \text{if } X \text{ is continuous.} \end{cases} \tag{5.4}$$

Note that the variance exists only if the corresponding sum or integral is absolutely convergent.

An alternative and, in general, more convenient way to compute the variance is via the relation

$$\text{Var} X = E X^2 - (E X)^2, \tag{5.5}$$

which is obtained by expanding the square in (5.3). As for the analogy with a physical body, the variance is related to the moment of inertia.

We close this section by defining *moments* and *central moments*. The former are

$$E X^n, \quad n = 1, 2, \ldots, \tag{5.6}$$

and the latter are

$$E(X - E X)^n, \quad n = 1, 2, \ldots, \tag{5.7}$$

provided they exist. In particular, the mean is the first moment ($n = 1$) and the variance is the second central moment ($n = 2$). The *absolute moments* and *absolute central moments* are

$$E|X|^r, \quad r > 0, \tag{5.8}$$

and

$$E|X - E X|^r, \quad r > 0, \tag{5.9}$$

respectively, provided they exist.

6 Joint Distributions and Independence

Let X and Y be random variables of the same kind (discrete or continuous). A complete description of the pair (X, Y) is given by the *joint distribution function*

$$F_{X,Y}(x, y) = P(X \leq x, Y \leq y), \quad -\infty < x, y < \infty. \tag{6.1}$$

In the discrete case there exists a *joint probability function*:

$$p_{X,Y}(x, y) = P(X = x, Y = y), \quad -\infty < x, y < \infty. \tag{6.2}$$

In the continuous case there exists a *joint density*:

$$f_{X,Y}(x, y) = \frac{\partial^2 F_{X,Y}(x, y)}{\partial x \partial y}, \quad -\infty < x, y < \infty. \tag{6.3}$$

The joint distribution function can be expressed in terms of the joint probability function and the joint density function, respectively, in the obvious way.

Next we turn to the concept of *independence*. Intuitively, we would require that $P(\{X \in A\} \cap \{Y \in B\}) = P(\{X \in A\}) \cdot P(\{Y \in B\})$ for all $A \subset \mathbb{R}$ and $B \subset \mathbb{R}$ in order for X and Y to be independent. However, just as in the definition of distribution functions, it suffices that this relation hold for sets $A = (-\infty, x]$ for all x and $B = (-\infty, y]$ for all y. Thus, X and Y are independent iff

$$P(\{X \le x\} \cap \{Y \le y\}) = P(\{X \le x\}) \cdot P(\{Y \le y\}), \quad (6.4)$$

for $-\infty < x, y < \infty$, that is, iff

$$F_{X,Y}(x,y) = F_X(x) \cdot F_Y(y), \quad (6.5)$$

for all x and y. In the discrete case this is equivalent to

$$p_{X,Y}(x,y) = p_X(x) \cdot p_Y(y), \quad (6.6)$$

for all x and y, and in the continuous case it is equivalent to

$$f_{X,Y}(x,y) = f_X(x) \cdot f_Y(y), \quad (6.7)$$

for $-\infty < x, y < \infty$.

The general case with sets of more than two random variables will be considered in Chapter 1.

7 Sums of Random Variables, Covariance, Correlation

A central part of probability theory is the study of sums of (independent) random variables. Here we confine ourselves to sums of two random variables, X and Y.

Let (X, Y) be a discrete two-dimensional random variable. In order to compute the probability function of $X + Y$, we wish to find the probabilities of the events $\{\omega : X(\omega) + Y(\omega) = z\}$ for all $z \in \mathbb{R}$. Consider a fixed $z \in \mathbb{R}$. Since $X(\omega) + Y(\omega) = z$ exactly when $X(\omega) = x$ and $Y(\omega) = y$, where $x + y = z$, it follows that

$$p_{X+Y}(z) = \sum_{\{(x,y):x+y=z\}} \sum p_{X,Y}(x,y) = \sum_x p_{X,Y}(x, z-x), \quad z \in \mathbb{R}.$$

If, in addition, X and Y are independent, then

$$p_{X+Y}(z) = \sum_x p_X(x) p_Y(z-x), \quad z \in \mathbb{R}, \quad (7.1)$$

which we recognize as the *convolution formula*.

A similar computation in the continuous case is a little more complicated. It is, however, a reasonable guess that the convolution formula should be

$$f_{X+Y}(z) = \int_{-\infty}^{\infty} f_X(x) f_Y(z-x)\, dx, \quad z \in \mathbb{R}. \quad (7.2)$$

That this is indeed the case can be shown by first considering $F_{X+Y}(z)$ and then differentiating. Also, if X and Y are *not* independent, then

$$f_{X+Y}(z) = \int_{-\infty}^{\infty} f_{X,Y}(x, z - x)\,dx, \quad z \in \mathbb{R},$$

in analogy with the discrete case.

Next we consider the mean and variance of sums of (two) random variables; we shall, in fact, consider linear combinations of them. To this end, let a and b be constants. It is easy to check that

$$E(aX + bY) = aE\,X + bE\,Y; \tag{7.3}$$

in other words, expectation is linear.

Further, by rearranging $(aX + bY - E(aX + bY))^2$ into $(a(X - E\,X) + b(Y - E\,Y))^2$, we obtain

$$\mathrm{Var}(aX + bY) = a^2\mathrm{Var}X + b^2\mathrm{Var}Y + 2abE(X - E\,X)(Y - E\,Y). \tag{7.4}$$

Since the double product does not vanish in general, we do not have a Pythagorean-looking identity. In fact, (7.4) provides the motivation for the definition of *covariance*:

$$\mathrm{Cov}(X, Y) = E(X - E\,X)(Y - E\,Y) \quad (= E\,XY - E\,XE\,Y). \tag{7.5}$$

Covariance is a measure of the interdependence of X and Y in the sense that it becomes large and positive when X and Y are both large and of the same sign; it is large and negative if X and Y are both large and of opposite signs. Since $\mathrm{Cov}(aX, bY) = ab\mathrm{Cov}(X, Y)$, it follows that the covariance is not scale invariant. A better measure of dependence is the *correlation coefficient*, which is a scale-invariant real number:

$$\rho_{X,Y} = \frac{\mathrm{Cov}(X, Y)}{\sqrt{\mathrm{Var}X \cdot \mathrm{Var}Y}}. \tag{7.6}$$

Moreover, $|\rho_{X,Y}| \leq 1$. If $\rho_{X,Y} = 0$, we say that X and Y are *uncorrelated*. There is a famous result to the effect that two independent random variables are uncorrelated but that the converse does not necessarily hold.

We also note that if X and Y, for example, are independent, then (7.4) reduces to the Pythagorean form. Furthermore, with $a = b = 1$ it follows, in particular, that the variance of the sum equals the sum of the variances; and with $a = 1$ and $b = -1$ it follows that the variance of the difference (also) equals the sum of the variances.

8 Limit Theorems

The next part of a first course in probability usually contains a survey of the most important distributions and their properties, a brief introduction to some of the most important limit theorems such as *the law of large numbers* and *the central limit theorem*, and results such as the Poisson approximation and normal approximation of the binomial distribution, for appropriate values of n and p.

As for the most important distributions, we refer once more to Appendix B.

The law of large numbers is normally presented in the so-called weak form under the assumption of finite variance. A preliminary tool for the proof is *Chebyshev's inequality*, which states that for a random variable U with mean m and variance σ^2 both finite, one has

$$P(|U - m| > \varepsilon) \le \frac{\sigma^2}{\varepsilon^2} \quad \text{for all} \quad \varepsilon > 0. \tag{8.1}$$

This inequality is, in fact, a special case of *Markov's inequality*, according to which

$$P(V > \varepsilon) \le \frac{E V^r}{\varepsilon^r} \tag{8.2}$$

for positive random variables, V, with $E V^r < \infty$.

The law of large numbers and the central limit theorem (the former under the assumption of finite mean only) are stated and proved in Chapter 6, so we refrain from recalling them here. The other limit theorems mentioned above are, in part, special cases of the central limit theorem. Some of them are also reviewed in examples and problems in Chapter 6.

9 Stochastic Processes

A first course in probability theory frequently concludes with a small chapter on stochastic processes, which contains definitions and some introduction to the theory of Markov processes and/or the Poisson process.

A stochastic process is a family of random variables, $X = \{X(t), t \in T\}$, where T is some index set. Typical cases are $T =$ the nonnegative integers (in which case the process is said to have discrete time) and $T = [0, 1]$ or $T = [0, \infty)$ (in which case the process is said to have continuous time). A stochastic process is in itself called discrete or continuous depending on the state space, which is the set of values assumed by the process.

As argued earlier, this book is devoted to "pure" probability theory, so we shall not discuss the general theory of stochastic processes. The only process we will discuss in detail is the Poisson process, one definition of which is that it has independent, stationary, Poisson-distributed increments.

10 The Contents of the Book

The purpose of this introductory chapter so far has been to provide a background and skim through the contents of the typical first course in probability. In this last section we briefly describe the contents of the following chapters.

In Chapter 1 we discuss multivariate random variables (random vectors) and the connection between joint and marginal distributions. In addition, we prove an important transformation theorem for continuous multivariate random variables, permitting us to find the distribution of the (vector-valued) function of a random vector.

Chapter 2 is devoted to conditional distributions; starting with (3.2) one can establish relations for conditional probabilities in the discrete case, which can be extended, by definitions, to the continuous case. Typically, one is given two jointly distributed random variables X and Y and wishes to find the (conditional) distribution of Y given that X has some fixed value. Conditional expectations and conditional variances are defined, and some results and relations are proved. Distributions with random parameters are discussed. One such example is the following: Suppose that X has a Poisson distribution with a parameter that itself is a random variable. What, really, is the distribution of X? Two further sections provide some words on Bayesian statistics and prediction and regression.

A very important tool in mathematics as well as in probability theory is the transform. In mathematics one talks about Laplace and Fourier transforms. The commonly used transforms in probability theory are the *(probability) generating function*, the *moment generating function*, and the *characteristic function*. The important feature of the transform is that adding independent random variables (convolution) corresponds to multiplying transforms. In Chapter 3 we present uniqueness theorems, "multiplication theorems," and some inversion results. Most results are given without proofs, since these would require mathematical tools beyond the scope of this book. Remarks on "why" some of the theorems hold, as well as examples, are given. One section deals with distributions with random parameters from the perspective of transforms. Another one is devoted to *sums of a random number of* independent, identically distributed (i.i.d.) random variables, where the number of summands is independent of the summands themselves. We thus consider $X_1 + X_2 + \cdots + X_N$, where $X_1, X_2, \ldots$ are independent and identically distributed random variables and N is a nonnegative, integer-valued random variable independent of $X_1, X_2, \ldots$. An application to the simplest kind of branching process is given.

Two interesting objects of a sample, that is, a set of independent, identically distributed observations of a random variable X, are the largest observation and the smallest observation. More generally, one can order the observations in increasing order. In Chapter 4 we derive the distributions of the ordered random variables, joint distributions of the smallest and the largest observation, and, more generally, of the whole ordered sample—the *order statistic*—as well as some functions of these.

The normal distribution is well known to be one of the most important distributions. In Chapter 5 we provide a detailed account of the *multivariate normal* distribution. In particular, three definitions are presented (and a fourth one in the problem section); the first two are always equivalent, and all

of them are equivalent in the nonsingular case. A number of important results
are proved, such as the equivalence of the uncorrelatedness and independence
of components of jointly normal random variables, special properties of linear
transformations of normal vectors, the independence of the sample mean and
the sample variance, and Cochran's theorem.

Chapter 6 is devoted to another important part of probability theory (and
statistics): *limit theorems*, a particular case being the asymptotic behavior of
sums of random variables as the number of summands tends to infinity. We
begin by defining four modes of convergence—almost sure convergence, convergence in probability, mean convergence, and distributional convergence—
and show that the limiting random variable or distribution is (essentially)
unique. We then proceed to show how the convergence concepts are related
to each other.

A very useful tool for distributional convergence is found in the so-called
continuity theorems, that is, limit theorems for transforms. The idea is that it
suffices to show that the sequence of, say, characteristic functions converges in
order to conclude that the corresponding sequence of random variables converges in distribution (convergence of transforms is often easier to establish
than is proof of distributional convergence directly). Two important applications are the *law of large numbers* and the *central limit theorem*, which are
stated and proved.

Another problem that is investigated is whether or not the sum sequence
converges if the individual sequences do. More precisely, if $\{U_n,\ n \geq 1\}$ and
$\{V_n,\ n \geq 1\}$ are sequences of random variables, such that U_n and V_n both
converge in some mode as $n \to \infty$, is it then true that $U_n + V_n$ converges as
$n \to \infty$?

Probability theory is, of course, much more than what one will find in
this book. Chapter 7 contains an outlook into some extensions and further
areas and concepts, such as stable distributions and domains of attraction
(that is, limit theorems when the variance does not exist), extreme value
theory and records. We close with an introduction to one of the most central
tools in probability theory and the theory of stochastic processes, namely
the theory of *martingales*. Although one needs a basic knowledge of measure
theory to fully appreciate the concept, one still will get the basic flavor with
our more elementary approach. The chapter thus may serve as an introduction
and appetizer to the more advanced theory of probability. For more on these
and additional topics we refer the reader to the more advanced literature, a
selection of which is cited in Appendix A.

This concludes the "pure probability" part. As mentioned above, we have
included a final chapter on the Poisson process. The reason for this is that
it is an extremely important and useful process for applications. Moreover,
it is common practice to use properties of the Poisson process that have not
been properly demonstrated. For example, one of the main features of the
Poisson process is the lack of memory property, which states: given that we
have waited some *fixed* time for an occurrence, the remaining waiting time

follows the same (exponential) law as the original waiting time. Equivalently, objects that break down according to a Poisson process never age. The proof of this property is easy to provide. However, one of the first applications of the property is to say that the waiting time between, say, the first and second occurrences in the process is also exponential, the motivation being the same, namely, that everything starts from scratch. Now, in this latter case we claim that everything starts from scratch (also) at a (certain) *random* time point. This distinction is usually not mentioned or, maybe, mentioned and then quickly forgotten.

In Chapter 8 we prove a number of properties of the Poisson process with the aid of the results and methods acquired earlier in the book. We frequently present different proofs of the same result. It is our belief that this illustrates the applicability of the different approaches and provides a comparison between the various techniques and their efficiencies. For example, the proof via an elementary method may well be longer than that based on a more sophisticated idea. On the other hand, the latter has in reality been preceded (somewhere else) by results that may, in turn, require difficult proofs (or which have been stated without proof).

To summarize, Chapter 8 gives a detailed account of the important Poisson process with proofs and at the same time provides a nice application of the theory of "pure" probability as we will have encountered it earlier in the book. The chapter closes with a short outlook on extensions, such as nonhomogeneous Poisson processes, birth (and death) processes, and renewal processes.

Every chapter concludes with a problem section. Some of the problems are fairly easy applications of the results in earlier sections, some are a little harder. Answers to the problems can be found in Appendix C.

One purpose of this book, obviously, is to make the reader realize that probability theory is an interesting, important, and fascinating subject. As a starting point for those who wish to know more, Appendix A contains some remarks and suggestions for further reading.

Throughout we use abbreviations to denote many standard distributions. Appendix B contains a list of these abbreviations and some useful facts: the probability function or the density function, mean, variance, and the characteristic function.

Multivariate Random Variables

1 Introduction

One-dimensional random variables are introduced when the object of interest is a one-dimensional function of the events (in the probability space $(\Omega, \mathcal{F}, P)$); recall Section 4 of the Introduction. In an analogous manner we now define *multivariate random variables*, or random vectors, as multivariate functions.

Definition 1.1. *An n-dimensional random variable or vector $\mathbf{X}$ is a (measurable) function from the probability space Ω to $\mathbb{R}^n$, that is,*

$$\mathbf{X} : \Omega \to \mathbb{R}^n. \qquad \square$$

Remark 1.1. We remind the reader that this text does not presuppose any knowledge of measure theory. This is why we do not explicitly mention that functions and sets are supposed to be *measurable*.

Remark 1.2. Sometimes we call $\mathbf{X}$ a *random variable* and sometimes we call it a *random vector*, in which case we consider it a *column vector*:

$$\mathbf{X} = (X_1, X_2, \ldots, X_n)'. \qquad \square$$

A complete description of the distribution of the random variable is provided by the *joint distribution function*

$$F_{X_1, X_2, \ldots, X_n}(x_1, \ldots, x_n) = P(X_1 \leq x_1, X_2 \leq x_2, \ldots, X_n \leq x_n),$$

for $x_k \in \mathbb{R}$, $k = 1, 2, \ldots, n$.

A more compact way to express this is

$$F_{\mathbf{X}}(\mathbf{x}) = P(\mathbf{X} \leq \mathbf{x}), \quad \mathbf{x} \in \mathbb{R}^n,$$

where the event $\{\mathbf{X} \leq \mathbf{x}\}$ is to be interpreted componentwise, that is,

A. Gut, *An Intermediate course in Probabilty*, Springer Texts in Statistics,
DOI: 10.1007/978-1-4419-0162-0_1,
© Springer Science + Business Media, LLC 2009

$$\{\mathbf{X} \le \mathbf{x}\} = \{X_1 \le x_1, \ldots, X_n \le x_n\} = \bigcap_{k=1}^{n} \{X_k \le x_k\}.$$

In the discrete case we introduce the *joint probability function*

$$p_{\mathbf{X}}(\mathbf{x}) = P(\mathbf{X} = \mathbf{x}), \quad \mathbf{x} \in \mathbb{R}^n,$$

that is,

$$p_{X_1, X_2, \ldots, X_n}(x_1, x_2, \ldots, x_n) = P(X_1 = x_1, \ldots, X_n = x_n)$$

for $x_k \in \mathbb{R}$, $k = 1, 2, \ldots, n$.

It follows that

$$F_{\mathbf{X}}(\mathbf{x}) = \sum_{\mathbf{y} \le \mathbf{x}} p_{\mathbf{X}}(\mathbf{y}),$$

that is,

$$F_{X_1, X_2, \ldots, X_n}(x_1, x_2, \ldots, x_n) = \sum_{y_1 \le x_1} \cdots \sum_{y_n \le x_n} p_{X_1, X_2, \ldots, X_n}(y_1, y_2, \ldots, y_n).$$

In the (absolutely) continuous case we define the *joint density (function)*

$$f_{\mathbf{X}}(\mathbf{x}) = \frac{d^n F_{\mathbf{X}}(\mathbf{x})}{d\mathbf{x}^n}, \quad \mathbf{x} \in \mathbb{R}^n,$$

that is,

$$f_{X_1, X_2, \ldots, X_n}(x_1, x_2, \ldots, x_n) = \frac{\partial^n F_{X_1, X_2, \ldots, X_n}(x_1, x_2, \ldots, x_n)}{\partial x_1 \partial x_2 \ldots \partial x_n},$$

where, again, $x_k \in \mathbb{R}$, $k = 1, 2, \ldots, n$.

Remark 1.3. Throughout we assume that all components of a random vector are of the same kind, either all discrete or all continuous.   □

It may well happen that in an n-dimensional problem one is only interested in the distribution of $m < n$ of the coordinate variables. We illustrate this situation with an example where $n = 2$.

Example 1.1. Let (X, Y) be a point that is uniformly distributed on the unit disc; that is, the joint distribution of X and Y is

$$f_{X,Y}(x, y) = \begin{cases} \frac{1}{\pi}, & \text{for } x^2 + y^2 \le 1, \\ 0, & \text{otherwise.} \end{cases}$$

Determine the distribution of the x-coordinate.   □

Choosing a point in the plane is obviously a two-dimensional task. However, the object of interest is a one-dimensional quantity; the problem is formulated in terms of the joint distribution of X and Y, and we are interested in the distribution of X (the density $f_X(x)$).

Before we solve this problem we shall study the discrete case, which, in some respects, is easier to handle.

Thus, suppose that (X, Y) is a given two-dimensional random variable whose joint probability function is $p_{X,Y}(x, y)$ and that we are interested in finding $p_X(x)$. We have

$$p_X(x) = P(X = x) = P(\bigcup_y \{X = x, Y = y\})$$

$$= \sum_y P(X = x, Y = y) = \sum_y p_{X,Y}(x, y).$$

A similar computation yields $p_Y(y)$. The distributions thus obtained are called *marginal distributions* (of X and Y, respectively).

The *marginal probability functions* are

$$p_X(x) = \sum_y p_{X,Y}(x, y)$$

and

$$p_Y(y) = \sum_x p_{X,Y}(x, y).$$

Analogous formulas hold in higher dimensions. They show that the probability function of a marginal distribution is obtained by summing the joint probability function over the components that are not of interest.

The *marginal distribution function* is obtained in the usual way. In the two-dimensional case we have, for example,

$$F_{X_1}(x) = \sum_{x' \leq x} p_{X_1}(x') = \sum_{x' \leq x} \sum_y p_{X_1, X_2}(x', y).$$

A corresponding discussion for the continuous case cannot be made immediately, since all probabilities involved equal zero. We therefore make definitions that are analogous to the results in the discrete case. In the two-dimensional case we define the *marginal density functions* as follows:

$$f_X(x) = \int_{-\infty}^{\infty} f_{X,Y}(x, y) \, DD$$

and

$$f_Y(y) = \int_{-\infty}^{\infty} f_{X,Y}(x, y) \, dx.$$

The marginal distribution function of X is

$$F_X(x) = \int_{-\infty}^{x} f_X(u)\,du = \int_{-\infty}^{x} \left(\int_{-\infty}^{\infty} f_{X,Y}(u,y)\,dy \right) du.$$

We now return to Example 1.1. Recall that the joint density of X and Y is

$$f_{X,Y}(x,y) = \begin{cases} \frac{1}{\pi}, & \text{for } x^2 + y^2 \leq 1, \\ 0, & \text{otherwise,} \end{cases}$$

which yields

$$f_X(x) = \int_{-\infty}^{\infty} f_{X,Y}(x,y)\,dy = \int_{-\sqrt{1-x^2}}^{\sqrt{1-x^2}} \frac{1}{\pi}\,dy = \frac{2}{\pi}\sqrt{1-x^2}$$

for $-1 < x < 1$ (and $f_X(x) = 0$ for $|x| \geq 1$).

As an extra precaution one might check that $\int_{-1}^{1} \frac{2}{\pi}\sqrt{1-x^2}\,dx = 1$. Similarly (by symmetry), we have

$$f_Y(y) = \frac{2}{\pi}\sqrt{1-y^2}, \quad -1 < y < 1.$$

Exercise 1.1. Let (X, Y, Z) be a point chosen uniformly within the three-dimensional unit sphere. Determine the marginal distributions of (X, Y) and X. $\qquad\square$

We have now seen how a model might well be formulated in a higher dimension than the actual problem of interest. The converse is the problem of discovering to what extent the marginal distributions determine the joint distribution. There exist counterexamples showing that the joint distribution is not necessarily uniquely determined by the marginal ones. Interesting applications are computer tomography and satellite pictures; in both applications one makes two-dimensional pictures and wishes to make conclusions about three-dimensional objects (the brain and the Earth).

We close this section by introducing the concepts of independence and uncorrelatedness.

The components of a random vector $\mathbf{X}$ are *independent* iff, for the joint distribution, we have

$$F_{\mathbf{X}}(\mathbf{x}) = \prod_{k=1}^{n} F_{X_k}(x_k), \quad x_k \in \mathbb{R}, \quad k = 1, 2, \ldots, n,$$

that is, iff the joint distribution function equals the product of the marginal ones. In the discrete case this is equivalent to

$$p_{\mathbf{X}}(\mathbf{x}) = \prod_{k=1}^{n} p_{X_k}(x_k), \quad x_k \in \mathbb{R}, \quad k = 1, 2, \ldots, n.$$

In the continuous case it is equivalent to

$$f_{\mathbf{X}}(\mathbf{x}) = \prod_{k=1}^{n} f_{X_k}(x_k), \quad x_k \in \mathbb{R}, \quad k = 1, 2, \ldots, n.$$

The random variables X and Y are *uncorrelated* iff their *covariance* equals zero, that is, iff

$$\mathrm{Cov}\,(X, Y) = E(X - E\,X)(Y - E\,Y) = 0.$$

If the variances are nondegenerate (and finite), the situation is equivalent to the *correlation coefficient* being equal to zero, that is

$$\rho_{X,Y} = \frac{\mathrm{Cov}\,(X, Y)}{\sqrt{\mathrm{Var}\,X \cdot \mathrm{Var}\,Y}} = 0$$

(recall that the correlation coefficient ρ is a scale-invariant real number and that $|\rho| \leq 1$).

In particular, independent random variables are uncorrelated. The converse is not necessarily true.

The random variables $X_1, X_2, \ldots, X_n$ are pairwise uncorrelated if every pair is uncorrelated.

Exercise 1.2. Are X and Y independent in Example 1.1? Are they uncorrelated?

Exercise 1.3. Let (X, Y) be a point that is uniformly distributed on a square whose corners are $(\pm 1, \pm 1)$. Determine the distribution(s) of the x- and y-coordinates. Are X and Y independent? Are they uncorrelated?   □

2 Functions of Random Variables

Frequently, one is not primarily interested in the random variables themselves, but in functions of them. For example, the sum and the difference of two random variables X and Y are, in fact, functions of the two-dimensional random variable (X, Y).

As an introduction we consider one-dimensional functions of one-dimensional random variables.

Example 2.1. Let $X \in U(0, 1)$, and put $Y = X^2$. Then

$$F_Y(y) = P(Y \leq y) = P(X^2 \leq y) = P(X \leq \sqrt{y}) = F_X(\sqrt{y}).$$

Differentiation yields

$$f_Y(y) = f_X(\sqrt{y}) \frac{1}{2\sqrt{y}} = \frac{1}{2\sqrt{y}}, \quad 0 < y < 1,$$

(and $f_Y(y) = 0$ otherwise).

Example 2.2. Let $X \in U(0,1)$, and put $Y = -\log X$. Then

$$F_Y(y) = P(Y \le y) = P(-\log X \le y) = P(X \ge e^{-y})$$
$$= 1 - F_X(e^{-y}) = 1 - e^{-y}, \quad y > 0,$$

which we recognize as $F_{\mathrm{Exp}(1)}(y)$ (or else we obtain $f_Y(y) = e^{-y}$, for $y > 0$, by differentiation and again that $Y \in \mathrm{Exp}(1)$).

Example 2.3. Let X have an arbitrary continuous distribution, and suppose that g is a differentiable, strictly increasing function (whose inverse g^{-1} thus exists uniquely). Set $Y = g(X)$. Computations like those above yield

$$F_Y(y) = P(g(X) \le y) = P(X \le g^{-1}(y)) = F_X\big(g^{-1}(y)\big)$$

and

$$f_Y(y) = f_X\big(g^{-1}(y)\big) \cdot \frac{d}{dy} g^{-1}(y).$$

If g had been strictly decreasing, we would have obtained

$$f_Y(y) = -f_X\big(g^{-1}(y)\big) \cdot \frac{d}{dy} g^{-1}(y).$$

(Note that $f_Y(y) > 0$ since $dg^{-1}(y)/dy < 0$).

To summarize, we have shown that if g is strictly monotone, then

$$f_Y(y) = f_X\big(g^{-1}(y)\big) \cdot |\frac{d}{dy} g^{-1}(y)|. \qquad \square$$

Our next topic is a multivariate analog of this result.

2.1 The Transformation Theorem

Let $\mathbf{X}$ be an n-dimensional, continuous, random variable with density $f_\mathbf{X}(\mathbf{x})$, and suppose that $\mathbf{X}$ has its mass concentrated on a set $S \subset \mathbb{R}^n$. Let $g = (g_1, g_2, \ldots, g_n)$ be a bijection from S to some set $T \subset \mathbb{R}^n$, and consider the n-dimensional random variable

$$\mathbf{Y} = g(\mathbf{X}).$$

This means that we consider the n one-dimensional random variables

$$Y_1 = g_1(X_1, X_2, \ldots, X_n),$$
$$Y_2 = g_2(X_1, X_2, \ldots, X_n),$$
$$\vdots$$
$$Y_n = g_n(X_1, X_2, \ldots, X_n).$$

Finally, assume, say, that g and its inverse are both continuously differentiable (in order for the Jacobian $\mathbf{J} = |d(\mathbf{x})/d(\mathbf{y})|$ to be well defined).

Theorem 2.1. *The density of* $\mathbf{Y}$ *is*

$$f_{\mathbf{Y}}(\mathbf{y}) = \begin{cases} f_{\mathbf{X}}(h_1(\mathbf{y}), h_2(\mathbf{y}), \ldots, h_n(\mathbf{y})) \cdot |\,\mathbf{J}\,|, & \text{for } \mathbf{y} \in T, \\ 0, & \text{otherwise}, \end{cases}$$

where h *is the (unique) inverse of* g *and where*

$$\mathbf{J} = \left| \frac{d(\mathbf{x})}{d(\mathbf{y})} \right| = \begin{vmatrix} \frac{\partial x_1}{\partial y_1} & \frac{\partial x_1}{\partial y_2} & \cdots & \frac{\partial x_1}{\partial y_n} \\ \frac{\partial x_2}{\partial y_1} & \frac{\partial x_2}{\partial y_2} & \cdots & \frac{\partial x_2}{\partial y_n} \\ \vdots & \vdots & \ddots & \vdots \\ \frac{\partial x_n}{\partial y_1} & \frac{\partial x_n}{\partial y_2} & \cdots & \frac{\partial x_n}{\partial y_n} \end{vmatrix};$$

that is, $\mathbf{J}$ *is the Jacobian.*

Proof. We first introduce the following piece of notation:

$$h(B) = \{\mathbf{x} : g(\mathbf{x}) \in B\}, \quad \text{for } B \subset \mathbb{R}^n.$$

Now,

$$P(\mathbf{Y} \in B) = P(\mathbf{X} \in h(B)) = \int_{h(B)} f_{\mathbf{X}}(\mathbf{x})d\mathbf{x}.$$

The change of variable $\mathbf{y} = g(\mathbf{x})$ yields

$$P(\mathbf{Y} \in B) = \int_B f_{\mathbf{X}}(h_1(\mathbf{y}), h_2(\mathbf{y}), \ldots, h_n(\mathbf{y})) \cdot |\,\mathbf{J}\,| \, d\mathbf{y},$$

according to the formula for changing variables in multiple integrals. The claim now follows in view of the following result:

Lemma 2.1. *Let* $\mathbf{Z}$ *be an* n-*dimensional continuous random variable. If, for every* $B \subset \mathbb{R}^n$,

$$P(\mathbf{Z} \in B) = \int_B h(\mathbf{x})\,d\mathbf{x},$$

then h *is the density of* $\mathbf{Z}$. $\square$

Remark 2.1. Note that the Jacobian in Theorem 2.1 reduces to the derivative of the inverse in Example 2.3 when $n = 1$. $\square$

Example 2.4. Let X and Y be independent $N(0, 1)$-distributed random variables. Show that $X+Y$ and $X-Y$ are independent $N(0, 2)$-distributed random variables.

We put $U = X + Y$ and $V = X - Y$. Inversion yields $X = (U + V)/2$ and $Y = (U - V)/2$, which implies that

$$\mathbf{J} = \begin{vmatrix} \frac{1}{2} & \frac{1}{2} \\ \frac{1}{2} & -\frac{1}{2} \end{vmatrix} = -\frac{1}{2}.$$

By Theorem 2.1 and independence, we now obtain

$$
\begin{aligned}
f_{U,V}(u,v) &= f_{X,Y}\Big(\frac{u+v}{2}, \frac{u-v}{2}\Big)\cdot |\,\mathbf{J}\,| \\
&= f_X\Big(\frac{u+v}{2}\Big)\cdot f_Y\Big(\frac{u-v}{2}\Big)\cdot |\,\mathbf{J}\,| \\
&= \frac{1}{\sqrt{2\pi}}e^{-\frac{1}{2}(\frac{u+v}{2})^2}\cdot \frac{1}{\sqrt{2\pi}}e^{-\frac{1}{2}(\frac{u-v}{2})^2}\cdot \frac{1}{2} \\
&= \frac{1}{\sqrt{2\pi\cdot 2}}e^{-\frac{1}{2}\frac{u^2}{2}}\cdot \frac{1}{\sqrt{2\pi\cdot 2}}e^{-\frac{1}{2}\frac{v^2}{2}},
\end{aligned}
$$

for $-\infty < u,\, v < \infty$. $\qquad\square$

Remark 2.2. That $X+Y$ and $X-Y$ are $N(0,2)$-distributed might be known from before; or it can easily be verified via the convolution formula. The important point here is that with the aid of Theorem 2.1 we may, in addition, prove independence.

Remark 2.3. We shall return to this example in Chapter 5 and provide a solution that exploits special properties of the multivariate normal distribution; see Examples 5.7.1 and 5.8.1. $\qquad\square$

Example 2.5. Let X and Y be independent $\mathrm{Exp}(1)$-distributed random variables. Show that $X/(X+Y)$ and $X+Y$ are independent, and find their distributions.

We put $U = X/(X+Y)$ and $V = X+Y$. Inversion yields $X = U\cdot V$, $Y = V - UV$, and

$$
\mathbf{J} = \begin{vmatrix} v & , & u \\ -v & & 1-u \end{vmatrix} = v.
$$

Theorem 2.1 and independence yield

$$
\begin{aligned}
f_{U,V}(u,v) &= f_{X,Y}(uv, v-uv)\cdot |\,\mathbf{J}\,| = f_X(uv)\cdot f_Y(v(1-u))\cdot |\,\mathbf{J}\,| \\
&= e^{-uv}\cdot e^{-v(1-u)}\cdot v = ve^{-v}
\end{aligned}
$$

for $0 < u < 1$ and $v > 0$, and $f_{U,V}(u,v) = 0$ otherwise, that is,

$$
f_{U,V}(u,v) = \begin{cases} 1\cdot ve^{-v}, & \text{for } 0 < u < 1,\ v > 0, \\ 0, & \text{otherwise.} \end{cases}
$$

This shows that $U \in U(0,1)$, that $V \in \Gamma(2,1)$, and that U and V are independent. $\qquad\square$

As a further application of Theorem 2.1 we prove the convolution formula (in the continuous case); recall formula (7.2) of the Introduction. We are thus given the continuous, independent random variables X and Y, and we seek the distribution of $X+Y$.

A first observation is that we start with *two* variables but seek the distribution of just *one* new one. The trick is to put $U = X + Y$ and to introduce an auxiliary variable V, which may be arbitrarily (that is, suitably) defined. With the aid of Theorem 2.1, we then obtain $f_{U,V}(u, v)$ and, finally, $f_U(u)$ by integrating over v.

Toward that end, set $U = X + Y$ and $V = X$. Inversion yields $X = V$, $Y = U - V$, and

$$\mathbf{J} = \begin{vmatrix} 0 & 1 \\ 1 & -1 \end{vmatrix} = -1,$$

from which we obtain

$$f_{U,V}(u, v) = f_{X,Y}(v, u - v) \cdot |\, \mathbf{J}\, | = f_X(v) \cdot f_Y(u - v) \cdot 1$$

and, finally,

$$f_U(u) = \int_{-\infty}^{\infty} f_X(v) f_Y(u - v)\, dv,$$

which is the desired formula.

Exercise 2.1. Derive the density for the difference, product, and ratio, respectively, of two independent, continuous random variables.   □

2.2 Many-to-One

A natural question is the following: What if g is not injective? Let us again begin with the case $n = 1$.

Example 2.6. A simple one-dimensional example is $y = x^2$. If X is a continuous, one-dimensional, random variable and $Y = X^2$, then

$$f_Y(y) = f_X(\sqrt{y}) \frac{1}{2\sqrt{y}} + f_X(-\sqrt{y}) \frac{1}{2\sqrt{y}}.$$

Note that the function is 2-to-1 and that we obtain *two terms*.   □

Now consider the general case. Suppose that the set $S \subset \mathbb{R}^n$ can be partitioned into m disjoint subsets $S_1, S_2, \ldots, S_m$ in $\mathbb{R}^n$, such that $g : S_k \to T$ is 1 to 1 and satisfies the assumptions of Theorem 2.1 for each k. Then

$$P(\mathbf{Y} \in T) = P(\mathbf{X} \in S) = P\left(\mathbf{X} \in \bigcup_{k=1}^{m} S_k\right) = \sum_{k=1}^{m} P(\mathbf{X} \in S_k), \qquad (2.1)$$

which, by Theorem 2.1 applied m times, yields

$$f_{\mathbf{Y}}(\mathbf{y}) = \sum_{k=1}^{m} f_{\mathbf{X}}(h_{1k}(\mathbf{y}), h_{2k}(\mathbf{y}), \ldots, h_{nk}(\mathbf{y})) \cdot |\, \mathbf{J}_k\, |, \qquad (2.2)$$

where, for $k = 1, 2, \ldots, m$, $(h_{1k}, h_{2k}, \ldots, h_{nk})$ is the inverse corresponding to the mapping from S_k to T and $\mathbf{J}_k$ is the Jacobian.

A reconsideration of Example 2.6 in light of this formula shows that the result there corresponds to the partition $S = (\mathbb{R} =) S_1 \cup S_2 \cup \{0\}$, where $S_1 = (0, \infty)$ and $S_2 = (-\infty, 0)$ and also that the first term in the right-hand side there corresponds to S_1 and the second one to S_2. The fact that the value at a single point may be arbitrarily chosen takes care of $f_Y(0)$.

Example 2.7. Steven is a beginner at darts, which means that the points where his darts hit the board can be assumed to be uniformly spread over the board. Find the distribution of the distance from one hitting point to the center of the board.

We assume, without restriction, that the radius of the board is 1 foot (this is only a matter of scaling). Let (X, Y) be the hitting point. We know from Example 1.1 that

$$f_{X,Y}(x, y) = \begin{cases} \frac{1}{\pi}, & \text{for } x^2 + y^2 \leq 1, \\ 0, & \text{otherwise.} \end{cases}$$

We wish to determine the distribution of $U = \sqrt{X^2 + Y^2}$, that is, the distribution of the distance from the hitting point to the origin. To this end we introduce the auxiliary random variable $V = \arctan(Y/X)$ and note that the range of the arctan function is $(-\pi/2, \pi/2)$. This means that we have a 2-to-1 mapping, since the points (X, Y) and $(-X, -Y)$ correspond to the same (U, V). By symmetry and since the Jacobian equals u, we obtain

$$f_{U,V}(u, v) = \begin{cases} 2 \cdot \frac{1}{\pi} \cdot u, & \text{for } 0 < u < 1,\ -\frac{\pi}{2} < v < \frac{\pi}{2}, \\ 0, & \text{otherwise.} \end{cases}$$

It follows that $f_U(u) = 2u$ for $0 < u < 1$ (and 0 otherwise), that $V \in U(-\pi/2, \pi/2)$, and that U and V are independent. □

3 Problems

1. Show that if $X \in C(0, 1)$, then so is $1/X$.
2. Let $X \in C(m, a)$. Determine the distribution of $1/X$.
3. Show that if $T \in t(n)$, then $T^2 \in F(1, n)$.
4. Show that if $F \in F(m, n)$, then $1/F \in F(n, m)$.
5. Show that if $X \in C(0, 1)$, then $X^2 \in F(1, 1)$.
6. Show that $\beta(1, 1) = U(0, 1)$.
7. Show that if $F \in F(m, n)$, then $1/(1 + \frac{m}{n}F) \in \beta(n/2, m/2)$.
8. Show that if X and Y are independent $N(0, 1)$-distributed random variables, then $X/Y \in C(0, 1)$.
9. Show that if $X \in N(0, 1)$ and $Y \in \chi^2(n)$ are independent random variables, then $X/\sqrt{Y/n} \in t(n)$.

10. Show that if $X \in \chi^2(m)$ and $Y \in \chi^2(n)$ are independent random variables, then $(X/m)/(Y/n) \in F(m, n)$.

11. Show that if X and Y are independent $\text{Exp}(a)$-distributed random variables, then $X/Y \in F(2, 2)$.

12. Let X and Y be independent random variables such that $X \in U(0, 1)$ and $Y \in U(0, \alpha)$. Find the density function of $Z = X + Y$.
    Remark. Note that there are two cases: $\alpha \geq 1$ and $\alpha < 1$.

13. Let X and Y have a joint density function given by

$$f(x, y) = \begin{cases} 1, & \text{for} \quad 0 \leq x \leq 2, \ \max(0, x - 1) \leq y \leq \min(1, x), \\ 0, & \text{otherwise.} \end{cases}$$

Determine the marginal density functions and the joint and marginal distribution functions.

14. Suppose that $X \in \text{Exp}(1)$, let Y be the integer part and Z the fractional part, that is, let
$$Y = [X] \quad \text{and} \quad Z = X - [X].$$
Show that Y and Z are independent and find their distributions.

15. Ottar jogs regularly. One day he started his run at 5:31 p.m. and returned at 5:46 p.m. The following day he started at 5:31 p.m. and returned at 5:47 p.m. His watch shows only hours and minutes (not seconds). What is the probability that the run the first day lasted longer than the run the second day?

16. A certain chemistry problem involves the numerical study of a lognormal random variable X. Suppose that the software package used requires the input of EY and $\text{Var}\,Y$ into the computer (where Y is normal and such that $X = e^Y$), but that one knows only the values of EX and $\text{Var}\,X$. Find expressions for the former mean and variance in terms of the latter.

17. Let X and Y be independent $\text{Exp}(a)$-distributed random variables. Find the density function of the random variable $Z = X/(1 + Y)$.

18. Let $X \in \text{Exp}(1)$ and $Y \in U(0, 1)$ be independent random variables. Determine the distribution (density) of $X + Y$.

19. The random vector $\mathbf{X} = (X_1, X_2, X_3)'$ has density function

$$f_{\mathbf{X}}(\mathbf{x}) = \begin{cases} \frac{2}{2e-5} \cdot x_1^2 \cdot x_2 \cdot e^{x_1 \cdot x_2 \cdot x_3}, & \text{for} \quad 0 < x_1, x_2, x_3 < 1, \\ 0, & \text{otherwise.} \end{cases}$$

Determine the distribution of $X_1 \cdot X_2 \cdot X_3$.

20. The random variables X_1 and X_2 are independent and equidistributed with density function

$$f(x) = \begin{cases} 4x^3, & \text{for} \quad 0 \leq x \leq 1, \\ 0, & \text{otherwise.} \end{cases}$$

Set $Y_1 = X_1\sqrt{X_2}$ and $Y_2 = X_2\sqrt{X_1}$.

(a) Determine the joint density function of Y_1 and Y_2.

(b) Are Y_1 and Y_2 independent?

21. Let $(X, Y)'$ have density

$$f(x, y) = \begin{cases} \frac{x}{(1+x)^2 \cdot (1+xy)^2}, & \text{for } x, y > 0, \\ 0, & \text{otherwise.} \end{cases}$$

Show that X and $X \cdot Y$ are independent, equidistriduted random variables and determine their distribution.

22. Let X and Y have joint density

$$f(x, y) = \begin{cases} cx(1 - y), & \text{when } 0 < x < y < 1, \\ 0, & \text{otherwise.} \end{cases}$$

Determine the distribution of $Y - X$.

23. Suppose that $(X, Y)'$ has a density function given by

$$f(x, y) = \begin{cases} e^{-x^2 y}, & \text{for } x \geq 1, y > 0, \\ 0, & \text{otherwise.} \end{cases}$$

Determine the distribution of $X^2 Y$.

24. Let X and Y have the following joint density function:

$$f(x, y) = \begin{cases} \lambda^2 e^{-\lambda y}, & \text{for } 0 < x < y, \\ 0, & \text{otherwise.} \end{cases}$$

Show that Y and $X/(Y-X)$ are independent, and find their distributions.

25. Let X and Y have joint density

$$f(x, y) = \begin{cases} cx, & \text{when } 0 < x^2 < y < \sqrt{x} < 1, \\ 0, & \text{otherwise.} \end{cases}$$

Determine the distribution of XY.

26. Suppose that X and Y are random variables with a joint density

$$f(x, y) = \begin{cases} \frac{1}{y} e^{-x/y} e^{-y}, & \text{when } 0 < x, y < \infty, \\ 0, & \text{otherwise.} \end{cases}$$

Show that X/Y and Y are independent standard exponential random variables and exploit this fact in order to compute $E X$ and $\text{Var } X$.

27. Let X and Y have joint density

$$f(x, y) = \begin{cases} cx, & \text{when } 0 < x^3 < y < \sqrt{x} < 1, \\ 0, & \text{otherwise.} \end{cases}$$

Determine the distribution of XY.

28. Let X and Y have joint density

$$f(x,y) = \begin{cases} cx, & \text{when } 0 < x^2 < y < \sqrt{x} < 1, \\ 0, & \text{otherwise.} \end{cases}$$

Determine the distribution of X^2/Y.

29. Suppose that $(X,Y)'$ has density

$$f(x,y) = \begin{cases} \frac{2}{(1+x+y)^3}, & \text{for } x, y > 0, \\ 0, & \text{otherwise.} \end{cases}$$

Determine the distribution of
(a) $X + Y$,
(b) $X - Y$.

30. Suppose that X and Y are random variables with a joint density

$$f(x,y) = \begin{cases} \frac{2}{5}(2x + 3y), & \text{when } 0 < x, y < 1, \\ 0, & \text{otherwise.} \end{cases}$$

Determine the distribution of $2X + 3Y$.

31. Suppose that X and Y are random variables with a joint density

$$f(x,y) = \begin{cases} xe^{-x-xy}, & \text{when } x > 0, y > 0, \\ 0, & \text{otherwise.} \end{cases}$$

Determine the distribution of $X(1 + Y)$.

32. Suppose that X and Y are random variables with a joint density

$$f(x,y) = \begin{cases} c\frac{x}{(1+y)^2}, & \text{when } 0 < y < x < 1, \\ 0, & \text{otherwise.} \end{cases}$$

Determine the distribution of $X/(1 + Y)^2$.

33. Suppose that X, Y, and Z are random variables with a joint density

$$f(x,y,z) = \begin{cases} \frac{6}{(1+x+y+z)^4}, & \text{when } x, y, z > 0, \\ 0, & \text{otherwise.} \end{cases}$$

Determine the distribution of $X + Y + Z$.

34. Suppose that X, Y, and Z are random variables with a joint density

$$f(x,y,z) = \begin{cases} ce^{-(x+y)^2}, & \text{for } -\infty < x < \infty, 0 < y < 1, \\ 0, & \text{otherwise.} \end{cases}$$

Determine the distribution of $X + Y$.

35. Suppose that X and Y are random variables with a joint density

$$f(x,y) = \begin{cases} \frac{c}{(1+x-y)^2}, & \text{when } 0 < y < x < 1, \\ 0, & \text{otherwise.} \end{cases}$$

Determine the distribution of $X - Y$.

36. Suppose that X and Y are random variables with a joint density

$$f(x,y) = \begin{cases} c \cdot \cos x, & \text{when } 0 < y < x < \frac{\pi}{2}, \\ 0, & \text{otherwise.} \end{cases}$$

Determine the distribution of Y/X.

37. Suppose that X and Y are independent $\mathrm{Pa}(1,1)$-distributed random variables. Determine the distributions of XY and X/Y.

38. Suppose that X and Y are random variables with a joint density

$$f(x,y) = \begin{cases} c \cdot \log y, & \text{when } 0 < y < x < 1, \\ 0, & \text{otherwise.} \end{cases}$$

Determine the distribution (density) of $Z = -\log(Y/X)$.

39. Let $X_1 \in \Gamma(a_1, b)$ and $X_2 \in \Gamma(a_2, b)$ be independent random variables. Show that X_1/X_2 and $X_1 + X_2$ are independent random variables, and determine their distributions.

40. Let $X \in \Gamma(r, 1)$ and $Y \in \Gamma(s, 1)$ be independent random variables.
    (a) Show that $X/(X + Y)$ and $X + Y$ are independent.
    (b) Show that $X/(X + Y) \in \beta(r, s)$.
    (c) Use (a) and (b) and the relation

$$X = (X + Y) \cdot \frac{X}{X + Y}$$

in order to compute the mean and the variance of the beta distribution.

41. Let X_1, X_2, and X_3 be independent random variables, and suppose that $X_i \in \Gamma(r_i, 1)$, $i = 1, 2, 3$. Set

$$Y_1 = \frac{X_1}{X_1 + X_2},$$

$$Y_2 = \frac{X_1 + X_2}{X_1 + X_2 + X_3},$$

$$Y_3 = X_1 + X_2 + X_3.$$

Determine the joint distribution of Y_1, Y_2, and Y_3. Conclusions?

42. Let X and Y be independent $N(0,1)$-distributed random variables.
    (a) What is the distribution of $X^2 + Y^2$?
    (b) Are $X^2 + Y^2$ and X/Y independent?
    (c) Determine the distribution of X/Y.

43. Let X and Y be independent random variables. Determine the distribution of $(X - Y)/(X + Y)$ if
    (a) $X, Y \in \text{Exp}(1)$,
    (b) $X, Y \in N(0, 1)$ (see also Problem 5.10.9(c)).
44. A random vector in $\mathbb{R}^2$ is chosen as follows: Its length, Z, and its angle, Θ, with the positive x-axis, are independent random variables, Z has density

$$f(z) = ze^{-z^2/2}, \quad z > 0,$$

and $\Theta \in U(0, 2\pi)$. Let Q denote the point of the vector. Determine the joint distribution of the Cartesian coordinates of Q.
45. Show that the following procedure generates $N(0, 1)$-distributed random numbers: Pick two independent $U(0, 1)$-distributed numbers U_1 and U_2 and set $X = \sqrt{-2 \log U_1} \cdot \cos(2\pi U_2)$ and $Y = \sqrt{-2 \log U_1} \cdot \sin(2\pi U_2)$. Show that X and Y are independent $N(0, 1)$-distributed random variables.

2

Conditioning

1 Conditional Distributions

Let A and B be events, and suppose that $P(B) > 0$. We recall from Section 3 of the Introduction that the conditional probability of A given B is defined as $P(A \mid B) = P(A \cap B)/P(B)$ and that $P(A \mid B) = P(A)$ if A and B are independent.

Now, let (X, Y) be a two-dimensional random variable whose components are discrete.

Example 1.1. A symmetric die is thrown twice. Let U_1 be a random variable denoting the number of dots on the first throw, let U_2 be a random variable denoting the number of dots on the second throw, and set $X = U_1 + U_2$ and $Y = \min\{U_1, U_2\}$.

Suppose we wish to find the distribution of Y for some given value of X, for example, $P(Y = 2 \mid X = 7)$.

Set $A = \{Y = 2\}$ and $B = \{X = 7\}$. From the definition of conditional probabilities we obtain

$$P(Y = 2 \mid X = 7) = P(A \mid B) = \frac{P(A \cap B)}{P(B)} = \frac{\frac{2}{36}}{\frac{1}{6}} = \frac{1}{3}. \qquad \square$$

With this method one may compute $P(Y = y \mid X = x)$ for any fixed value of x as y varies for arbitrary, discrete, jointly distributed random variables. This leads to the following definition.

Definition 1.1. *Let X and Y be discrete, jointly distributed random variables. For $P(X = x) > 0$ the conditional probability function of Y given that $X = x$ is*

$$p_{Y\mid X=x}(y) = P(Y = y \mid X = x) = \frac{p_{X,Y}(x, y)}{p_X(x)},$$

and the conditional distribution function of Y given that $X = x$ is

A. Gut, *An Intermediate course in Probabilty*, Springer Texts in Statistics,
DOI: 10.1007/978-1-4419-0162-0_2,
© Springer Science + Business Media, LLC 2009

$$F_{Y|X=x}(y) = \sum_{z \le y} p_{Y|X=x}(z).$$ □

Exercise 1.1. Show that $p_{Y|X=x}(y)$ is a probability function of a true probability distribution.   □

It follows immediately (please check) that

$$p_{Y|X=x}(y) = \frac{p_{X,Y}(x,y)}{p_X(x)} = \frac{p_{X,Y}(x,y)}{\sum_z p_{X,Y}(x,z)}$$

and that

$$F_{Y|X=x}(y) = \frac{\sum_{z \le y} p_{X,Y}(x,z)}{p_X(x)} = \frac{\sum_{z \le y} p_{X,Y}(x,z)}{\sum_z p_{X,Y}(x,z)}.$$

Exercise 1.2. Compute the conditional probability function $p_{Y|X=x}(y)$ and the conditional distribution function $F_{Y|X=x}(y)$ in Example 1.1.   □

Now let X and Y have a joint continuous distribution. Expressions like $P(Y = y \mid X = x)$ have no meaning in this case, since the probability that a fixed value is assumed equals zero. However, an examination of how the preceding conditional probabilities are computed makes the following definition very natural.

Definition 1.2. *Let X and Y have a joint continuous distribution. For $f_X(x) > 0$, the conditional density function of Y given that $X = x$ is*

$$f_{Y|X=x}(y) = \frac{f_{X,Y}(x,y)}{f_X(x)},$$

and the conditional distribution function of Y given that $X = x$ is

$$F_{Y|X=x}(y) = \int_{-\infty}^{y} f_{Y|X=x}(z)\,dz.$$ □

In analogy with the discrete case, we further have

$$f_{Y|X=x}(y) = \frac{f_{X,Y}(x,y)}{\int_{-\infty}^{\infty} f_{X,Y}(x,z)\,dz}$$

and

$$F_{Y|X=x}(y) = \frac{\int_{-\infty}^{y} f_{X,Y}(x,z)\,dz}{\int_{-\infty}^{\infty} f_{X,Y}(x,z)\,dz}.$$

Exercise 1.3. Show that $f_{Y|X=x}(y)$ is a density function of a true probability distribution.

Exercise 1.4. Find the conditional distribution of Y given that $X = x$ in Example 1.1.1 and Exercise 1.1.3.

Exercise 1.5. Prove that if X and Y are independent then the conditional distributions and the unconditional distributions are the same. Explain why this is reasonable.   □

Remark 1.1. Definitions 1.1 and 1.2 can be extended to situations with more than two random variables. How?   □

2 Conditional Expectation and Conditional Variance

In the same vein as the concepts of expected value and variance are introduced as convenient location and dispersion measures for (ordinary) random variables or distributions, it is natural to introduce analogs to these concepts for conditional distributions. The following example shows how such notions enter naturally.

Example 2.1. A stick of length one is broken at a random point, uniformly distributed over the stick. The remaining piece is broken once more. Find the expected value and variance of the piece that now remains.

In order to solve this problem we let $X \in U(0,1)$ be the first remaining piece. The second remaining piece Y is uniformly distributed on the interval $(0, X)$. This is to be interpreted as follows: Given that $X = x$, the random variable Y is uniformly distributed on the interval $(0, x)$:

$$Y \mid X = x \in U(0, x),$$

that is, $f_{Y|X=x}(y) = 1/x$ for $0 < y < x$ and 0, otherwise. Clearly, $EX = 1/2$ and $\operatorname{Var} X = 1/12$. Furthermore, intuition suggests that

$$E(Y \mid X = x) = \frac{x}{2} \quad \text{and} \quad \operatorname{Var}(Y \mid X = x) = \frac{x^2}{12}. \tag{2.1}$$

We wish to determine EY and $\operatorname{Var} Y$ somehow with the aid of the preceding relations.   □

We are now ready to state our first definition.

Definition 2.1. *Let X and Y be jointly distributed random variables. The conditional expectation of Y given that $X = x$ is*

$$E(Y \mid X = x) = \begin{cases} \displaystyle\sum_y y\, p_{Y|X=x}(y) & \text{in the discrete case,} \\ \displaystyle\int_{-\infty}^{\infty} y\, f_{Y|X=x}(y)\, dy & \text{in the continuous case,} \end{cases}$$

provided the relevant sum or integral is absolutely convergent.   □

Exercise 2.1. Let X, Y, Y_1, and Y_2 be random variables, let g be a function, and c a constant. Show that

(a) $E(c \mid X = x) = c$,
(b) $E(Y_1 + Y_2 \mid X = x) = E(Y_1 \mid X = x) + E(Y_2 \mid X = x)$,
(c) $E(cY \mid X = x) = c \cdot E(Y \mid X = x)$,
(d) $E(g(X,Y) \mid X = x) = E(g(x,Y) \mid X = x)$,
(e) $E(Y \mid X = x) = EY$ if X and Y are independent. $\square$

The conditional distribution of Y given that $X = x$ depends on the value of x (unless X and Y are independent). This implies that the conditional expectation $E(Y \mid X = x)$ is a function of x, that is,

$$E(Y \mid X = x) = h(x) \tag{2.2}$$

for some function h. (If X and Y are independent, then check that $h(x) = EY$, a constant.)

An object of considerable interest and importance is the random variable $h(X)$, which we denote by

$$h(X) = E(Y \mid X). \tag{2.3}$$

This random variable is of interest not only in the context of probability theory (as we shall see later) but also in statistics in connection with estimation. Loosely speaking, it turns out that if Y is a "good" estimator and X is "suitably" chosen, then $E(Y \mid X)$ is a "better" estimator. Technically, given a so-called unbiased estimator U of a parameter θ, it is possible to construct another unbiased estimator V by considering the conditional expectation of U with respect to what is called a sufficient statistic T; that is, $V = E(U \mid T)$. The point is that $EU = EV = \theta$ (unbiasedness) and that $\operatorname{Var} V \leq \operatorname{Var} U$ (this follows essentially from the sufficiency and Theorem 2.3 ahead). For details, we refer to the statistics literature provided in Appendix A.

A natural question at this point is: What is the expected value of the random variable $E(Y \mid X)$?

Theorem 2.1. *Suppose that $E|Y| < \infty$. Then*

$$E\big(E(Y \mid X)\big) = EY.$$

Proof. We prove the theorem for the continuous case and leave the (completely analogous) proof for the discrete case as an exercise.

$$E\big(E(Y \mid X)\big) = E\,h(X) = \int_{-\infty}^{\infty} h(x)\,f_X(x)\,dx$$

$$= \int_{-\infty}^{\infty} E(Y \mid X = x)\,f_X(x)\,dx$$

$$= \int_{-\infty}^{\infty} \left(\int_{-\infty}^{\infty} y\,f_{Y\mid X=x}(y)\,dy \right) f_X(x)\,dx$$

$$= \int_{-\infty}^{\infty} \int_{-\infty}^{\infty} y \frac{f_{X,Y}(x,y)}{f_X(x)} f_X(x) \, dy \, dx = \int_{-\infty}^{\infty} \int_{-\infty}^{\infty} y f_{X,Y}(x,y) \, dy \, dx$$

$$= \int_{-\infty}^{\infty} y \left(\int_{-\infty}^{\infty} f_{X,Y}(x,y) \, dx \right) dy = \int_{-\infty}^{\infty} y f_Y(y) \, dy = EY. \qquad \square$$

Remark 2.1. Theorem 2.1 can be interpreted as an "expectation version" of the law of total probability.

Remark 2.2. Clearly, EY must exist in order for Theorem 2.1 to make sense, that is, the corresponding sum or integral must be absolutely convergent. Now, given this assumption, one can show that $E(E(Y \mid X))$ exists and is finite and that the computations in the proof, such as reversing orders of integration, are permitted. We shall, in the sequel, permit ourselves at times to be somewhat sloppy about such verifications. Analogous remarks apply to further results ahead.

We close this remark by pointing out that the conclusion always holds in case Y is nonnegative, in the sense that if one of the members is infinite, then so is the other. $\qquad \square$

Exercise 2.2. The object of this exercise is to show that if we do not assume that $E|Y| < \infty$ in Theorem 2.1, then the conclusion does not necessarily hold. Namely, suppose that $X \in \Gamma(1/2, 2) \, (= \chi^2(1))$ and that

$$f_{Y|X=x}(y) = \frac{1}{\sqrt{2\pi}} x^{\frac{1}{2}} e^{-\frac{1}{2}xy^2}, \quad -\infty < y < \infty.$$

(a) Compute $E(Y|X = x)$, $E(Y|X)$, and, finally, $E(E(Y|X))$.
(b) Show that $Y \in C(0,1)$.
(c) What about EY? $\qquad \square$

We are now able to find EY in Example 2.1.

Example 2.1 (continued). It follows from the definition that the first part of (2.1) holds:

$$E(Y \mid X = x) = \frac{x}{2}, \quad \text{that is,} \quad h(x) = \frac{x}{2}.$$

An application of Theorem 2.1 now yields

$$EY = E(E(Y \mid X)) = E\,h(X) = E\left(\frac{1}{2}X\right) = \frac{1}{2}EX = \frac{1}{2} \cdot \frac{1}{2} = \frac{1}{4}.$$

We have thus determined EY without prior knowledge about the distribution of Y. $\qquad \square$

Exercise 2.3. Find the expectation of the remaining piece after it has been broken off n times. $\qquad \square$

Remark 2.3. That the result $EY = 1/4$ is reasonable can intuitively be seen from the fact that X on average equals $1/2$ and that Y on average equals half the value of X, that is $1/2$ of $1/2$. The proof of Theorem 2.1 consists, in fact, of a stringent version of this kind of argument.   □

Theorem 2.2. *Let X and Y be random variables and g be a function. We have*

(a) $E\big(g(X)Y \mid X\big) = g(X) \cdot E(Y \mid X)$, *and*
(b) $E(Y \mid X) = EY$ *if X and Y are independent.*   □

Exercise 2.4. Prove Theorem 2.2.   □

Remark 2.4. Conditioning with respect to X means that X should be interpreted as known, and, hence, $g(X)$ as a constant that thus may be moved in front of the expectation (recall Exercise 2.1(a)). This explains why Theorem 2.2(a) should hold. Part (b) follows from the fact that the conditional distribution and the unconditional distribution coincide if X and Y are independent; in particular, this should remain true for the conditional expectation and the unconditional expectation (recall Exercises 1.5 and 2.1(e)).   □

A natural problem is to find the variance of the remaining piece Y in Example 2.1, which, in turn, suggests the introduction of the concept of conditional variance.

Definition 2.2. *Let X and Y have a joint distribution. The conditional variance of Y given that $X = x$ is*

$$\operatorname{Var}(Y \mid X = x) = E\big((Y - E(Y \mid X = x))^2 \mid X = x\big),$$

provided the corresponding sum or integral is absolutely convergent.   □

The conditional variance is (also) a function of x; call it $v(x)$. The corresponding random variable is

$$v(X) = \operatorname{Var}(Y \mid X). \tag{2.4}$$

The following result is fundamental.

Theorem 2.3. *Let X and Y be random variables and g a real-valued function. If $EY^2 < \infty$ and $E\big(g(X)\big)^2 < \infty$, then*

$$E\big(Y - g(X)\big)^2 = E\operatorname{Var}(Y \mid X) + E\big(E(Y \mid X) - g(X)\big)^2.$$

Proof. An expansion of the left-hand side yields

$$
\begin{aligned}
E\big(Y - g(X)\big)^2 \\
&= E\big(Y - E(Y \mid X) + E(Y \mid X) - g(X)\big)^2 \\
&= E\big(Y - E(Y \mid X)\big)^2 + 2E\big(Y - E(Y \mid X)\big)\big(E(Y \mid X) - g(X)\big) \\
&\quad + E\big(E(Y \mid X) - g(X)\big)^2.
\end{aligned}
$$

Using Theorem 2.1, the right-hand side becomes

$$E\,E\big((Y - E(Y \mid X))^2 \mid X\big) + 2E\,E\big((Y - E(Y \mid X))$$
$$\times (E(Y \mid X) - g(X)) \mid X\big) + E\big(E(Y \mid X) - g(X)\big)^2$$
$$= E\,\mathrm{Var}(Y \mid X) + 2E\{(E(Y \mid X) - g(X))\,E(Y - E(Y \mid X) \mid X)\}$$
$$+ E\big(E(Y \mid X) - g(X)\big)^2$$

by Theorem 2.2(a). Finally, since $E(Y - E(Y \mid X) \mid X) = 0$, this equals

$$E\,\mathrm{Var}(Y \mid X) + 2E\{(E(Y \mid X) - g(X)) \cdot 0\} + E\big(E(Y \mid X) - g(X)\big)^2,$$

which was to be proved.   □

The particular choice $g(X) = EY$, together with an application of Theorem 2.1, yields the following corollary:

Corollary 2.3.1. *Suppose that $EY^2 < \infty$. Then*

$$\mathrm{Var}\,Y = E\,\mathrm{Var}\,(Y \mid X) + \mathrm{Var}\,(E(Y \mid X)).$$
□

Example 2.1 (continued). Let us determine $\mathrm{Var}\,Y$ with the aid of Corollary 2.3.1.

It follows from second part of formula (2.1) that

$$\mathrm{Var}(Y \mid X = x) = \frac{1}{12}x^2, \quad \text{and hence,} \quad v(X) = \frac{1}{12}X^2,$$

so that

$$E\,\mathrm{Var}(Y \mid X) = E\,v(X) = E\Big(\frac{1}{12}X^2\Big) = \frac{1}{12} \cdot \frac{1}{3} = \frac{1}{36}.$$

Furthermore,

$$\mathrm{Var}\big(E(Y \mid X)\big) = \mathrm{Var}(h(X)) = \mathrm{Var}\Big(\frac{1}{2}X\Big) = \frac{1}{4}\mathrm{Var}(X) = \frac{1}{4} \cdot \frac{1}{12} = \frac{1}{48}.$$

An application of Corollary 2.3.1 finally yields $\mathrm{Var}\,Y = 1/36 + 1/48 = 7/144$.

We have thus computed $\mathrm{Var}\,Y$ without knowing the distribution of Y.   □

Exercise 2.5. Find the distribution of Y in Example 2.1, and verify the values of EY and $\mathrm{Var}\,Y$ obtained above.   □

A discrete variant of Example 2.1 is the following: Let X be uniformly distributed over the numbers $1, 2, \ldots, 6$ (that is, throw a symmetric die) and let Y be uniformly distributed over the numbers $1, 2, \ldots, X$ (that is, then throw a symmetric die with X faces). In this case,

$$h(x) = E(Y \mid X = x) = \frac{1+x}{2},$$

from which it follows that

$$EY = E\,h(X) = E\Big(\frac{1+X}{2}\Big) = \frac{1}{2}(1 + E\,X) = \frac{1}{2}(1 + 3.5) = 2.25.$$

The computation of $\mathrm{Var}\,Y$ is somewhat more elaborate. We leave the details to the reader.   □

3 Distributions with Random Parameters

We begin with two examples:

Example 3.1. Suppose that the density X of red blood corpuscles in humans follows a Poisson distribution whose parameter depends on the observed individual. This means that for Jürg we have $X \in \mathrm{Po}(m_J)$, where m_J is Jürg's parameter value, while for Alice we have $X \in \mathrm{Po}(m_A)$, where m_A is Alice's parameter value. For a person selected at random we may consider the parameter value M as a random variable such that, given that $M = m$, we have $X \in \mathrm{Po}(m)$; namely,

$$P(X = k \mid M = m) = e^{-m} \cdot \frac{m^k}{k!}, \quad k = 0, 1, 2, \ldots . \tag{3.1}$$

Thus, if we *know* that Alice was chosen, then $P(X = k \mid M = m_A) = e^{-m_A} \cdot m_A^k / k!$, for $k = 0, 1, 2, \ldots$, as before. We shall soon see that X itself (unconditioned) need not follow a Poisson distribution.

Example 3.2. A radioactive substance emits α-particles in such a way that the number of emitted particles during an hour, N, follows a $\mathrm{Po}(\lambda)$-distribution. The particle counter, however, is somewhat unreliable in the sense that an emitted particle is registered with probability p $(0 < p < 1)$, whereas it remains unregistered with probability $q = 1 - p$. All particles are registered independently of each other. This means that if we *know* that n particles were emitted during a specific hour, then the number of registered particles $X \in \mathrm{Bin}(n, p)$, that is,

$$P(X = k \mid N = n) = \binom{n}{k} p^k q^{n-k}, \quad k = 0, 1, \ldots, n \tag{3.2}$$

(and $N \in \mathrm{Po}(\lambda)$). If, however, we observe the process during an arbitrarily chosen hour, it follows, as will be seen below, that the number of registered particles does not follow a binomial distribution (but instead a Poisson distribution). □

The common feature in these examples is that the random variable under consideration, X, has a known distribution but with a parameter that is a random variable. Somewhat imprecisely, we might say that in Example 3.1 we have $X \in \mathrm{Po}(M)$, where M follows some distribution, and that in Example 3.2 we have $X \in \mathrm{Bin}(N, p)$, where $N \in \mathrm{Po}(\lambda)$. We prefer, however, to describe these cases as

$$X \mid M = m \in \mathrm{Po}(m) \quad \text{with} \quad M \in F, \tag{3.3}$$

where F is some distribution, and

$$X \mid N = n \in \mathrm{Bin}(n, p) \quad \text{with} \quad N \in \mathrm{Po}(\lambda), \tag{3.4}$$

respectively.

Let us now determine the (unconditional) distributions of X in our examples, where, in Example 3.1, we assume that $M \in \mathrm{Exp}(1)$.

Example 3.1 (continued). We thus have

$$X \mid M = m \in \mathrm{Po}(m) \quad \text{with} \quad M \in \mathrm{Exp}(1). \tag{3.5}$$

By (the continuous version of) the law of total probability, we obtain, for $k = 0, 1, 2, \ldots,$

$$
\begin{aligned}
P(X = k) &= \int_0^\infty P(X = k \mid M = x) \cdot f_M(x)\, dx \\
&= \int_0^\infty e^{-x} \frac{x^k}{k!} \cdot e^{-x}\, dx = \int_0^\infty \frac{x^k}{k!} e^{-2x}\, dx \\
&= \frac{1}{2^{k+1}} \cdot \int_0^\infty \frac{1}{\Gamma(k+1)} 2^{k+1} x^{k+1-1} e^{-2x}\, dx \\
&= \frac{1}{2^{k+1}} \cdot 1 = \frac{1}{2} \cdot \left(\frac{1}{2}\right)^k,
\end{aligned}
$$

that is, $X \in \mathrm{Ge}(1/2)$. The unconditional distribution in this case thus is not a Poisson distribution; it is a geometric distribution. $\square$

Exercise 3.1. Determine the distribution of X if M has

(a) an $\mathrm{Exp}(a)$-distribution,
(b) a $\Gamma(p, a)$-distribution. $\square$

Note also that we may use the formulas from Section 2 to compute EX and $\mathrm{Var}\,X$ without knowing the distribution of X. Namely, since $E(X \mid M = m) = m$ (i.e., $h(M) = E(X \mid M) = M$), Theorem 2.1 yields

$$EX = E\big(E(X \mid M)\big) = EM = 1,$$

and Corollary 2.3.1 yields

$$\mathrm{Var}\,X = E\,\mathrm{Var}(X \mid M) + \mathrm{Var}\big(E(X \mid M)\big) = EM + \mathrm{Var}\,M = 1 + 1 = 2.$$

If, however, the distribution has been determined (as above), the formulas from Section 2 may be used for checking.

If applied to Exercise 3.1(a), the latter formulas yield $EX = a$ and $\mathrm{Var}\,X = a + a^2$. Since this situation differs from Example 3.1 only by a rescaling of M, one might perhaps guess that the solution is another geometric distribution. If this were true, we would have

$$EX = a = \frac{q}{p} = \frac{1-p}{p} = \frac{1}{p} - 1; \quad p = \frac{1}{a+1}.$$

This value of p inserted in the expression for the variance yields

$$\frac{q}{p^2} = \frac{1-p}{p^2} = \frac{1}{p^2} - \frac{1}{p} = (a+1)^2 - (a+1) = a^2 + a,$$

which coincides with our computations above and provides the guess that $X \in \text{Ge}(1/(a+1))$.

Remark 3.1. In Example 3.1 we used the results of Section 2.2 to confirm our *result*. In Exercise 3.1(a) they were used to confirm (provide) a *guess*.   □

We now turn to the α-particles.

Example 3.2 (continued). Intuitively, the deficiency of the particle counter implies that the radiation actually measured is, on average, a fraction p of the original Poisson stream of particles. We might therefore expect that the number of registered particles during one hour should be a $\text{Po}(\lambda p)$-distributed random variable. That this is actually correct is verified next.

The model implies that

$$X \mid N = n \in \text{Bin}(n, p) \quad \text{with} \quad N \in \text{Po}(\lambda).$$

The law of total probability yields, for $k = 0, 1, 2, \ldots$,

$$
\begin{aligned}
P(X = k) &= \sum_{n=0}^{\infty} P(X = k \mid N = n) \cdot P(N = n) \\
&= \sum_{n=k}^{\infty} \binom{n}{k} p^k q^{n-k} \cdot e^{-\lambda} \frac{\lambda^n}{n!} \\
&= \frac{p^k}{k!} e^{-\lambda} \sum_{n=k}^{\infty} \frac{\lambda^n}{(n-k)!} q^{n-k} = \frac{(\lambda p)^k}{k!} e^{-\lambda} \sum_{n=k}^{\infty} \frac{(\lambda q)^{n-k}}{(n-k)!} \\
&= \frac{(\lambda p)^k}{k!} e^{-\lambda} \sum_{j=0}^{\infty} \frac{(\lambda q)^j}{j!} = \frac{(\lambda p)^k}{k!} e^{-\lambda} \cdot e^{\lambda q} = e^{-\lambda p} \frac{(\lambda p)^k}{k!},
\end{aligned}
$$

that is, $X \in \text{Po}(\lambda p)$. The unconditional distribution thus is not a binomial distribution; it is a Poisson distribution.   □

Remark 3.2. This is an example of a so-called thinned Poisson process. For more details, we refer to Section 8.6.   □

Exercise 3.2. Use Theorem 2.1 and Corollary 2.3.1 to check the values of $E\,X$ and $\text{Var}\,X$.   □

A family of distributions that is of special interest is the family of mixed normal, or mixed Gaussian, distributions. These are normal distributions with a random variance, namely,

$$X \mid \Sigma^2 = y \in N(\mu, y) \quad \text{with} \quad \Sigma^2 \in F, \tag{3.6}$$

where F is some distribution (on $(0, \infty)$).

For simplicity we assume in the following that $\mu = 0$.

As an example, consider normally distributed observations with rare disturbances. More specifically, the observations might be $N(0, 1)$-distributed with probability 0.99 and $N(0, 100)$-distributed with probability 0.01. We may write this as

$$X \in N(0, \Sigma^2), \quad \text{where} \quad P(\Sigma^2 = 1) = 0.99 \quad \text{and} \quad P(\Sigma^2 = 100) = 0.01.$$

By Theorem 2.1 it follows immediately that $E X = 0$. As for the variance, Corollary 2.3.1 tells us that

$$\text{Var } X = E \text{Var}\left(X \mid \Sigma^2\right) + \text{Var}\left(E(X \mid \Sigma^2)\right)$$
$$= E \Sigma^2 = 0.99 \cdot 1 + 100 \cdot 0.01 = 1.99.$$

If Σ^2 has a continuous distribution, computations such as those above yield

$$F_X(x) = \int_0^\infty \Phi\left(\frac{x}{\sqrt{y}}\right) f_{\Sigma^2}(y)\, dy,$$

from which the density function of X is obtained by differentiation:

$$f_X(x) = \int_0^\infty \frac{1}{\sqrt{y}} \phi\left(\frac{x}{\sqrt{y}}\right) f_{\Sigma^2}(y)\, dy = \int_0^\infty \frac{1}{\sqrt{2\pi y}} e^{-x^2/2y} f_{\Sigma^2}(y)\, dy. \quad (3.7)$$

Mean and variance can be found via the results of Section 2:

$$E X = E\left(E(X \mid \Sigma^2)\right) = 0,$$
$$\text{Var } X = E \text{Var}\left(X \mid \Sigma^2\right) + \text{Var}\left(E(X \mid \Sigma^2)\right) = E \Sigma^2.$$

Next, we determine the distribution of X under the particular assumption that $\Sigma^2 \in \text{Exp}(1)$. We are thus faced with the situation

$$X \mid \Sigma^2 = y \in N(0, y) \quad \text{with} \quad \Sigma^2 \in \text{Exp}(1) \qquad (3.8)$$

By (3.7),

$$f_X(x) = \int_0^\infty \frac{1}{\sqrt{2\pi y}} e^{-x^2/2y} e^{-y}\, dy = \left[\, \text{set } y = u^2 \,\right]$$
$$= \int_0^\infty \frac{1}{\sqrt{2\pi}} e^{-x^2/2u^2} e^{-u^2} \cdot 2\, du = \sqrt{\frac{2}{\pi}} \int_0^\infty \exp\left\{-\frac{x^2}{2u^2} - u^2\right\} du.$$

In order to solve this integral, the following device may be of use: Let $x > 0$, set

$$I(x) = \int_0^\infty \exp\left\{-\frac{x^2}{2u^2} - u^2\right\} du,$$

differentiate (differentiation and integration may be interchanged), and make the change of variable $y = x/u\sqrt{2}$. This yields

$$I'(x) = \int_0^\infty \left(-\frac{x}{u^2}\right) \exp\left\{-\frac{x^2}{2u^2} - u^2\right\} du = -\sqrt{2} \int_0^\infty \exp\left\{-y^2 - \frac{x^2}{2y^2}\right\} dy.$$

It follows that I satisfies the differential equation

$$I'(x) = -\sqrt{2}I(x)$$

with the initial condition

$$I(0) = \int_0^\infty e^{-u^2} du = \frac{\sqrt{\pi}}{2},$$

the solution of which is

$$I(x) = \frac{\sqrt{\pi}}{2} e^{-x\sqrt{2}}, \quad x > 0. \tag{3.9}$$

By inserting (3.9) into the expression for $f_X(x)$, and noting that the density is symmetric around $x = 0$, we finally obtain

$$f_X(x) = \sqrt{\frac{2}{\pi}} \frac{\sqrt{\pi}}{2} e^{-|x|\sqrt{2}} = \frac{1}{\sqrt{2}} e^{-|x|\sqrt{2}} = \frac{1}{2}\sqrt{2} e^{-|x|\sqrt{2}}, \quad -\infty < x < \infty,$$

that is, $X \in L(\frac{1}{\sqrt{2}})$; a Laplace distribution.

An extra check yields $E X = 0$ and $\operatorname{Var} X = E \Sigma^2 = 1 \ (= 2 \cdot (\frac{1}{\sqrt{2}})^2)$, as desired.

Exercise 3.3. Show that if X has a normal distribution such that the mean is zero and the inverse of the variance is Γ-distributed, viz.,

$$X \mid \Sigma^2 = \lambda \in N(0, 1/\lambda) \quad \text{with} \quad \Sigma^2 \in \Gamma\left(\frac{n}{2}, \frac{2}{n}\right),$$

then $X \in t(n)$.

Exercise 3.4. Sheila has a coin with $P(\text{head}) = p_1$ and Betty has a coin with $P(\text{head}) = p_2$. Sheila tosses her coin m times. Each time she obtains "heads," Betty tosses her coin (otherwise not). Find the distribution of the total number of heads obtained by Betty.

Further, check that mean and variance coincide with the values obtained by Theorem 2.1 and Corollary 2.3.1. Alternatively, find mean and variance first and try to guess the desired distribution (and check if your guess was correct).

As a hint, observe that the game can be modeled as follows: Let N be the number of heads obtained by Sheila and X be the number of heads obtained by Betty. We thus wish to find the distribution of X, where

$$X \mid N = n \in \operatorname{Bin}(n, p_2) \quad \text{with} \quad N \in \operatorname{Bin}(m, p_1), \quad 0 < p_1, p_2 < 1. \qquad \square$$

We shall return to the topic of this section in Section 3.5.

4 The Bayesian Approach

A typical problem in probability theory begins with assumptions such as "let $X \in Po(m)$," "let $Y \in N(\mu, \sigma^2)$," "toss a symmetric coin 15 times," and so forth. In the computations that follow, one tacitly assumes that all parameters are known, that the coin is *exactly* symmetric, and so on.

In statistics one assumes (certain) parameters to be unknown, for example, that the coin might be asymmetric, and one searches for methods, devices, and rules to decide whether or not one should believe in certain hypotheses. Two typical illustrations in the Gaussian approach are "μ unknown and σ known" and "μ and σ unknown."

The Bayesian approach is a kind of compromise. One claims, for example, that parameters are never *completely* unknown; one always has *some* prior opinion or knowledge about them.

A probabilistic model describing this approach was given in Example 3.1. The opening statement there was that the density of red blood corpuscles follows a Poisson distribution. One interpretation of that statement could have been that whenever we are faced with a blood sample the density of red blood corpuscles in the sample is Poissonian. The Bayesian approach taken in Example 3.1 is that whenever we know from whom the blood sample has been taken, the density of red blood corpuscles in the sample is Poissonian, however, with a parameter depending on the individual. If we do not know from whom the sample has been taken, then the parameter is unknown; it is a random variable following some distribution. We also found that if this distribution is the standard exponential, then the density of red blood corpuscles is geometric (and hence not Poissonian).

The prior knowledge about the parameters in this approach is expressed in such a way that the parameters are assumed to follow some probability distribution, called the *prior* (or a priori) distribution. If one wishes to assume that a parameter is "completely unknown," one might solve the situation by attributing some uniform distribution to the parameter.

In this terminology we may formulate our findings in Example 3.1 as follows: If the parameter in a Poisson distribution has a standard exponential prior distribution, then the random variable under consideration follows a $Ge(1/2)$-distribution.

Frequently, one performs random experiments in order to estimate (unknown) parameters. The estimates are based on observations from some probability distribution. The Bayesian analog is to determine the conditional distribution of the parameter given the result of the random experiment. Such a distribution is called the *posterior* (or a posteriori) distribution.

Next we determine the posterior distribution in Example 3.1.

Example 4.1. The model in the example was

$$X \mid M = m \in Po(m) \quad \text{with} \quad M \in Exp(1). \tag{4.1}$$

We further had found that $X \in \mathrm{Ge}(1/2)$. Now we wish to determine the conditional distribution of M given the value of X.

For $x > 0$, we have

$$F_{M|X=k}(x) = P(M \leq x \mid X = k) = \frac{P(\{M \leq x\} \cap \{X = k\})}{P(X = k)}$$

$$= \frac{\int_0^x P(X = k \mid M = y) \cdot f_M(y)\, dy}{P(X = k)}$$

$$= \frac{\int_0^x e^{-y} \frac{y^k}{k!} \cdot e^{-y}\, dy}{(\frac{1}{2})^{k+1}} = \int_0^x \frac{1}{\Gamma(k+1)} y^k 2^{k+1} e^{-2y}\, dy\,,$$

which, after differentiation, yields

$$f_{M|X=k}(x) = \frac{1}{\Gamma(k+1)} x^k 2^{k+1} e^{-2x}, \quad x > 0.$$

Thus, $M \mid X = k \in \Gamma(k+1, \frac{1}{2})$ or, in our new terminology, the posterior distribution of M given that X equals k is $\Gamma(k+1, \frac{1}{2})$. $\qquad\square$

Remark 4.1. Note that, starting from the distribution of X given M (and from that of M), we have determined the distribution of M given X and that the solution of the problem, in fact, amounted to applying a continuous version of Bayes' formula. $\qquad\square$

Exercise 4.1. Check that $E\,M$ and $\mathrm{Var}\,M$ are what they are supposed to be by applying Theorem 2.1 and Corollary 2.3.1 to the posterior distribution. $\square$

We conclude this section by studying coin tossing from the Bayesian point of view under the assumption that nothing is known about $p = P(\text{heads})$.

Let X_n be the number of heads after n coin tosses. *One* possible model is

$$X_n \mid P = p \in \mathrm{Bin}(n, p) \quad \text{with} \quad P \in U(0, 1). \tag{4.2}$$

The prior distribution of P, thus, is the $U(0, 1)$-distribution. Models of this kind are called *mixed binomial models*.

For $k = 0, 1, 2, \ldots, n$, we now obtain (via some facts about the beta distribution)

$$P(X_n = k) = \int_0^1 \binom{n}{k} x^k (1 - x)^{n-k} \cdot 1\, dx$$

$$= \binom{n}{k} \int_0^1 x^{(k+1)-1} (1 - x)^{(n+1-k)-1}\, dx$$

$$= \binom{n}{k} \frac{\Gamma(k+1)\Gamma(n+1-k)}{\Gamma(k+1+n+1-k)}$$

$$= \frac{n!\, k!\, (n-k)!}{k!\, (n-k)!\, (n+1)!} = \frac{1}{n+1}\,.$$

This means that X_n is uniformly distributed over the integers $0, 1, \ldots, n$.

A second thought reveals that this is a very reasonable conclusion. Since *nothing* is known about the coin (in the sense of relation (4.2)), there is nothing that favors a specific outcome, that is, all outcomes should be equally probable.

If p is known, we know that the results in different tosses are independent and that the probability of heads given that we obtained 100 heads in a row (still) equals p. What about these facts in the Bayesian model?

$$P(X_{n+1} = n+1 \mid X_n = n) = \frac{P(\{X_{n+1} = n+1\} \cap \{X_n = n\})}{P(X_n = n)}$$

$$= \frac{P(X_{n+1} = n+1)}{P(X_n = n)}$$

$$= \frac{\frac{1}{n+2}}{\frac{1}{n+1}} = \frac{n+1}{n+2} \to 1 \quad \text{as} \quad n \to \infty.$$

This means that if we know that there were many heads in a row then the (conditional) probability of another head is very large; the results in different tosses are not at all independent.

Why is this the case? Let us find the posterior distribution of P.

$$P(P \le x \mid X_n = k) = \frac{\int_0^x P(X_n = k \mid P = y) \cdot f_P(y) \, dy}{P(X_n = k)}$$

$$= \frac{\int_0^x \binom{n}{k} y^k (1-y)^{n-k} \cdot 1 \, dy}{\frac{1}{n+1}}$$

$$= (n+1) \binom{n}{k} \int_0^x y^k (1-y)^{n-k} \, dy.$$

Differentiation yields

$$f_{P \mid X_n = k}(x) = \frac{\Gamma(n+2)}{\Gamma(k+1)\Gamma(n+1-k)} x^k (1-x)^{n-k}, \quad 0 < x < 1,$$

viz., a $\beta(k+1, n+1-k)$-distribution.

For $k = n$ we obtain in particular (or, by direct computation)

$$f_{P \mid X_n = n}(x) = (n+1)x^n, \quad 0 < x < 1.$$

It follows that

$$P(P > 1 - \varepsilon \mid X_n = n) = 1 - (1 - \varepsilon)^{n+1} \to 1 \quad \text{as} \quad n \to \infty$$

for all $\varepsilon > 0$. This means that if we know that there were many heads in a row then we also know that p is close to 1 and thus that it is very likely that the next toss will yield another head.

Remark 4.2. It is, of course, possible to consider the posterior distribution as a prior distribution for a further random experiment, and so on. □

5 Regression and Prediction

A common statistics problem is to analyze how different (levels of) treatments or treatment combinations affect the outcome of an experiment. The yield of a crop, for example, may depend on variability in watering, fertilization, climate, and other factors in the various areas where the experiment is performed. One problem is that one cannot predict the outcome y exactly, meaning without error, even if the levels of the treatments $x_1, x_2, \ldots, x_n$ are known exactly. An important function for predicting the outcome is the conditional expectation of the (random) outcome Y given the (random) levels of treatment $X_1, X_2, \ldots, X_n$.

Let $X_1, X_2, \ldots, X_n$ and Y be jointly distributed random variables, and set

$$h(\mathbf{x}) = h(x_1, \ldots, x_n) = E(Y \mid X_1 = x_1, \ldots, X_n = x_n) = E(Y \mid \mathbf{X} = \mathbf{x}).$$

Definition 5.1. *The function h is called the* regression function Y on $\mathbf{X}$. □

Remark 5.1. For $n = 1$ we have $h(x) = E(Y \mid X = x)$, which is the ordinary conditional expectation. □

Definition 5.2. *A* predictor *(for Y) based on $\mathbf{X}$ is a function, $d(\mathbf{X})$. The predictor is called* linear *if d is linear, that is, if $d(\mathbf{X}) = a_0 + a_1 X_1 + \cdots + a_n X_n$, where $a_0, a_1, \ldots, a_n$ are constants.* □

Predictors are used to predict (as the name suggests). The prediction error is given by the random variable

$$Y - d(\mathbf{X}). \tag{5.1}$$

There are several ways to compare different predictors. One suitable measure is defined as follows:

Definition 5.3. *The* expected quadratic prediction error *is*

$$E\big(Y - d(\mathbf{X})\big)^2.$$

Moreover, if d_1 and d_2 are predictors, we say that d_1 is better than d_2 if $E(Y - d_1(\mathbf{X}))^2 \leq E(Y - d_2(\mathbf{X}))^2$. □

In the following we confine ourselves to considering the case $n = 1$. A predictor is thus a function of X, $d(X)$, and the expected quadratic prediction error is $E(Y - d(X))^2$. If the predictor is linear, that is, if $d(X) = a + bX$, where a and b are constants, the expected quadratic prediction error is $E(Y - (a + bX))^2$.

Example 5.1. Pick a point uniformly distributed in the triangle $x, y \geq 0$, $x + y \leq 1$. We wish to determine the regression functions $E(Y \mid X = x)$ and $E(X \mid Y = y)$.

To solve this problem we first note that the joint density of X and Y is

$$f_{X,Y}(x, y) = \begin{cases} c, & \text{for} \quad x, y \geq 0, x + y \leq 1, \\ 0, & \text{otherwise}, \end{cases}$$

where c is some constant, which is found by noticing that the total mass equals 1. We thus have

$$1 = \int_{-\infty}^{\infty} \int_{-\infty}^{\infty} f_{X,Y}(x, y) \, dx dy = \int_0^1 \left(\int_0^{1-x} c \, dy \right) dx$$

$$= c \int_0^1 (1 - x) \, dx = c \left[-\frac{(1-x)^2}{2} \right]_0^1 = \frac{c}{2},$$

from which it follows that $c = 2$.

In order to determine the conditional densities we first compute the marginal ones:

$$f_X(x) = \int_{-\infty}^{\infty} f_{X,Y}(x, y) \, dy = \int_0^{1-x} 2 \, dy = 2(1 - x), \quad 0 < x < 1,$$

$$f_Y(y) = \int_{-\infty}^{\infty} f_{X,Y}(x, y) \, dx = \int_0^{1-y} 2 \, dx = 2(1 - y), \quad 0 < y < 1.$$

Incidentally, X and Y have the same distribution for reasons of symmetry. Finally,

$$f_{Y \mid X = x}(y) = \frac{f_{X,Y}(x, y)}{f_X(x)} = \frac{2}{2(1 - x)} = \frac{1}{1 - x}, \quad 0 < y < 1 - x,$$

and so

$$E(Y \mid X = x) = \int_0^{1-x} y \cdot \frac{1}{1 - x} \, dy = \frac{1}{1 - x} \left[\frac{y^2}{2} \right]_0^{1-x} = \frac{(1 - x)^2}{2(1 - x)} = \frac{1 - x}{2}$$

and, by symmetry,

$$E(X \mid Y = y) = \frac{1 - y}{2}. \qquad \square$$

Remark 5.2. Note also, for example, that $Y \mid X = x \in U(0, 1 - x)$ in the example, that is, the density is, for x fixed, a constant (which is the inverse of the length of the interval $(0, 1 - x)$). This implies that $E(Y \mid X = x) = (1-x)/2$, which agrees with the previous results. It also provides an alternative solution to the last part of the problem. In this case the gain is marginal, but in a more technically complicated situation it might be more substantial. $\square$

Exercise 5.1. Solve the same problem when

$$f_{X,Y}(x,y) = \begin{cases} cx, & \text{for } 0 < x, y < 1, \\ 0, & \text{otherwise.} \end{cases}$$

Exercise 5.2. Solve the same problem when

$$f_{X,Y}(x,y) = \begin{cases} e^{-y}, & \text{for } 0 < x < y, \\ 0, & \text{otherwise.} \end{cases} \qquad \square$$

Theorem 5.1. *Suppose that* $EY^2 < \infty$. *Then* $h(X) = E(Y \mid X)$ *(i.e., the regression function* Y *on* X*) is the best predictor of* Y *based on* X.

Proof. By Theorem 2.3 we know that for an arbitrary predictor $d(X)$,

$$E(Y - d(X))^2 = E \operatorname{Var}(Y \mid X) + E(h(X) - d(X))^2 \geq E \operatorname{Var}(Y \mid X),$$

where equality holds iff $d(X) = h(X)$ (more precisely, iff $P(d(X) = h(X)) = 1$). The choice $d(x) = h(x)$ thus yields minimal expected quadratic prediction error. $\qquad \square$

Example 5.2. In Example 5.1 we found the regression function of Y based on X to be $(1 - X)/2$. By Theorem 5.1 it is the best predictor of Y based on X. A simple calculation shows that the expected quadratic prediction error is $E(Y - (1 - X)/2)^2 = 1/48$.

We also noted that X and Y have the same marginal distribution. A (very) naive suggestion for another predictor therefore might be X itself. The expected quadratic prediction error for this predictor is $E(Y - X)^2 = 1/4 > 1/48$, which shows that the regression function is indeed a better predictor. $\square$

Sometimes it is difficult to determine regression functions explicitly. In such cases one might be satisfied with the best *linear* predictor. This means that one wishes to minimize $E(Y - (a + bX))^2$ as a function of a and b, which leads to the well-known method of least squares. The solution of this problem is given in the following result.

Theorem 5.2. *Suppose that* $EX^2 < \infty$ *and* $EY^2 < \infty$. *Set* $\mu_x = EX$, $\mu_y = EY$, $\sigma_x^2 = \operatorname{Var} X$, $\sigma_y^2 = \operatorname{Var} Y$, $\sigma_{xy} = \operatorname{Cov}(X, Y)$, *and* $\rho = \sigma_{xy}/\sigma_x\sigma_y$. *The best linear predictor of* Y *based on* X *is*

$$L(X) = \alpha + \beta X,$$

where
$$\alpha = \mu_y - \frac{\sigma_{xy}}{\sigma_x^2}\mu_x = \mu_y - \rho\frac{\sigma_y}{\sigma_x}\mu_x \quad \text{and} \quad \beta = \frac{\sigma_{xy}}{\sigma_x^2} = \rho\frac{\sigma_y}{\sigma_x}. \qquad \square$$

The best linear predictor thus is

$$\mu_y + \rho \frac{\sigma_y}{\sigma_x}(X - \mu_x).\tag{5.2}$$

Definition 5.4. *The line* $y = \mu_y + \rho\frac{\sigma_y}{\sigma_x}(x - \mu_x)$ *is called the* regression line Y *on* X. *The slope,* $\rho\frac{\sigma_y}{\sigma_x}$, *of the line is called the* regression coefficient. $\square$

Remark 5.3. Note that $y = L(x)$, where $L(X)$ is the best linear predictor of Y based on X.

Remark 5.4. If, in particular, (X, Y) has a joint Gaussian distribution, it turns out that the regression function is linear, that is, for this very important case the best linear predictor is, in fact, the best predictor. For details, we refer the reader to Section 5.6. $\square$

Example 5.1 (continued). The regression function Y on X turned out to be linear in this example; $y = (1-x)/2$. It follows in particular that the regression function coincides with the regression line Y on X. The regression coefficient equals $-1/2$. $\square$

The expected quadratic prediction error of the best linear predictor of Y based on X is obtained as follows:

Theorem 5.3. $E\big(Y - L(X)\big)^2 = \sigma_y^2(1 - \rho^2)$.

Proof.

$$E\big(Y - L(X)\big)^2 = E\big(Y - \mu_y - \rho\frac{\sigma_y}{\sigma_x}(X - \mu_x)\big)^2 = E(Y - \mu_y)^2$$
$$+ \rho^2\frac{\sigma_y^2}{\sigma_x^2}E(X - \mu_x)^2 - 2\rho\frac{\sigma_y}{\sigma_x}E(Y - \mu_y)(X - \mu_x)$$
$$= \sigma_y^2 + \rho^2 \cdot \sigma_y^2 - 2\rho\frac{\sigma_y}{\sigma_x}\sigma_{xy} = \sigma_y^2(1 - \rho^2). \qquad \square$$

Definition 5.5. *The quantity* $\sigma_y^2(1 - \rho^2)$ *is called* residual variance. $\square$

Exercise 5.3. Check via Theorem 5.3 that the residual variance in Example 5.1 equals $1/48$ as was claimed in Example 5.2. $\square$

The regression line X on Y is determined similarly. It is

$$x = \mu_x + \rho\frac{\sigma_x}{\sigma_y}(y - \mu_y),$$

which can be rewritten as

$$y = \mu_y + \frac{1}{\rho} \cdot \frac{\sigma_y}{\sigma_x}(x - \mu_x)$$

if $\rho \neq 0$. The regression lines Y on X and X on Y are thus, in general, different. They coincide iff they have the same slope—iff

$$\rho \cdot \frac{\sigma_y}{\sigma_x} = \frac{1}{\rho} \cdot \frac{\sigma_y}{\sigma_x} \quad \Longleftrightarrow \quad |\rho| = 1,$$

that is, iff there exists a linear relation between X and Y. $\qquad\square$

Example 5.1 (continued). The regression function X on Y was also linear (and coincides with the regression line X on Y). The line has the form $x = (1-y)/2$, that is, $y = 1 - 2x$. In particular, we note that the slopes of the regression lines are $-1/2$ and -2, respectively. $\qquad\square$

6 Problems

1. Let X and Y be independent $\mathrm{Exp}(1)$-distributed random variables. Find the conditional distribution of X given that $X + Y = c$ (c is a positive constant).
2. Let X and Y be independent $\Gamma(2, a)$-distributed random variables. Find the conditional distribution of X given that $X + Y = 2$.
3. The life of a repairing device is $\mathrm{Exp}(1/a)$-distributed. Peter wishes to use it on n different, independent, $\mathrm{Exp}(1/na)$-distributed occasions.
   (a) Compute the probability P_n that this is possible.
   (b) Determine the limit of P_n as $n \to \infty$.
4. The life T (hours) of the lightbulb in an overhead projector follows an $\mathrm{Exp}(10)$-distribution. During a normal week it is used a $\mathrm{Po}(12)$-distributed number of lectures lasting exactly one hour each. Find the probability that a projector with a newly installed lightbulb functions throughout a normal week (without replacing the lightbulb).
5. The random variables $N, X_1, X_2, \ldots$ are independent, $N \in \mathrm{Po}(\lambda)$, and $X_k \in \mathrm{Be}(1/2)$, $k \geq 1$. Set

$$Y_1 = \sum_{k=1}^{N} X_k \quad \text{and} \quad Y_2 = N - Y_1$$

   ($Y_1 = 0$ for $N = 0$). Show that Y_1 and Y_2 are independent, and determine their distributions.
6. Suppose that $X \in N(0,1)$ and $Y \in \mathrm{Exp}(1)$ are independent random variables. Prove that $X\sqrt{2Y}$ has a standard Laplace distribution.
7. Let $N \in \mathrm{Ge}(p)$ and set $X = (-1)^N$. Compute
   (a) $E\,X$ and $\mathrm{Var}\,X$,
   (b) the distribution (probability function) of X.
8. The density function of the two-dimensional random variable (X, Y) is

$$f_{X,Y}(x,y) = \begin{cases} \dfrac{x^2}{2 \cdot y^3} \cdot e^{-\frac{x}{y}}, & \text{for } 0 < x < \infty, \quad 0 < y < 1, \\ 0, & \text{otherwise.} \end{cases}$$

(a) Determine the distribution of Y.

(b) Find the conditional distribution of X given that $Y = y$.

(c) Use the results from (a) and (b) to compute EX and $\operatorname{Var} X$.

9. The density of the random vector $(X, Y)'$ is

$$f_{X,Y}(x, y) = \begin{cases} cx, & \text{for } x \geq 0, \quad y \geq 0, \quad x + y \leq 1, \\ 0, & \text{otherwise.} \end{cases}$$

Compute

(a) c,

(b) the conditional expectations $E(Y \mid X = x)$ and $E(X \mid Y = y)$.

10. Suppose X and Y have a joint density function given by

$$f(x, y) = \begin{cases} cx^2, & \text{for } 0 < x < y < 1, \\ 0, & \text{otherwise.} \end{cases}$$

Find c, the marginal density functions, EX, EY, and the conditional expectations $E(Y \mid X = x)$ and $E(X \mid Y = y)$.

11. Suppose X and Y have a joint density function given by

$$f(x, y) = \begin{cases} c \cdot x^2 y, & \text{for } 0 < y < x < 1, \\ 0, & \text{otherwise.} \end{cases}$$

Compute c, the marginal densities, EX, EY, and the conditional expectations $E(Y \mid X = x)$ and $E(X \mid Y = y)$.

12. Let X and Y have joint density

$$f(x, y) = \begin{cases} cxy, & \text{when } 0 < y < x < 1, \\ 0, & \text{otherwise.} \end{cases}$$

Compute the conditional expectations $E(Y \mid X = x)$ and $E(X \mid Y = y)$.

13. Let X and Y have joint density

$$f(x, y) = \begin{cases} cy, & \text{when } 0 < y < x < 2, \\ 0, & \text{otherwise.} \end{cases}$$

Compute the conditional expectations $E(Y \mid X = x)$ and $E(X \mid Y = y)$.

14. Suppose that X and Y are random variables with joint density

$$f(x, y) = \begin{cases} c(x + 2y), & \text{when } 0 < x < y < 1, \\ 0, & \text{otherwise.} \end{cases}$$

Compute the regression functions $E(Y \mid X = x)$ and $E(X \mid Y = y)$.

15. Suppose that X and Y are random variables with a joint density

$$f(x,y) = \begin{cases} \frac{2}{5}(2x+3y), & \text{when } 0 < x, y < 1, \\ 0, & \text{otherwise.} \end{cases}$$

Compute the conditional expectations $E(Y \mid X = x)$ and $E(X \mid Y = y)$.

16. Let X and Y be random variables with a joint density

$$f(x,y) = \begin{cases} \frac{4}{5}(x+3y)e^{-x-2y}, & \text{when } x, y > 0, \\ 0, & \text{otherwise.} \end{cases}$$

Compute the regression functions $E(Y \mid X = x)$ and $E(X \mid Y = y)$.

17. Suppose that the joint density of X and Y is given by

$$f(x,y) = \begin{cases} xe^{-x-xy}, & \text{when } x > 0, \, y > 0, \\ 0, & \text{otherwise.} \end{cases}$$

Determine the regression functions $E(Y \mid X = x)$ and $E(X \mid Y = y)$.

18. Let the joint density function of X and Y be given by

$$f(x,y) = \begin{cases} c(x+y), & \text{for } 0 < x < y < 1, \\ 0, & \text{otherwise.} \end{cases}$$

Determine c, the marginal densities, EX, EY, and the conditional expectations $E(Y \mid X = x)$ and $E(X \mid Y = y)$.

19. Let the joint density of X and Y be given by

$$f_{X,Y}(x,y) = \begin{cases} c, & \text{for } 0 \le x \le 1, \quad x^2 \le y \le x, \\ 0, & \text{otherwise.} \end{cases}$$

Compute c, the marginal densities, and the conditional expectations $E(Y \mid X = x)$ and $E(X \mid Y = y)$.

20. Suppose that X and Y are random variables with joint density

$$f(x,y) = \begin{cases} cx, & \text{when } 0 < x < 1, \, x^3 < y < x^{1/3}, \\ 0, & \text{otherwise.} \end{cases}$$

Compute the conditional expectations $E(Y \mid X = x)$ and $E(X \mid Y = y)$.

21. Suppose that X and Y are random variables with joint density

$$f(x,y) = \begin{cases} cy, & \text{when } 0 < x < 1, \, x^4 < y < x^{1/4}, \\ 0, & \text{otherwise.} \end{cases}$$

Compute the conditional expectations $E(Y \mid X = x)$ and $E(X \mid Y = y)$.

22. Let the joint density function of X and Y be given by

$$f(x, y) = \begin{cases} c \cdot x^3 y, & \text{for} \quad x, y > 0, \quad x^2 + y^2 \leq 1, \\ 0, & \text{otherwise.} \end{cases}$$

Compute c, the marginal densities, and the conditional expectations $E(Y \mid X = x)$ and $E(X \mid Y = y)$.

23. The joint density function of X and Y is given by

$$f(x, y) = \begin{cases} c \cdot xy, & \text{for} \quad x, y > 0, \quad 4x^2 + y^2 \leq 1, \\ 0, & \text{otherwise.} \end{cases}$$

Compute c, the marginal densities, and the conditional expectations $E(Y \mid X = x)$ and $E(X \mid Y = y)$.

24. Let X and Y have joint density

$$f(x, y) = \begin{cases} \dfrac{c}{x^3 y}, & \text{when} \quad 1 < y < x, \\ 0, & \text{otherwise.} \end{cases}$$

Compute the conditional expectations $E(Y \mid X = x)$ and $E(X \mid Y = y)$.

25. Let X and Y have joint density

$$f(x, y) = \begin{cases} \dfrac{c}{x^4 y}, & \text{when} \quad 1 < y < x, \\ 0, & \text{otherwise.} \end{cases}$$

Compute the conditional expectations $E(Y \mid X = x)$ and $E(X \mid Y = y)$.

26. Suppose that X and Y are random variables with a joint density

$$f(x, y) = \begin{cases} \dfrac{c}{(1 + x - y)^2}, & \text{when} \quad 0 < y < x < 1, \\ 0, & \text{otherwise.} \end{cases}$$

Compute the conditional expectations $E(Y \mid X = x)$ and $E(X \mid Y = y)$.

27. Suppose that X and Y are random variables with a joint density

$$f(x, y) = \begin{cases} c \cdot \cos x, & \text{when} \quad 0 < y < x < \frac{\pi}{2}, \\ 0, & \text{otherwise.} \end{cases}$$

Compute the conditional expectations $E(Y \mid X = x)$ and $E(X \mid Y = y)$.

28. Let X and Y have joint density

$$f(x, y) = \begin{cases} c \log y, & \text{when} \quad 0 < y < x < 1, \\ 0, & \text{otherwise.} \end{cases}$$

Compute the conditional expectations $E(Y \mid X = x)$ and $E(X \mid Y = y)$.

29. The random vector $(X, Y)'$ has the following joint distribution:
$$P(X = m, Y = n) = \binom{m}{n} \frac{1}{2^m} \frac{m}{15},$$
where $m = 1, 2, \ldots, 5$ and $n = 0, 1, \ldots, m$. Compute $E(Y \mid X = m)$.

30. Show that a suitable power of a Weibull-distributed random variable whose parameter is gamma-distributed is Pareto-distributed. More precisely, show that if
$$X \mid A = a \in W(\tfrac{1}{a}, \tfrac{1}{b}) \quad \text{with} \quad A \in \Gamma(p, \theta),$$
then X^b has a (translated) Pareto distribution.

31. Show that an exponential random variable such that the inverse of the parameter is gamma-distributed is Pareto-distributed. More precisely, show that if
$$X \mid M = m \in \text{Exp}(m) \quad \text{with} \quad M^{-1} \in \Gamma(p, a),$$
then X has a (translated) Pareto distribution.

32. Let X and Y be random variables such that
$$Y \mid X = x \in \text{Exp}(1/x) \quad \text{with} \quad X \in \Gamma(2, 1).$$
(a) Show that Y has a translated Pareto distribution.
(b) Compute EY.
(c) Check the value in (b) by recomputing it via our favorite formula for conditional means.

33. Suppose that the random variable X is uniformly distributed symmetrically around zero, but in such a way that the parameter is uniform on $(0, 1)$; that is, suppose that
$$X \mid A = a \in U(-a, a) \quad \text{with} \quad A \in U(0, 1).$$
Find the distribution of X, EX, and $\text{Var } X$.

34. In Section 4 we studied the situation when a coin, such that $p = P(\text{head})$ is considered to be a $U(0, 1)$-distributed random variable, is tossed, and found (i.a.) that if $X_n = \#$ heads after n tosses, then X_n is uniformly distributed over the integers $0, 1, \ldots, n$.
Suppose instead that p is considered to be $\beta(2, 2)$-distributed. What then? More precisely, consider the following model:
$$X_n \mid Y = y \in \text{Bin}(n, y) \quad \text{with} \quad f_Y(y) = 6y(1 - y), \ 0 < y < 1.$$
(a) Compute $E X_n$ and $\text{Var } X_n$.
(b) Determine the distribution of X_n.

35. Let X and Y be jointly distributed random variables such that
$$Y \mid X = x \in \text{Bin}(n, x) \quad \text{with} \quad X \in U(0, 1).$$
Compute EY, $\text{Var } Y$, and $\text{Cov}(X, Y)$ (without using what is known from Section 4 about the distribution of Y).

36. Let X and Y be jointly distributed random variables such that

$$Y \mid X = x \in \mathrm{Fs}(x) \quad \text{with} \quad f_X(x) = 3x^2, \quad 0 \le x \le 1.$$

Compute EY, $\mathrm{Var}\, Y$, $\mathrm{Cov}\,(X, Y)$, and the distribution of Y.

37. Let X be the number of coin tosses until heads is obtained. Suppose that the probability of heads is unknown in the sense that we consider it to be a random variable $Y \in U(0, 1)$.
   (a) Find the distribution of X (cf. Problem 3.8.48).
   (b) The expected value of an Fs-distributed random variable exists, as is well known. What about $E\,X$?
   (c) Suppose that the value $X = n$ has been observed. Find the posterior distribution of Y, that is, the distribution of $Y \mid X = n$.

38. Let p be the probability that the tip points downward after a person throws a drawing pin once. Annika throws a drawing pin until it points downward for the first time. Let X be the number of throws for this to happen. She then throws the drawing pin another X times. Let Y be the number of times the drawing pin points downward in the latter series of throws. Find the distribution of Y (cf. Problem 3.8.31).

39. A point P is chosen uniformly in an n-dimensional sphere of radius 1. Next, a point Q is chosen uniformly within the concentric sphere, centered at the origin, going through P. Let X and Y be the distances of P and Q, respectively, to the common center. Find the joint density function of X and Y and the conditional expectations $E(Y \mid X = x)$ and $E(X \mid Y = y)$.
   Hint 1. Begin by trying the case $n = 2$.
   Hint 2. The volume of an n-dimensional sphere of radius r is equal to $c_n r^n$, where c_n is some constant (which is of no interest for the problem).
   Remark. For $n = 1$ we rediscover the stick from Example 2.1.

40. Let X and Y be independent random variables. The conditional distribution of Y given that $X = x$ then does not depend on x. Moreover, $E(Y \mid X = x)$ is independent of x; recall Theorem 2.2(b) and Remark 2.4. Now, suppose instead that $E(Y \mid X = x)$ is independent of x (i.e., that $E(Y \mid X) = EY$). We say that Y *has constant regression with respect to X*. However, it does not necessarily follow that X and Y are independent. Namely, let the joint density of X and Y be given by

$$f(x, y) = \begin{cases} \frac{1}{2}, & \text{for} \quad |x| + |y| \le 1, \\ 0, & \text{otherwise.} \end{cases}$$

Show that Y has constant regression with respect to X and/but that X and Y are not independent.

3

Transforms

1 Introduction

In Chapter 1 we learned how to handle transformations in order to find the distribution of new (constructed) random variables. Since the arithmetic mean or average of a set of (independent) random variables is a very important object in probability theory as well as in statistics, we focus in this chapter on sums of independent random variables (from which one easily finds corresponding results for the average). We know from earlier work that the convolution formula may be used but also that the sums or integrals involved may be difficult or even impossible to compute. In particular, this is the case if the number of summands is "large." In that case, however, the central limit theorem is (frequently) applicable. This result will be proved in the chapter on convergence; see Theorem 6.5.2.

Exercise 1.1. Let X_1, X_2, ... be independent $U(0,1)$-distributed random variables.

(a) Find the distribution of $X_1 + X_2$.
(b) Find the distribution of $X_1 + X_2 + X_3$.
(c) Show that the distribution of $S_n = X_1 + X_2 + \cdots + X_n$ is given by

$$F_{S_n}(x) = \frac{1}{n!} \sum_{k=0}^{n-1} (-1)^k \binom{n}{k} (x-k)_+^n, \quad 0 \le x \le n,$$

where $x_+ = \max\{x, 0\}$. □

Even if, in theory, we have solved this problem, we face new problems if we actually wish to compute $P(S_n \le x)$ for some given x already for moderately sized values of n; for example, what is $P(S_5 \le \pi)$?

In this chapter we shall learn how such problems can be transformed into new problems, how the new (simpler) problems are solved, and finally that these solutions can be retransformed or inverted to provide a solution to the original problems.

A. Gut, *An Intermediate course in Probabilty*, Springer Texts in Statistics,
DOI: 10.1007/978-1-4419-0162-0_3,
© Springer Science + Business Media, LLC 2009

Remark 1.1. In order to determine the distribution of sums of independent random variables we mentioned the convolution formula. From analysis we recall that the problem of convolving functions can be transformed to the problem of multiplying their Laplace transforms or Fourier transforms (which is a simpler task). □

We begin, however, with an example from a different area.

Example 1.1. Let $a_1, a_2, \ldots, a_n$ be positive reals. We want to know their product.

This is a "difficult" problem. We therefore find the logarithms of the numbers, add them to yield $\sum_{k=1}^{n} \log a_k$, and then invert. □

Figure 1.1 illustrates the procedure.

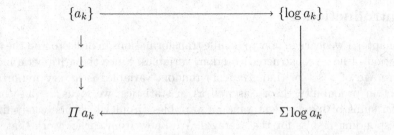

Figure 1.1

We obtained the correct answer since $\exp\{\sum_{k=1}^{n} \log a_k\} = \prod_{k=1}^{n} a_k$.
The central ideas of the solution thus are

(a) addition is easier to perform than multiplication;
(b) the logarithm has a unique inverse (i.e., if $\log x = \log y$, then $x = y$), namely, the exponential function.

As for sums of independent random variables, the topic of this chapter, we shall introduce three transforms: the (probability) generating function, the moment generating function, and the characteristic function. Two common features of these transforms are that

(a) summation of independent random variables (convolution) corresponds to multiplication of the transforms;
(b) the transformation is 1-to-1, namely, there is a uniqueness theorem to the effect that if two random variables have the same transform then they also have the same distribution.

Notation: The notation
$$X \overset{d}{=} Y$$
means that the random variables X and Y are equidistributed. □

Remark 1.2. It is worth pointing out that two random variables, X and Y, may well have the property $X \overset{d}{=} Y$ and yet $X(\omega) \neq Y(\omega)$ for *all* ω. A very simple example is the following: Toss a fair coin once and set

$$X = \begin{cases} 1, & \text{if the outcome is heads,} \\ 0, & \text{if the outcome is tails,} \end{cases}$$

and

$$Y = \begin{cases} 1, & \text{if the outcome is tails,} \\ 0, & \text{if the outcome is heads.} \end{cases}$$

Clearly, $X \in \text{Be}(1/2)$ and $Y \in \text{Be}(1/2)$, in particular, $X \overset{d}{=} Y$. But $X(\omega)$ and $Y(\omega)$ differ for every ω. $\qquad\square$

2 The Probability Generating Function

Definition 2.1. *Let X be a nonnegative, integer-valued random variable. The* (probability) generating function *of X is*

$$g_X(t) = E\,t^X = \sum_{n=0}^{\infty} t^n \cdot P(X = n).$$
$\qquad\square$

Remark 2.1. The generating function is defined at least for $|t| \leq 1$, since it is a power series with coefficients in $[0, 1]$. Note also that $g_X(1) = \sum_{n=0}^{\infty} P(X = n) = 1$. $\qquad\square$

Theorem 2.1. *Let X and Y be nonnegative, integer-valued random variables. If $g_X = g_Y$, then $p_X = p_Y$.* $\qquad\square$

The theorem states that if two nonnegative, integer-valued random variables have the same generating function then they follow the same probability law. It is thus the uniqueness theorem mentioned in the previous section. The result is a special case of the uniqueness theorem for power series. We refer to the literature cited in Appendix A for a complete proof.

Theorem 2.2. *Let $X_1, X_2, \ldots, X_n$ be independent, nonnegative, integer-valued random variables, and set $S_n = X_1 + X_2 + \cdots + X_n$. Then*

$$g_{S_n}(t) = \prod_{k=1}^{n} g_{X_k}(t).$$

Proof. Since $X_1, X_2, \ldots, X_n$ are independent, the same is true for $t^{X_1}, t^{X_2}, \ldots, t^{X_n}$, which yields

$$g_{S_n}(t) = E\,t^{X_1+X_2+\cdots+X_n} = E \prod_{k=1}^{n} t^{X_k} = \prod_{k=1}^{n} E\,t^{X_k} = \prod_{k=1}^{n} g_{X_k}(t). \qquad\square$$

This result asserts that adding independent, nonnegative, integer-valued random variables corresponds to multiplying their generating functions (recall Example 1.1(a)).

A case of particular importance is given next.

Corollary 2.2.1. *If, in addition, $X_1, X_2, \ldots, X_n$ are equidistributed, then*

$$g_{S_n}(t) = \big(g_X(t)\big)^n. \qquad \qquad \Box$$

Termwise differentiation of the generating function (this is permitted (at least) for $|t| < 1$) yields

$$g'_X(t) = \sum_{n=1}^{\infty} nt^{n-1}P(X = n), \qquad (2.1)$$

$$g''_X(t) = \sum_{n=2}^{\infty} n(n-1)t^{n-2}P(X = n), \qquad (2.2)$$

and, in general, for $k = 1, 2, \ldots,$

$$g_X^{(k)}(t) = \sum_{n=k}^{\infty} n(n-1)\cdots(n-k+1)t^{n-k}P(X = n). \qquad (2.3)$$

By putting $t = 0$ in (2.1)–(2.3), we obtain $g_X^{(n)}(0) = n! \cdot P(X = n)$, that is,

$$P(X = n) = \frac{g_X^{(n)}(0)}{n!}. \qquad (2.4)$$

The probability generating function thus generates the probabilities; hence the name of the transform.

By letting $t \nearrow 1$ in (2.1)–(2.3) (this requires a little more care), the following result is obtained.

Theorem 2.3. *Let X be a nonnegative, integer-valued random variable, and suppose that $E|X|^k < \infty$ for some $k = 1, 2, \ldots$. Then*

$$E\,X(X - 1)\cdots(X - k + 1) = g_X^{(k)}(1). \qquad \qquad \Box$$

Remark 2.2. Derivatives at $t = 1$ are throughout to be interpreted as limits as $t \nearrow 1$. For simplicity, however, we use the simpler notation $g'(1)$, $g''(1)$, and so on. $\qquad \Box$

The following example illustrates the relevance of this remark.

Example 2.1. Suppose that X has the probability function

$$p(k) = \frac{C}{k^2}, \quad k = 1, 2, 3, \ldots,$$

(where, to be precise, $C^{-1} = \sum_{k=1}^{\infty} 1/k^2 = \pi^2/6$). The divergence of the harmonic series tells us that the distribution does not have a finite mean.

Now, the generating function is

$$g(t) = \frac{6}{\pi^2} \sum_{k=1}^{\infty} \frac{t^k}{k^2}, \quad \text{for} \quad |t| \leq 1,$$

so that

$$g'(t) = \frac{6}{\pi^2} \sum_{k=1}^{\infty} \frac{t^{k-1}}{k} = -\frac{6}{\pi^2} \cdot \frac{\log(1-t)}{t} \nearrow +\infty \quad \text{as} \quad t \nearrow 1.$$

This shows that although the generating function itself exists for $t = 1$, the derivative only exists for all t *strictly smaller* than 1, but *not* for the boundary value $t = 1$. $\square$

For $k = 1$ and $k = 2$ we have, in particular, the following result:

Corollary 2.3.1 *Let X be as before. Then*

(a) $E|X| < \infty \implies E X = g'_X(1)$, *and*

(b) $E X^2 < \infty \implies \text{Var } X = g''_X(1) + g'_X(1) - (g'_X(1))^2$. $\square$

Exercise 2.1. Prove Corollary 2.3.1. $\square$

Next we consider some special distributions:

The Bernoulli distribution. Let $X \in \text{Be}(p)$. Then

$$g_X(t) = q \cdot t^0 + p \cdot t^1 = q + pt, \quad \text{for all} \quad t,$$
$$g'_X(t) = p, \quad \text{and} \quad g''_X(t) = 0,$$

which yields

$$E X = g'_X(1) = p$$

and

$$\text{Var } X = g''_X(1) + g'_X(1) - (g'_X(1))^2 = 0 + p - p^2 = p(1-p) = pq.$$

The binomial distribution. Let $X \in \text{Bin}(n, p)$. Then

$$g_X(t) = \sum_{k=0}^{n} t^k \binom{n}{k} p^k q^{n-k} = \sum_{k=0}^{n} \binom{n}{k} (pt)^k q^{n-k} = (q + pt)^n,$$

for all t. Furthermore,

$$g'_X(t) = n(q + pt)^{n-1} \cdot p \quad \text{and} \quad g''_X(t) = n(n-1)(q + pt)^{n-2} \cdot p^2,$$

which yields

$$E X = np \quad \text{and} \quad \text{Var } X = n(n-1)p^2 + np - (np)^2 = npq.$$

We further observe that

$$g_{\text{Bin}(n,p)}(t) = \big(g_{\text{Be}(p)}(t)\big)^n,$$

which, according to Corollary 2.2.1, shows that if $Y_1, Y_2, \ldots, Y_n$ are independent, Be(p)-distributed random variables, and $X_n = Y_1 + Y_2 + \cdots + Y_n$, then

$$g_{X_n}(t) = g_{\text{Bin}(n,p)}(t).$$

By Theorem 2.1 (uniqueness) it follows that $X_n \in \text{Bin}(n, p)$, a conclusion that, alternatively, could be proved by the convolution formula and induction.

Similarly, if $X_1 \in \text{Bin}(n_1, p)$ and $X_2 \in \text{Bin}(n_2, p)$ are independent, then, by Theorem 2.2,

$$g_{X_1+X_2}(t) = (q + pt)^{n_1+n_2} = g_{\text{Bin}(n_1+n_2,p)}(t),$$

which proves that $X_1 + X_2 \in \text{Bin}(n_1 + n_2, p)$ and hence establishes, in a simple manner, the addition theorem for the binomial distribution.

Remark 2.3. It is instructive to reprove the last results by actually using the convolution formula. We stress, however, that the simplicity of the method of generating functions is illusory, since it in fact exploits various results on generating functions and their derivatives. □

The geometric distribution. Let $X \in \text{Ge}(p)$. Then

$$g_X(t) = \sum_{k=0}^{\infty} t^k pq^k = p \sum_{k=0}^{\infty} (tq)^k = \frac{p}{1 - qt}, \quad |t| < \frac{1}{q}.$$

Moreover,

$$g'_X(t) = -\frac{p}{(1 - qt)^2} \cdot (-q) = \frac{pq}{(1 - qt)^2}$$

and

$$g''_X(t) = -\frac{2pq}{(1 - qt)^3} \cdot (-q) = \frac{2pq^2}{(1 - qt)^3},$$

from which it follows that $E X = q/p$ and $\text{Var } X = q/p^2$.

Exercise 2.2. Let $X_1, X_2, \ldots, X_n$ be independent Ge(p)-distributed random variables. Determine the distribution of $X_1 + X_2 + \cdots + X_n$. □

The Poisson distribution. Let $X \in \text{Po}(m)$. Then

$$g_X(t) = \sum_{k=0}^{\infty} t^k e^{-m} \frac{m^k}{k!} = e^{-m} \sum_{k=0}^{\infty} \frac{(mt)^k}{k!} = e^{m(t-1)}.$$

Exercise 2.3. (a) Let $X \in \text{Po}(m)$. Show that $E\,X = \text{Var}\,X = m$.
(b) Let $X_1 \in \text{Po}(m_1)$ and $X_2 \in \text{Po}(m_2)$ be independent random variables.
Show that $X_1 + X_2 \in \text{Po}(m_1 + m_2)$. □

3 The Moment Generating Function

In spite of their usefulness, probability generating functions are of limited use
in that they are only defined for nonnegative, integer-valued random variables.
Important distributions, such as the normal distribution and the exponential
distribution, cannot be handled with this transform. This inconvenience is
overcome as follows:

Definition 3.1. *The* moment generating function *of a random variable X is*

$$\psi_X(t) = E\,e^{tX},$$

*provided there exists $h > 0$, such that the expectation exists and is finite for
$|t| < h$.* □

Remark 3.1. As a first observation we mention the close connection between
moment generating functions and Laplace transforms of real-valued functions.
Indeed, for a nonnegative random variable X, one may define the Laplace
transform

$$E\,e^{-sX} \quad \text{for} \quad s \geq 0,$$

which thus always exist (why?). Analogously, one may view the moment generating function as a two-sided Laplace transform.

Remark 3.2. Note that for nonnegative, integer-valued random variables we
have $\psi(t) = g(e^t)$, for $|t| < h$, provided the moment generating function exists
(for $|t| < h$). □

The uniqueness and multiplication theorems are presented next. The
proofs are analogous to those for the generating function.

Theorem 3.1. *Let X and Y be random variables. If there exists $h > 0$, such
that $\psi_X(t) = \psi_Y(t)$ for $|t| < h$, then $X \stackrel{d}{=} Y$.* □

Theorem 3.2. *Let $X_1, X_2, \ldots, X_n$ be independent random variables whose
moment generating functions exist for $|t| < h$ for some $h > 0$, and set $S_n =
X_1 + X_2 + \cdots + X_n$. Then*

$$\psi_{S_n}(t) = \prod_{k=1}^{n} \psi_{X_k}(t), \quad |t| < h.$$ □

Corollary 3.2.1. *If, in addition,* X_1, X_2, ..., X_n *are equidistributed, then*

$$\psi_{S_n}(t) = \left(\psi_X(t)\right)^n, \quad |t| < h. \qquad \square$$

For the probability generating function we found that the derivatives at zero produced the probabilities (which motivated the name of the transform). The derivatives at 0 of the moment generating function produce the moments (hence the name of the transform).

Theorem 3.3. *Let X be a random variable whose moment generating function $\psi_X(t)$, exists for $|t| < h$ for some $h > 0$. Then*

(a) *all moments exist, that is,* $E\,|X|^r < \infty$ *for all* $r > 0$;
(b) $E\,X^n = \psi_X^{(n)}(0)$ *for* $n = 1, 2, \ldots$.

Proof. We prove the theorem in the continuous case, leaving the completely analogous proof in the discrete case as an exercise.

By assumption,

$$\int_{-\infty}^{\infty} e^{tx} f_X(x)\,dx < \infty \quad \text{for} \quad |t| < h.$$

Let t, $0 < t < h$, be given. The assumption implies that, for every $x_1 > 0$,

$$\int_{x_1}^{\infty} e^{tx} f_X(x)\,dx < \infty \quad \text{and} \quad \int_{-\infty}^{-x_1} e^{-tx} f_X(x)\,dx < \infty. \qquad (3.1)$$

Since $|x|^r / e^{|tx|} \to 0$ as $x \to \infty$ for all $r > 0$, we further have

$$|x|^r \le e^{|tx|} \quad \text{for} \quad |x| > x_2. \qquad (3.2)$$

Now, let $x_0 > x_2$. It follows from (3.1) and (3.2) that

$$\int_{-\infty}^{\infty} |x|^r f_X(x)\,dx$$

$$= \int_{-\infty}^{-x_0} |x|^r f_X(x)\,dx + \int_{-x_0}^{x_0} |x|^r f_X(x)\,dx + \int_{x_0}^{\infty} |x|^r f_X(x)\,dx$$

$$\le \int_{-\infty}^{-x_0} e^{-tx} f_X(x)\,dx + |x_0|^r \cdot P(|X| \le x_0) + \int_{x_0}^{\infty} e^{tx} f_X(x)\,dx < \infty.$$

This proves (a), from which (b) follows by differentiation:

$$\psi_X^{(n)}(t) = \int_{-\infty}^{\infty} x^n e^{tx} f_X(x)\,dx$$

and, hence,

$$\psi_X^{(n)}(0) = \int_{-\infty}^{\infty} x^n f_X(x)\,dx = E\,X^n. \qquad \square$$

Remark 3.3. The idea in part (a) is that the exponential function grows more rapidly than every polynomial. As a consequence, $|x|^r \leq e^{|tx|}$ as soon as $|x| > x_2$ (say). On the other hand, for $|x| < x_2$ we trivially have $|x|^r \leq Ce^{|tx|}$ for some constant C. It follows that for all x

$$|x|^r \leq (C+1)e^{|tx|},$$

and hence that

$$E\,|X|^r \leq (C+1)E\,e^{|tX|} < \infty \quad \text{for} \quad |t| < h.$$

Note that this, in fact, proves Theorem 3.2(a) in the continuous case as well as in the discrete case.

Remark 3.4. Taylor expansion of the exponential function yields

$$e^{tX} = 1 + \sum_{n=1}^{\infty} \frac{t^n X^n}{n!} \quad \text{for} \quad |t| < h.$$

By taking expectation termwise (this is permitted), we obtain

$$\psi_X(t) = E\,e^{tX} = 1 + \sum_{n=1}^{\infty} \frac{t^n}{n!} E\,X^n \quad \text{for} \quad |t| < h.$$

Termwise differentiation (which is also permitted) yields the result of part (b). A special feature with the series expansion is that if the moment generating function is given in that form we may simply read off the moments; $E\,X^n$ is the coefficient of $t^n/n!$, $n = 1, 2, \ldots$, in the series expansion. □

Let us now, as in the previous section, study some known distributions. First, some discrete ones:

The Bernoulli distribution. Let $X \in \text{Be}(p)$. Then $\psi_X(t) = q + pe^t$. Differentiation yields $E\,X = p$ and $\text{Var}\,X = pq$. Taylor expansion of e^t leads to

$$\psi_X(t) = q + p\sum_{n=0}^{\infty} \frac{t^n}{n!} = 1 + \sum_{n=1}^{\infty} \frac{t^n}{n!} \cdot p,$$

from which it follows that $E\,X^n = p$, $n = 1, 2, \ldots$. In particular, $E\,X = p$ and $\text{Var}\,X = p - p^2 = pq$.

The binomial distribution. Let $X \in \text{Bin}(n,p)$. Then

$$\psi_X(t) = \sum_{k=0}^{n} e^{tk} \binom{n}{k} p^k q^{n-k} = \sum_{k=0}^{n} \binom{n}{k} (pe^t)^k q^{n-k} = (q + pe^t)^n.$$

Differentiation yields $E\,X = np$ and $\text{Var}\,X = npq$.

Taylor expansion can also be performed in this case, but it is more cumbersome. If, however, we only wish to find EX and $\operatorname{Var}X$ it is not too hard:

$$\psi_X(t) = \left(q + pe^t\right)^n = \left(q + p\sum_{k=0}^{\infty}\frac{t^k}{k!}\right)^n = \left(1 + pt + p\frac{t^2}{2!} + \cdots\right)^n$$

$$= 1 + npt + \binom{n}{2}p^2t^2 + np\frac{t^2}{2} + \cdots$$

$$= 1 + npt + \left(n(n-1)p^2 + np\right)\frac{t^2}{2} + \cdots .$$

Here the ellipses mean that the following terms contain t raised to at least the third degree. By identifying the coefficients we find that $EX = np$ and that $EX^2 = n(n-1)p^2 + np$, which yields $\operatorname{Var}X = npq$.

Remark 3.5. Let us immediately point out that in this particular case this is not a very convenient procedure for determining EX and $\operatorname{Var}X$; the purpose was merely to illustrate the method.   □

Exercise 3.1. Prove, with the aid of moment generating functions, that if $Y_1, Y_2, \ldots, Y_n$ are independent $\mathrm{Be}(p)$-distributed random variables, then $Y_1 + Y_2 + \cdots + Y_n \in \mathrm{Bin}(n,p)$.

Exercise 3.2. Prove, similarly, that if $X_1 \in \mathrm{Bin}(n_1, p)$ and $X_2 \in \mathrm{Bin}(n_2, p)$ are independent, then $X_1 + X_2 \in \mathrm{Bin}(n_1 + n_2, p)$.   □

The geometric distribution. For $X \in \mathrm{Ge}(p)$ computations like those made for the generating function yield $\psi_X(t) = p/(1 - qe^t)$ (for $qe^t < 1$). Differentiation yields EX and $\operatorname{Var}X$.

The Poisson distribution. For $X \in \mathrm{Po}(m)$ we obtain $\psi_X(t) = e^{m(e^t - 1)}$ for all t, and so forth.

Next we compute the moment generating function for some continuous distributions.

The uniform (rectangular) distribution. Let $X \in U(a, b)$. Then

$$\psi_X(t) = \int_a^b e^{tx}\frac{1}{b-a}\,dx = \frac{1}{b-a}\left[\frac{1}{t}e^{tx}\right]_a^b = \frac{e^{tb} - e^{ta}}{t(b-a)}$$

for all t. In particular,

$$\psi_{U(0,1)}(t) = \frac{e^t - 1}{t} \quad \text{and} \quad \psi_{U(-1,1)}(t) = \frac{e^t - e^{-t}}{2t} = \frac{\sinh t}{t}.$$

The moments can be obtained by differentiation. If, instead, we use Taylor expansion, then

$$\psi_X(t) = \frac{1}{t(b-a)}\left[1 + \sum_{n=1}^{\infty}\frac{(tb)^n}{n!} - \left(1 + \sum_{n=1}^{\infty}\frac{(ta)^n}{n!}\right)\right]$$

$$= \frac{1}{t(b-a)}\sum_{n=1}^{\infty}\left(\frac{(tb)^n}{n!} - \frac{(ta)^n}{n!}\right) = \frac{1}{b-a}\sum_{n=1}^{\infty}\frac{b^n - a^n}{n!}t^{n-1}$$

$$= 1 + \sum_{n=1}^{\infty}\frac{b^{n+1} - a^{n+1}}{(b-a)(n+1)!}t^n = 1 + \sum_{n=1}^{\infty}\frac{b^{n+1} - a^{n+1}}{(b-a)(n+1)}\cdot\frac{t^n}{n!},$$

from which we conclude that

$$E\,X^n = \frac{b^{n+1} - a^{n+1}}{(b-a)(n+1)} \quad\text{for}\quad n = 1, 2, \ldots,$$

and thus, in particular, the known expressions for mean and variance, via

$$E\,X = \frac{b^2 - a^2}{2(b-a)} = \frac{a+b}{2},$$

$$E\,X^2 = \frac{b^3 - a^3}{3(b-a)} = \frac{b^2 + ab + a^2}{3},$$

$$\text{Var}\,X = \frac{b^2 + ab + a^2}{3} - \left(\frac{a+b}{2}\right)^2 = \frac{(b-a)^2}{12}.$$

The exponential distribution. Let $X \in \text{Exp}(a)$. Then

$$\psi_X(t) = \int_0^{\infty} e^{tx}\frac{1}{a}e^{-x/a}\,dx = \frac{1}{a}\int_0^{\infty}e^{-x(\frac{1}{a}-t)}\,dx$$

$$= \frac{1}{a}\cdot\frac{1}{\frac{1}{a}-t} = \frac{1}{1-at} \quad\text{for}\quad t < \frac{1}{a}.$$

Furthermore, $\psi_X'(t) = a/(1-at)^2$, $\psi_X''(t) = 2a^2/(1-at)^3$, and, in general, $\psi_X^{(n)}(t) = n!a^n/(1-at)^{n+1}$. It follows that $E\,X^n = n!a^n$, $n = 1, 2, \ldots$, and, in particular, that $E\,X = a$ and $\text{Var}\,X = a^2$.

Exercise 3.3. Perform a Taylor expansion of the moment generating function, and verify the expressions for the moments. $\quad\quad\square$

The gamma distribution. For $X \in \Gamma(p, a)$, we have

$$\psi_X(t) = \int_0^{\infty}e^{tx}\frac{1}{\Gamma(p)}x^{p-1}\frac{1}{a^p}e^{-x/a}\,dx$$

$$= \frac{1}{a^p}\cdot\frac{1}{(\frac{1}{a}-t)^p}\int_0^{\infty}\frac{1}{\Gamma(p)}x^{p-1}\left(\frac{1}{a}-t\right)^p e^{-x(\frac{1}{a}-t)}\,dx$$

$$= \frac{1}{a^p}\frac{1}{(\frac{1}{a}-t)^p}\cdot 1 = \frac{1}{(1-at)^p} \quad\text{for}\quad t < \frac{1}{a}.$$

As is standard by now, the moments may be obtained via differentiation. Note also that $\psi(t) = (\psi_{\text{Exp}(a)}(t))^p$. Thus, for $p = 1, 2, \ldots$, we conclude from Corollary 3.2.1 and Theorem 3.1 that if $Y_1, Y_2, \ldots, Y_p$ are independent, Exp(a)-distributed random variables then $Y_1 + Y_2 + \cdots + Y_p \in \Gamma(p, a)$.

Exercise 3.4. (a) Check the details of the last statement.
(b) Show that if $X_1 \in \Gamma(p_1, a)$ and $X_2 \in \Gamma(p_2, a)$ are independent random variables then $X_1 + X_2 \in \Gamma(p_1 + p_2, a)$.   □

The standard normal distribution. Suppose that $X \in N(0, 1)$. Then

$$\psi_X(t) = \int_{-\infty}^{\infty} e^{tx} \frac{1}{\sqrt{2\pi}} \exp\{-x^2/2\}\, dx$$

$$= e^{t^2/2} \int_{-\infty}^{\infty} \frac{1}{\sqrt{2\pi}} \exp\{-(x-t)^2/2\}\, dx = e^{t^2/2}, \quad -\infty < t < \infty.$$

The general normal (Gaussian) distribution. Suppose that $X \in N(\mu, \sigma^2)$. Then

$$\psi_X(t) = \int_{-\infty}^{\infty} e^{tx} \frac{1}{\sigma\sqrt{2\pi}} \exp\left\{-\frac{(x-\mu)^2}{2\sigma^2}\right\} dx$$

$$= e^{t\mu + \sigma^2 t^2/2} \int_{-\infty}^{\infty} \frac{1}{\sigma\sqrt{2\pi}} \exp\left\{-\frac{(x - \mu - \sigma^2 t)^2}{2\sigma^2}\right\} dx$$

$$= e^{t\mu + \sigma^2 t^2/2}, \quad -\infty < t < \infty.$$

The computations in the special case and the general case are essentially the same; it is a matter of completing squares. However, this is a bit more technical in the general case.

This leads to the following useful result, which shows how to derive the moment generating function of a linear transformation of a random variable.

Theorem 3.4. *Let X be a random variable and a and b be real numbers. Then*

$$\psi_{aX+b}(t) = e^{tb} \psi_X(at).$$

Proof. $\psi_{aX+b}(t) = E\, e^{t(aX+b)} = e^{tb} \cdot E\, e^{(at)X} = e^{tb} \cdot \psi_X(at).$ □

As an illustration we show how the moment generating function for a general normal distribution can be derived from the moment generating function of the standard normal one.

Thus, suppose that $X \in N(\mu, \sigma^2)$. We then know that $X \stackrel{d}{=} \sigma Y + \mu$, where $Y \in N(0, 1)$. An application of Theorem 3.4 thus tells us that

$$\psi_X(t) = e^{t\mu} \psi_Y(\sigma t) = e^{t\mu + \sigma^2 t^2/2},$$

as expected.

Exercise 3.5. (a) Show that if $X \in N(\mu, \sigma^2)$ then $E\,X = \mu$ and $\operatorname{Var} X = \sigma^2$.
(b) Let $X_1 \in N(\mu_1, \sigma_1^2)$ and $X_2 \in N(\mu_2, \sigma_2^2)$ be independent random variables. Show that $X_1 + X_2$ is normally distributed, and find the parameters.
(c) Let $X \in N(0, \sigma^2)$. Show that $E\,X^{2n+1} = 0$ for $n = 0, 1, 2, \ldots$, and that $E\,X^{2n} = [(2n)!/2^n n!] \cdot \sigma^{2n} = (2n-1)!!\sigma^{2n} = 1 \cdot 3 \cdots (2n-1)\sigma^{2n}$ for $n = 1, 2, \ldots$.

Exercise 3.6. (a) Show that if $X \in N(0,1)$ then $X^2 \in \chi^2(1)$ by computing the moment generating function of X^2, that is, by showing that

$$\psi_{X^2}(t) = E \exp\{tX^2\} = \frac{1}{\sqrt{1-2t}} \quad \text{for} \quad t < \frac{1}{2}.$$

(b) Show that if $X_1 \in N(0,1)$ and $X_2 \in N(0,1)$ are independent then $X_1^2 + X_2^2 \in \chi^2(2) \quad (= \operatorname{Exp}(2))$. $\square$

For two-dimensional analogs to Exercise 3.6, see Problems 5.10.36 and 37.

The Cauchy distribution. The moment generating function does not exist for the Cauchy distribution, since $\int [e^{tx}/(1+x^2)]\,dx$ is divergent for all $t \neq 0$. Note also that the nonexistence of the moment generating function follows from Theorem 3.3(a), since no moments of order 1 and above exist.

According to Theorem 3.3(a), it is conceivable that there might exist distributions with moments of all orders and, yet, the moment generating function does not exist in any neighborhood around zero. In fact, the *log-normal distribution* is one such example. To see this we first note that if $X \in LN(\mu, \sigma^2)$, then $X \overset{d}{=} e^Y$, where $Y \in N(\mu, \sigma^2)$, which implies that

$$f_X(x) = \begin{cases} \dfrac{1}{\sigma x \sqrt{2\pi}} \exp\{-\dfrac{(\log x - \mu)^2}{2\sigma^2}\}, & \text{for} \quad x > 0, \\ 0, & \text{otherwise.} \end{cases}$$

It follows that

$$E\,X^r = E\,e^{rY} = \psi_Y(r) = \exp\{r\mu + \tfrac{1}{2}\sigma^2 r^2\},$$

for any $r > 0$, that is, all moments exist.

However, since $e^x \geq x^n/n!$ for any n, it follows that, for any $t > 0$,

$$\begin{aligned} E \exp\{tX\} = E \exp\{te^Y\} &\geq E \frac{(te^Y)^n}{n!} = \frac{t^n}{n!} E\,e^{nY} \\ &= \frac{t^n}{n!}\psi_Y(n) = \frac{t^n}{n!}\exp\{n\mu + \tfrac{1}{2}\sigma^2 n^2\} \\ &= \frac{1}{n!}\exp\{n(\log t + \mu + \tfrac{1}{2}\sigma^2 n)\}, \end{aligned}$$

which can be made arbitrarily large by choosing n sufficiently large, since $\log t + \mu + \tfrac{1}{2}\sigma^2 n \geq \tfrac{1}{4}\sigma^2 n$ for any fixed $t > 0$ as $n \to \infty$ and $\exp\{cn^2\}/n! \to \infty$

as $n \to \infty$ for any positive constant c. The moment generating function thus does not exist for any positive t.

Another class of distributions that possesses moments of all orders but not a moment generating function is the class of *generalized gamma distributions* whose densities are

$$f(x) = Cx^{\beta-1}e^{-x^{\alpha}}, \quad x > 0,$$

where $\beta > -1$, $0 < \alpha < 1$, and C is a normalizing constant (that is chosen such that the total mass equals 1).

It is clear that all moments exist, but, since $\alpha < 1$, we have

$$\int_{-\infty}^{\infty} e^{tx}x^{\beta-1}e^{-x^{\alpha}}\,dx = +\infty$$

for *all* $t > 0$, so that the moment generating function does not exist.

Remark 3.6. The fact that the integral is finite for all $t < 0$ is no contradiction, since for a moment generating function to exist we require finiteness of the integral in a neighborhood of zero, that is, for $|t| < h$ for some $h > 0$.   □

We close this section by defining the moment generating function for random vectors.

Definition 3.2. *Let* $\mathbf{X} = (X_1, X_2, \ldots, X_n)'$ *be a random vector. The moment generating function of* $\mathbf{X}$ *is*

$$\psi_{X_1,\ldots,X_n}(t_1,\ldots,t_n) = E\,e^{t_1X_1+\cdots+t_nX_n},$$

provided there exist $h_1, h_2, \ldots, h_n > 0$ *such that the expectation exists for* $|t_k| < h_k$, $k = 1, 2, \ldots, n$.   □

Remark 3.7. In vector notation (where, thus, $\mathbf{X}, \mathbf{t}$, and $\mathbf{h}$ are column vectors) the definition may be rewritten in the more compact form

$$\psi_{\mathbf{X}}(\mathbf{t}) = E\,e^{\mathbf{t}'\mathbf{X}},$$

provided there exists $\mathbf{h} > \mathbf{0}$, such that the expectation exists for $|\mathbf{t}| < \mathbf{h}$ (the inequalities being interpreted componentwise).   □

4 The Characteristic Function

So far we have introduced two transforms: the generating function and the moment generating function. The advantage of moment generating functions over generating functions is that they can be defined for all kinds of random variables. However, the moment generating function does not exist for all distributions; the Cauchy and the log-normal distributions are two such examples. In this section we introduce a third transform, the characteristic function, which exists for *all* distributions. A minor technical complication, however, is that this transform is complex-valued and therefore requires somewhat more sophisticated mathematics in order to be dealt with stringently.

Definition 4.1. *The* characteristic function *of a* random variable X *is*

$$\varphi_X(t) = E\, e^{itX} = E(\cos tX + i \sin tX).\qquad\qquad \square$$

As mentioned above, the characteristic function is complex-valued. Since

$$|E\, e^{itX}| \leq E\, |e^{itX}| = E\,1 = 1, \qquad\qquad (4.1)$$

it follows that the characteristic function exists for all t and for all random variables.

Remark 4.1. Apart from a minus sign in the exponent (and, possibly, a factor $\sqrt{1/2\pi}$), characteristic functions coincide with Fourier transforms in the continuous case and with Fourier series in the discrete case. $\square$

We begin with some basic facts and properties.

Theorem 4.1. *Let X be a random variable. Then*

(a) $|\varphi_X(t)| \leq \varphi_X(0) = 1;$
(b) $\overline{\varphi_X(t)} = \varphi_X(-t);$
(c) $\varphi_X(t)$ *is (uniformly) continuous.*

Proof. (a) $\varphi_X(0) = E\, e^{i\cdot 0 \cdot X} = 1$. This, together with (4.1), proves (a).
(b) We have

$$\overline{\varphi_X(t)} = E(\cos tX - i \sin tX) = E(\cos(-t)X + i\sin(-t)X)$$
$$= E\, e^{i(-t)X} = \varphi_X(-t).$$

(c) Let t be arbitrary and $h > 0$ (a similar argument works for $h < 0$). Then

$$|\varphi_X(t+h) - \varphi_X(t)| = |E\, e^{i(t+h)X} - E\, e^{itX}|$$
$$= |E\, e^{itX}(e^{ihX} - 1)| \leq E|e^{itX}(e^{ihX} - 1)|$$
$$= E\,|e^{ihX} - 1|. \qquad\qquad (4.2)$$

Now, suppose that X has a continuous distribution; the discrete case is treated analogously.

For the function e^{ix} we have the trivial estimate $|e^{ix} - 1| \leq 2$, but also the more delicate one $|e^{ix} - 1| \leq |x|$. With the aid of these estimates we obtain, for $A > 0$,

$$E|e^{ihX} - 1| = \int_{-\infty}^{-A} |e^{ihx} - 1| f_X(x)\, dx + \int_{-A}^{A} |e^{ihx} - 1| f_X(x)\, dx$$
$$+ \int_{A}^{\infty} |e^{ihx} - 1| f_X(x)\, dx$$
$$\leq \int_{-\infty}^{-A} 2 f_X(x)\, dx + \int_{-A}^{A} |hx| f_X(x)\, dx + \int_{A}^{\infty} 2 f_X(x)\, dx$$
$$\leq 2P(|X| \geq A) + hAP(|X| \leq A)$$
$$\leq 2P(|X| \geq A) + hA. \qquad\qquad (4.3)$$

Let $\varepsilon > 0$ be arbitrarily small. It follows from (4.2) and (4.3) that

$$|\varphi_X(t+h) - \varphi_X(t)| \leq 2P(|X| \geq A) + hA < \varepsilon, \qquad (4.4)$$

provided we first choose A so large that $2P(|X| \geq A) < \varepsilon/2$, and then h so small that $hA < \varepsilon/2$. This proves the continuity of φ_X. Since the estimate in (4.4) does not depend on t, we have, in fact, shown that φ_X is uniformly continuous. $\qquad \square$

Theorem 4.2. *Let X and Y be random variables. If $\varphi_X = \varphi_Y$, then $X \stackrel{d}{=} Y$.* $\square$

This is the uniqueness theorem for characteristic functions. Next we present, without proof, some inversion theorems.

Theorem 4.3. *Let X be a random variable with distribution function F and characteristic function φ. If F is continuous at a and b, then*

$$F(b) - F(a) = \lim_{T \to \infty} \frac{1}{2\pi} \int_{-T}^{T} \frac{e^{-itb} - e^{-ita}}{-it} \cdot \varphi(t)\,dt. \qquad \square$$

Remark 4.2. Observe that Theorem 4.2 is an immediate corollary of Theorem 4.3. This is due to the fact that the former theorem is an existence result (only), whereas the latter provides a formula for explicitly computing the distribution function in terms of the characteristic function. $\qquad \square$

Theorem 4.4. *If, in addition, $\int_{-\infty}^{\infty} |\varphi(t)|\,dt < \infty$, then X has a continuous distribution with density*

$$f(x) = \frac{1}{2\pi} \int_{-\infty}^{\infty} e^{-itx} \cdot \varphi(t)\,dt. \qquad \square$$

Theorem 4.5. *If the distribution of X is discrete, then*

$$P(X = x) = \lim_{T \to \infty} \frac{1}{2T} \int_{-T}^{T} e^{-itx} \cdot \varphi(t)\,dt. \qquad \square$$

As for the name of the transform, we have just seen that every random variable possesses a unique characteristic function; the characteristic function characterizes the distribution uniquely.

The proof of the following result, the multiplication theorem for characteristic functions, is similar to those for the other transforms and is therefore omitted.

Theorem 4.6. *Let X_1, X_2, $\ldots$, X_n be independent random variables, and set $S_n = X_1 + X_2 + \cdots + X_n$. Then*

$$\varphi_{S_n}(t) = \prod_{k=1}^{n} \varphi_{X_k}(t). \qquad \square$$

Corollary 4.6.1. *If, in addition, X_1, X_2, ..., X_n are equidistributed, then*

$$\varphi_{S_n}(t) = \left(\varphi_X(t)\right)^n.$$
□

Since we have derived the transform of several known distributions in the two previous sections, we leave some of them as exercises in this section.

Exercise 4.1. Show that $\varphi_{\mathrm{Be}(p)}(t) = q + pe^{it}$, $\varphi_{\mathrm{Bin}(n,p)}(t) = (q + pe^{it})^n$, $\varphi_{\mathrm{Ge}(p)}(t) = p/(1 - qe^{it})$, and $\varphi_{\mathrm{Po}(m)}(t) = \exp\{m(e^{it} - 1)\}$. □

Note that for the computation of these characteristic functions one seems to perform the same work as for the computation of the corresponding moment generating function, the only difference being that t is replaced by it. In fact, in the discrete cases we considered in the previous sections, the computations are really completely analogous. The binomial theorem, convergence of geometric series, and Taylor expansion of the exponential function hold unchanged in the complex case.

The situation is somewhat more complicated for continuous distributions.

The uniform (rectangular) distribution. Let $X \in U(a,b)$. Then

$$
\begin{aligned}
\varphi_X(t) &= \int_a^b e^{itx} \frac{1}{b-a}\, dx = \frac{1}{b-a} \int_a^b (\cos tx + i \sin tx)\, dx \\
&= \frac{1}{b-a} \cdot \left[\frac{1}{t} \sin tx - i\frac{1}{t} \cos tx\right]_a^b \\
&= \frac{1}{b-a} \cdot \frac{1}{t}(\sin bt - \sin at - i\cos bt + i\cos at) \\
&= \frac{1}{it(b-a)}(i\sin bt - i\sin at + \cos bt - \cos at) \\
&= \frac{e^{itb} - e^{ita}}{it(b-a)} \quad (= \psi_X(it)).
\end{aligned}
$$

In particular,

$$
\varphi_{U(0,1)}(t) = \frac{e^{it}-1}{it} \quad \text{and} \quad \varphi_{U(-1,1)}(t) = \frac{e^{it}-e^{-it}}{2it} = \frac{\sin t}{t}. \tag{4.5}
$$

The (mathematical) complication is that we cannot integrate as easily as we could before. However, in this case we observe that the derivative of e^{ix} equals ie^{ix}, which justifies the integration and hence implies that the computations here are "the same" as for the moment generating function.

For the exponential and gamma distributions, the complication arises in the following manner:

The exponential distribution. Let $X \in \mathrm{Exp}(a)$. Then

$$\varphi_X(t) = \int_0^\infty e^{itx} \frac{1}{a} e^{-x/a}\, dx = \frac{1}{a} \int_0^\infty e^{-x(\frac{1}{a} - it)}\, dx$$

$$= \frac{1}{a} \cdot \frac{1}{\frac{1}{a} - it} = \frac{1}{1 - ait}.$$

The gamma distribution. Let $X \in \Gamma(p, a)$. We are faced with the same problems as for the exponential distribution. The conclusion is that $\varphi_{\Gamma(p,a)}(t) = (1 - ait)^{-p}$.

The standard normal (Gaussian) distribution. Let $X \in N(0, 1)$. Then

$$\varphi_X(t) = \int_{-\infty}^\infty e^{itx} \frac{1}{\sqrt{2\pi}} e^{-\frac{1}{2}x^2}\, dx$$

$$= e^{-t^2/2} \int_{-\infty}^\infty \frac{1}{\sqrt{2\pi}} e^{-\frac{1}{2}(x - it)^2}\, dx = e^{-t^2/2}.$$

In this case one cannot argue as before, since there is no primitive function. Instead we observe that the moment generating function can be extended into a function that is analytic in the complex plane. The characteristic function equals the thus extended function along the imaginary axis, from which we conclude that $\varphi_X(t) = \psi_X(it)\ (= e^{(it)^2/2} = e^{-t^2/2})$.

It is now possible to prove the addition theorems for the various distributions just as for generating functions and moment generating functions.

Exercise 4.2. Prove the addition theorems for the binomial, Poisson, and gamma distributions.   □

In Remark 3.4 we gave a series expansion of the moment generating function. Following is the counterpart for characteristic functions:

Theorem 4.7. *Let X be a random variable. If $E|X|^n < \infty$ for some $n = 1, 2, \ldots$, then*

(a) $\varphi_X^{(k)}(0) = i^k \cdot E X^k$ *for* $k = 1, 2, \ldots, n$;
(b) $\varphi_X(t) = 1 + \sum_{k=1}^n E X^k \cdot (it)^k/k! + o(|t|^n)$ *as* $t \to 0$.   □

Remark 4.3. For $n = 2$ we obtain, in particular,

$$\varphi_X(t) = 1 + itE X - \frac{t^2}{2} E X^2 + o(t^2) \quad \text{as} \quad t \to 0.$$

If, moreover, $E X = 0$ and $\text{Var } X = \sigma^2$, then

$$\varphi_X(t) = 1 - \tfrac{1}{2} t^2 \sigma^2 + o(t^2) \quad \text{as} \quad t \to 0.$$ □

Exercise 4.3. Find the mean and variance of the binomial, Poisson, uniform, exponential, and standard normal distributions.   □

The conclusion of Theorem 4.7 is rather natural in view of Theorem 3.3 and Remark 3.4. Note, however, that a random variable whose moment generating function exists has moments of all orders (Theorem 3.3(a)), which implies that the series expansion can be carried out as an infinite sum. Since, however, all random variables (in particular, those without (higher order) moments) possess a characteristic function, it is reasonable to expect that the expansion here can only be carried out as long as moments exist. The order of magnitude of the remainder follows from estimating the difference of e^{ix} and the first part of its (complex) Taylor expansion.

Furthermore, a comparison between Theorems 3.3(b) and 4.7(a) tempts one to guess that these results could be derived from one another; once again the relation $\varphi_X(t) = \psi_X(it)$ seems plausible. This relation is, however, not true in general—recall that there are random variables, such as the Cauchy distribution, for which the moment generating function does not exist. In short, the validity of the relation depends on to what extent (if at all) the function $E\,e^{izX}$, where z is *complex-valued*, is an analytic function of z, a problem that will not be considered here (recall, however, the earlier arguments for the standard normal distribution).

Theorem 4.7 states that if the moment of a given order exists, then the characteristic function is differentiable, and the moments up to that order can be computed via the derivatives of the characteristic function as stated in the theorem. A natural question is whether a converse holds. The answer is yes, but only for moments of *even* order.

Theorem 4.8. *Let X be a random variable. If, for some $n = 0, 1, 2, \ldots$, the characteristic function φ has a finite derivative of order $2n$ at $t = 0$, then $E|X|^{2n} < \infty$ (and the conclusions of Theorem 4.7 hold).*

The "problem" with the converse is that if we want to apply Theorem 4.8 to show that the mean is finite we must first show that the second derivative of the characteristic function exists. Since there exist distributions with finite mean whose characteristic functions are not twice differentiable (such as the so-called stable distributions with index between 1 and 2), the theorem is not always applicable.

Next we present the analog of Theorem 3.4 on how to find the transform of a linearly transformed random variable.

Theorem 4.9. *Let X be a random variable and a and b be real numbers. Then*

$$\varphi_{aX+b}(t) = e^{ibt} \cdot \varphi_X(at).$$

Proof. $\varphi_{aX+b}(t) = E\,e^{it(aX+b)} = e^{itb} \cdot E\,e^{i(at)X} = e^{itb} \cdot \varphi_X(at).$ □

Exercise 4.4. Let $X \in N(\mu, \sigma^2)$. Use the expression above for the characteristic function of the standard normal distribution and Theorem 4.9 to show that $\varphi_X(t) = e^{it\mu - \sigma^2 t^2/2}$.

Exercise 4.5. Prove the addition theorem for the normal distribution.   □

The Cauchy distribution. For $X \in C(0,1)$, one can show that

$$\varphi_X(t) = \int_{-\infty}^{\infty} e^{itx} \cdot \frac{1}{\pi} \frac{1}{1+x^2} \, dx = e^{-|t|}.$$

A device for doing this is the following: If we "already happen to know" that the difference between two independent, Exp(1)-distributed random variables is $L(1)$-distributed, then we know that

$$\varphi_{L(1)}(t) = \frac{1}{1-it} \cdot \frac{1}{1+it} = \frac{1}{1+t^2}$$

(use Theorem 4.6 and Theorem 4.9 (with $a = -1$ and $b = 0$)). We thus have

$$\frac{1}{1+t^2} = \int_{-\infty}^{\infty} e^{itx} \tfrac{1}{2} e^{-|x|} \, dx.$$

A change of variables, such that $x \to t$ and $t \to x$, yields

$$\frac{1}{1+x^2} = \int_{-\infty}^{\infty} e^{itx} \tfrac{1}{2} e^{-|t|} \, dt,$$

and, by symmetry,

$$\frac{1}{1+x^2} = \int_{-\infty}^{\infty} e^{-itx} \tfrac{1}{2} e^{-|t|} \, dt,$$

which can be rewritten as

$$\frac{1}{\pi} \cdot \frac{1}{1+x^2} = \frac{1}{2\pi} \int_{-\infty}^{\infty} e^{-itx} e^{-|t|} \, dt. \tag{4.6}$$

A comparison with the inversion formula given in Theorem 4.4 shows that since the left-hand side of (4.6) is the density of the $C(0,1)$-distribution, it necessarily follows that $e^{-|t|}$ is the characteristic function of this distribution.

Exercise 4.6. Use Theorem 4.9 to show that $\varphi_{C(m,a)}(t) = e^{itm} \varphi_X(at) = e^{itm - a|t|}$.   □

Our final result in this section is a consequence of Theorems 4.9 and 4.1(b).

Theorem 4.10. *Let X be a random variable. Then*

$$\varphi_X \text{ is real} \iff X \overset{d}{=} -X$$

(i.e., iff the distribution of X is symmetric).

Proof. Theorem 4.9 (with $a = -1$ and $b = 0$) and Theorem 4.1(b) together yield

$$\varphi_{-X}(t) = \varphi_X(-t) = \overline{\varphi_X(t)}. \tag{4.7}$$

First suppose that φ_X is real-valued, that is, that $\overline{\varphi_X(t)} = \varphi_X(t)$. It follows that $\varphi_{-X}(t) = \varphi_X(t)$, or that X and $-X$ have the same characteristic function. By the uniqueness theorem they are equidistributed.

Now suppose that $X \overset{d}{=} -X$. Then $\varphi_X(t) = \varphi_{-X}(t)$, which, together with (4.7), yields $\varphi_X(t) = \overline{\varphi_X(t)}$, that is, φ_X is real-valued.   □

Exercise 4.7. Show that if X and Y are i.i.d. random variables then $X - Y$ has a symmetric distribution.

Exercise 4.8. Show that one cannot find i.i.d. random variables X and Y such that $X - Y \in U(-1, 1)$.   □

We conclude by defining the characteristic function for random vectors.

Definition 4.2. *Let* $\mathbf{X} = (X_1, X_2 \ldots, X_n)'$ *be a random vector. The characteristic function of* $\mathbf{X}$ *is*

$$\varphi_{X_1,\ldots,X_n}(t_1, \ldots, t_n) = E\, e^{i(t_1 X_1 + \cdots + t_n X_n)}.$$

In the more compact vector notation (cf. Remark 3.7) this may be rewritten as

$$\varphi_{\mathbf{X}}(\mathbf{t}) = E\, e^{i t' \mathbf{X}}.$$ □

In particular, the following special formulas, which are useful at times, can be obtained:

$$\varphi_{X_1,\ldots,X_n}(t, t, \ldots, t) = E\, e^{it(X_1 + \cdots + X_n)} = \varphi_{X_1 + \cdots + X_n}(t)$$

and

$$\varphi_{X_1,\ldots,X_n}(t, 0, \ldots, 0) = \varphi_{X_1}(t).$$

Characteristic functions of random vectors are an important tool in the treatment of the multivariate normal distribution in Chapter 5.

5 Distributions with Random Parameters

This topic was treated in Section 2.3 by conditioning methods. Here we show how Examples 2.3.1 and 2.3.2 (in the reverse order) can be tackled with the aid of transforms. Let us begin by saying that transforms are often easier to work with computationally than the conditioning methods. However, one reason for this is that behind the transform approach there are theorems that sometimes are rather sophisticated.

Example 2.3.2 (continued). Recall that the point of departure was

$$X \mid N = n \in \text{Bin}(n, p) \quad \text{with} \quad N \in \text{Po}(\lambda). \tag{5.1}$$

An application of Theorem 2.2.1 yields

$$g_X(t) = E\big(E(t^X \mid N)\big) = E\,h(N),$$

where

$$h(n) = E(t^X \mid N = n) = (q + pt)^n,$$

from which it follows that

$$g_X(t) = E(q + pt)^N = g_N(q + pt) = e^{\lambda((q+pt)-1)} = e^{\lambda p(t-1)},$$

that is, $X \in \text{Po}(\lambda p)$ (why?). Note also that $g_N(q + pt) = g_N(g_{\text{Be}(p)}(t))$.

Example 2.3.1 (continued). We had

$$X \mid M = m \in \text{Po}(m) \quad \text{with} \quad M \in \text{Exp}(1).$$

By using the moment generating function (for a change) and Theorem 2.2.1, we obtain

$$\psi_X(t) = E\,e^{tX} = E\big(E(e^{tX} \mid M)\big) = E\,h(M),$$

where

$$h(m) = E(e^{tX} \mid M = m) = \psi_{X \mid M=m}(t) = e^{m(e^t - 1)}.$$

Thus,

$$\psi_X(t) = E\,e^{M(e^t - 1)} = \psi_M(e^t - 1) = \frac{1}{1 - (e^t - 1)}$$

$$= \frac{1}{2 - e^t} = \frac{\frac{1}{2}}{1 - \frac{1}{2}e^t} = \psi_{\text{Ge}(1/2)}(t),$$

and we conclude that $X \in \text{Ge}(1/2)$.   □

Remark 5.1. It may be somewhat faster to use generating functions, but it is useful to practise another transform.   □

Exercise 5.1. Solve Exercise 2.3.1 using transforms.   □

In Section 2.3 we also considered the situation

$$X \mid \Sigma^2 = y \in N(0, y) \quad \text{with} \quad \Sigma^2 \in \text{Exp}(1),$$

which is the normal distribution with mean zero and an exponentially distributed variance. After hard work we found that $X \in L(1/\sqrt{2})$. The alternative, using characteristic functions and Theorem 2.2.1, yields

$$\varphi_X(t) = E\,e^{itX} = E\big(E(e^{itX} \mid \Sigma^2)\big) = E\,h(\Sigma^2),$$

where

$$h(y) = \varphi_{X\mid\Sigma^2=y}(t) = e^{-t^2 y/2},$$

and so

$$\varphi_X(t) = E\,e^{-t^2\Sigma^2/2} = \psi_{\Sigma^2}\Big(-\tfrac{t^2}{2}\Big)$$
$$= \frac{1}{1 - \big(-\tfrac{t^2}{2}\big)} = \frac{1}{1 + (\tfrac{1}{\sqrt{2}})^2 t^2} = \varphi_{L(1/\sqrt{2})}(t),$$

and the desired conclusion follows. At this point, however, let us stress once again that the price of the simpler computations here are some general theorems (Theorem 2.2.1 and the uniqueness theorem for characteristic functions), the proofs of which are all the more intricate.

Exercise 5.2. Solve Exercise 2.3.3 using transforms.   □

6 Sums of a Random Number of Random Variables

An important generalization of the theory of sums of independent random variables is the theory of sums of *a random number of* (independent) random variables. Apart from being a theory in its own right, it has several interesting and important applications. In this section we study this problem under the additional assumption that *the number of terms in the sum is independent of the summands*; in the following section we present an important application to branching processes (the interested reader might pause here for a moment and read the first few paragraphs of that section).

Before proceeding, however, here are some examples that will be solved after some theory has been presented.

Example 6.1. Consider a roulette wheel with the numbers $0, 1, \ldots, 36$. Charlie bets one dollar on number 13 until it appears. He then bets one dollar the same number of times on number 36. We wish to determine his expected loss in the second round (in which he bets on number 36).

Example 6.2. Let $X_1, X_2, \ldots$ be independent, Exp(1)-distributed random variables, and let $N \in \mathrm{Fs}(p)$ be independent of $X_1, X_2, \ldots$. We wish to find the distribution of $X_1 + X_2 + \cdots + X_N$.

In Section 5 we presented a solution of Example 2.3.2 based on transforms. Next we present another solution based on transforms where, instead, we consider the random variable in focus as a sum of a random number of Be(p)-distributed random variables.

Example 2.3.2 (continued). As before, let N be the number of emitted particles during a given hour. We introduce the following indicator random variables:

$$Y_k = \begin{cases} 1, & \text{if the } k\text{th particle is registered,} \\ 0, & \text{otherwise.} \end{cases}$$

Then

$$X = Y_1 + Y_2 + \cdots + Y_N$$

equals the number of registered particles during this particular hour.   □

Thus, the general idea is that we are given a set X_1, X_2, ... of i.i.d. random variables with partial sums $S_n = X_1 + X_2 + \cdots + X_n$, for $n \geq 1$. Furthermore, N is a nonnegative, integer-valued random variable that is independent of X_1, X_2, ... Our aim is to investigate the random variable

$$S_N = X_1 + X_2 + \cdots + X_N, \tag{6.1}$$

where $S_N = S_0 = 0$ when $N = 0$.

For $A \subset (-\infty, \infty)$, we have

$$P(S_N \in A \mid N = n) = P(S_n \in A \mid N = n) = P(S_n \in A), \tag{6.2}$$

where the last equality is due to the independence of N and X_1, X_2, ... The interpretation of (6.2) is that the distribution of S_N, given $N = n$, is the same as that of S_n.

Remark 6.1. Let $N = \min\{n : S_n > 0\}$. Clearly, $P(S_N > 0) = 1$. This implies that if the summands are allowed to assume negative values (with positive probability) then so will S_n, whereas S_N is always positive. However, in this case N is not independent of the summands; on the contrary, N is defined in terms of the summands.   □

In case the summands are nonnegative and integer-valued, the generating function of S_N can be derived as follows:

Theorem 6.1. *Let X_1, X_2, ... be i.i.d. nonnegative, integer-valued random variables, and let N be a nonnegative, integer-valued random variable, independent of X_1, X_2, ... Set $S_0 = 0$ and $S_n = X_1 + X_2 + \cdots + X_n$, for $n \geq 1$. Then*

$$g_{S_N}(t) = g_N(g_X(t)). \tag{6.3}$$

Proof. We have

$$g_{S_N}(t) = E t^{S_N} = \sum_{n=0}^{\infty} E(t^{S_N} \mid N = n) \cdot P(N = n)$$

$$= \sum_{n=0}^{\infty} E(t^{S_n} \mid N = n) \cdot P(N = n) = \sum_{n=0}^{\infty} E(t^{S_n}) \cdot P(N = n)$$

$$= \sum_{n=0}^{\infty} (g_X(t))^n \cdot P(N = n) = g_N(g_X(t)).   □$$

Remark 6.2. In the notation of Chapter 2 and with the aid of Theorem 2.2.1, we may alternatively write

$$g_{S_N}(t) = E\, t^{S_N} = E\left(E\left(t^{S_N} \mid N\right)\right) = E\, h(N),$$

where

$$h(n) = E\left(t^{S_N} \mid N = n\right) = \cdots = \left(g_X(t)\right)^n,$$

which yields

$$g_{S_N}(t) = E\left(g_X(t)\right)^N = g_N\left(g_X(t)\right). \qquad \square$$

Theorem 6.2. *Suppose that the conditions of Theorem 6.1 are satisfied.*

(a) *If, moreover,*

$$E\, N < \infty \quad and \quad E\, |X| < \infty,$$

then

$$E\, S_N = E\, N \cdot E\, X.$$

(b) *If, in addition,*

$$\mathrm{Var}\, N < \infty \quad and \quad \mathrm{Var}\, X < \infty,$$

then

$$\mathrm{Var}\, S_N = E\, N \cdot \mathrm{Var}\, X + (E\, X)^2 \cdot \mathrm{Var}\, N.$$

Proof. It follows from Corollary 2.3.1 that

$$E\, S_N = g_{S_N}'(1) \tag{6.4}$$

and that

$$\mathrm{Var}\, S_N = g_{S_N}''(1) + g_{S_N}'(1) - \left(g_{S_N}'(1)\right)^2. \tag{6.5}$$

Furthermore, by differentiating the right-hand side of (6.3), using the chain rule, we obtain

$$g_{S_N}'(t) = g_N'\left(g_X(t)\right) \cdot g_X'(t),$$

which, after letting $t \nearrow 1$, yields

$$E\, S_N = g_{S_N}'(1) = g_N'(1) \cdot g_X'(1) = E\, N \cdot E\, X.$$

This proves (a).

A further differentiation shows that

$$g_{S_N}''(t) = g_N''\left(g_X(t)\right) \cdot \left(g_X'(t)\right)^2 + g_N'\left(g_X(t)\right) \cdot g_X''(t),$$

which yields

$$\begin{aligned}
g_{S_N}''(1) &= g_N''(1) \cdot \left(g_X'(1)\right)^2 + g_N'(1) \cdot g_X''(1) \\
&= E\, N(N-1) \cdot (E\, X)^2 + E\, N \cdot E\, X(X-1).
\end{aligned}$$

It finally follows that

$$\operatorname{Var} S_N = g''_{S_N}(1) + g'_{S_N}(1) - \left(g'_{S_N}(1)\right)^2$$
$$= E\,N(N-1)\cdot(E\,X)^2 + E\,N\cdot E\,X(X-1)$$
$$+ E\,N\cdot E\,X - (E\,N\cdot E\,X)^2$$
$$= E\,N\cdot\operatorname{Var} X + (E\,X)^2\cdot\operatorname{Var} N. \qquad\square$$

Theorem 6.2 can also be proved directly by modifying the proof of Theorem 6.1 in the obvious manner. As for (a) we then have

$$E\,S_N = \sum_{n=0}^{\infty} E\left(S_N \mid N = n\right)\cdot P(N = n)$$
$$= \sum_{n=0}^{\infty} E\left(S_n \mid N = n\right)\cdot P(N = n)$$
$$= \sum_{n=0}^{\infty} E\left(S_n\right)\cdot P(N = n) = \sum_{n=0}^{\infty} nE\,X\cdot P(N = n)$$
$$= E\,X\cdot\sum_{n=0}^{\infty} nP(N = n) = E\,X\cdot E\,N.$$

Note in particular that this proof is valid for arbitrary $X_1, X_2, \ldots$ (some argument concerning the absolute convergence is needed).

Exercise 6.1. Compute $E\,S_N^2$ similarly and prove Theorem 6.2(b). $\qquad\square$

In the notation of Chapter 2 we have, for Theorem 6.2(a) (cf. Remark 6.2),

$$E\,S_N = E\big(E(S_N \mid N)\big) = E\,h(N),$$

where

$$h(n) = E(S_N \mid N = n) = E(S_n \mid N = n) = E\,S_n = nE\,X,$$

that is,

$$E\,S_N = E(N\,E\,X) = E\,X\cdot E\,N.$$

For an alternative proof of Theorem 6.2(b), we use Corollary 2.2.3.1, according to which

$$\operatorname{Var} S_N = E\operatorname{Var}\left(S_N \mid N\right) + \operatorname{Var}\left(E(S_N \mid N)\right).$$

Since (check!)

$$\operatorname{Var}(S_N \mid N = n) = \operatorname{Var}(S_n \mid N = n) = \operatorname{Var} S_n = n\operatorname{Var} X,$$

it follows that

$$E\operatorname{Var}(S_N \mid N) = E(N\operatorname{Var} X) = E\,N\cdot\operatorname{Var} X.$$

Furthermore, $E(S_N \mid N = n) = nE\,X$, which yields

$$\mathrm{Var}\big(E(S_N \mid N)\big) = \mathrm{Var}(N \cdot E\,X) = (E\,X)^2 \cdot \mathrm{Var}\,N,$$

and the desired conclusion follows.

Let us now use these results in order to obtain another solution of Example 2.3.2 and to solve the problem posed in Example 6.1.

Example 2.3.2 (continued). Recall that N was the number of emitted particles during a given hour, that we kept track of whether particles were registered or not by the indicator variables $Y_1, Y_2, \ldots$, and that the number of registered particles during this particular hour was given by $X = Y_1 + Y_2 + \cdots + Y_N$.

An application of Theorem 6.1 now yields

$$g_X(t) = g_N\big(g_Y(t)\big) = \exp\{\lambda(g_Y(t) - 1)\} = e^{\lambda(q+pt-1)} = e^{\lambda p(t-1)},$$

which is the generating function of a $\mathrm{Po}(\lambda p)$-distribution. It follows from the uniqueness theorem for generating functions that $X \in \mathrm{Po}(\lambda p)$.

Moreover, by Theorem 6.2,

$$E\,X = E\,N \cdot E\,Y = \lambda \cdot p,$$
$$\mathrm{Var}\,X = E\,N \cdot \mathrm{Var}\,Y + (E\,Y)^2 \mathrm{Var}\,N = \lambda \cdot pq + p^2 \cdot \lambda = \lambda p. \qquad \square$$

Remark 6.3. The answers here and in Section 5 are obviously the same, but they are obtained somewhat differently. Analogous arguments can be made in other examples. This provides a link between the two sections. $\qquad \square$

As for Example 6.1, let $N \in \mathrm{Fs}(1/37)$ equal the number of bets on number 13, and let $Y_1, Y_2, \ldots$ be the losses in the bets on number 36. Thus

$$Y_k = \begin{cases} 1, & \text{if number 36 does not appear,} \\ -35, & (\text{i.e., } -36+1) \quad \text{otherwise,} \end{cases}$$

and $Y_1, Y_2, \ldots$ are independent with $P(Y_k = 1) = 36/37$ and $P(Y_k = -35) = 1/37$ (note that a negative loss is a gain). With this notation Charlie's total loss in the second round equals $X = Y_1 + Y_2 + \cdots + Y_N$, and an application of Theorem 6.2(a) yields

$$E\,X = E\,N \cdot E\,Y = 37 \cdot \left(1 \cdot \frac{36}{37} - 35 \cdot \frac{1}{37}\right) = 1.$$

If we wish to determine his overall loss, we have to add $(N-1) \cdot 1 - 35$ (or $N \cdot 1 - 36$) to X, in which case we find that the expected overall loss equals 2.

Although this does not seem so terrible, we must remember that this game requires access to an infinite amount of money to start with.

Exercise 6.2. Find the generating function of his loss in the second round. Try also to find it for his overall loss. $\qquad \square$

If, as in Example 6.2, the summands have a continuous distribution, then Theorem 6.1 no longer applies, since the generating function is not defined for such random variables. However, the following result holds.

Theorem 6.3. *Let X_1, X_2, ... be i.i.d. random variables, whose moment generating function exists for $|t| < h$ for some $h > 0$. Furthermore, let N be a nonnegative, integer-valued random variable independent of X_1, X_2, ... Set $S_0 = 0$ and $S_n = X_1 + X_2 + \cdots + X_n$, for $n \geq 1$. Then*

$$\psi_{S_N}(t) = g_N(\psi_X(t)).\qquad\square$$

The proof is completely analogous to the proof of Theorem 6.1 and is therefore left as an exercise.

Exercise 6.3. Prove Theorem 6.2 by starting from Theorem 6.3. Note, however, that this requires the existence of the moment generating function of the summands, a restriction that we know from above is not necessary for Theorem 6.2 to hold. $\qquad\square$

Next we solve the problem posed in Example 6.2. Recall from there that we were given X_1, X_2, ... independent, Exp(1)-distributed random variables and $N \in \mathrm{Fs}(p)$ independent of X_1, X_2, ... and that we wish to find the distribution of $X_1 + X_2 + \cdots + X_N$.

With the (by now) usual notation we have, by Theorem 6.3, for $t < p$,

$$\psi_{S_N}(t) = g_N(\psi_X(t)) = \frac{p \cdot \frac{1}{1-t}}{1 - q\frac{1}{1-t}} = \frac{p}{1 - t - q} =$$

$$= \frac{p}{p - t} = \frac{1}{1 - \frac{t}{p}} = \psi_{\mathrm{Exp}(1/p)}(t),$$

which, by the uniqueness theorem for moment generating functions, shows that $S_N \in \mathrm{Exp}(1/p)$. $\qquad\square$

Remark 6.4. If in Example 6.2 we had assumed that $N \in \mathrm{Ge}(p)$, we would have obtained

$$\psi_{S_N}(t) = \frac{p}{1 - q\frac{1}{1-t}} = \frac{p(1 - t)}{p - t} = p + q\frac{1}{1 - \frac{t}{p}}.$$

This means that S_N is a mixture of a $\delta(0)$-distribution and an Exp(1/p)-distribution, the weights being p and q, respectively. An intuitive argument supporting this is that $P(S_N = 0) = P(N = 0) = p$. If $N \geq 1$, then S_N behaves as in Example 6.2. The distribution of S_N thus is neither discrete nor continuous; it is a mixture. Note also that a geometric random variable that is known to be positive is, in fact, Fs-distributed; if $Z \in \mathrm{Ge}(p)$, then $Z \mid Z > 0 \in \mathrm{Fs}(p)$. $\qquad\square$

Finally, if the summands do not possess a moment generating function, then characteristic functions can be used in the obvious way.

Theorem 6.4. *Let $X_1, X_2, \ldots$ be i.i.d. random variables, and let N be a nonnegative, integer-valued random variable independent of $X_1, X_2, \ldots$. Set $S_0 = 0$ and $S_n = X_1 + X_2 + \cdots + X_n$, for $n \geq 1$. Then*

$$\varphi_{S_N}(t) = g_N\big(\varphi_X(t)\big). \qquad \qquad \square$$

Exercise 6.4. Prove Theorem 6.4.

Exercise 6.5. Use Theorem 6.4 to prove Theorem 6.2. $\square$

7 Branching Processes

An important application for the results of the previous section is provided by the theory of branching processes, which is described by the following model:

At time $t = 0$ there exists an initial population (a group of ancestors or founding members) $X(0)$. During its lifespan, every individual gives birth to a random number of children. During their lifespans, these children give birth to a random number of children, and so on. The reproduction rules for the simplest case, which is the only one we shall consider, are

(a) all individuals give birth according to the same probability law, independently of each other;
(b) the number of children produced by an individual is independent of the number of individuals in their generation.

Such branching processes are called *Galton–Watson processes* after Sir Francis Galton (1822–1911)—a cousin of Charles Darwin—who studied the decay of English peerage and other family names of distinction (he contested the hypothesis that distinguished family names are more likely to become extinct than names of ordinary families) and Rev. Henry William Watson (1827–1903). They met via problem 4001 posed by Galton in the *Educational Times*, 1 April 1873, for which Watson proposed a solution in the same journal, 1 August 1873. Another of Galton's achievements was that he established the use of fingerprints in the police force.

In the sequel we also assume that $X(0) = 1$; this is a common assumption, made in order to simplify some nonsignificant matters. Furthermore, since individuals give birth, we attribute the female sex to them. Finally, to avoid certain trivialities, we exclude, throughout, the degenerate case—when each individual always gives birth to exactly one child.

Example 7.1. Family names. Assume that men and women who live together actually marry and that the woman changes her last name to that of her husband (as in the old days). A family name thus survives only through sons. If sons are born according to the rules above, the evolution of a family name may be described by a branching process. In particular, one might be interested in whether or not a family name will live on forever or become extinct.

Instead of family names, one might consider some mutant gene and its survival or otherwise.

Example 7.2. Nuclear reactions. The fission caused by colliding neutrons results in a (random) number of new neutrons, which, when they collide produce new neutrons, and so on.

Example 7.3. Waiting lines. A customer who arrives at an empty server (or a telephone call that arrives at a switchboard) may be viewed as an ancestor. The customers (or calls) arriving while he is being served are his children, and so on. The process continues as long as there are people waiting to be served.

Example 7.4. The laptometer. When the sprows burst in a laptometer we are faced with failures of the first kind. Now, every sprow that bursts causes failures of the second kind (independently of the number of failures of the first kind and of the other sprows). Suppose the number of failures of the first kind during one hour follows the $Po(\lambda)$-distribution and that the number of failures of the second kind caused by one sprow follows the $Bin(n, p)$-distribution. Find the mean and variance of the total number of failures during one hour.   □

We shall solve the problem posed in Example 7.4 later.

Now, let, for $n \geq 1$,

$$X(n) = \# \text{ individuals in generation } n,$$

let Y and $\{Y_k, k \geq 1\}$ be generic random variables denoting the number of children obtained by individuals, and set $p_k = P(Y = k)$, $k = 0, 1, 2, \ldots$. Recall that we exclude the case $P(Y = 1) = 1$.

Consider the initial population or the ancestor $X(0)$ $(= 1 = \text{Eve})$. Then $X(1)$ equals the number of children of the ancestor and $X(1) \stackrel{d}{=} Y$. Next, let $Y_1, Y_2, \ldots$ be the number of children obtained by the first, second, ... child. It follows from the assumptions that $Y_1, Y_2, \ldots$ are i.i.d. and, furthermore, independent of $X(1)$. Since

$$X(2) = Y_1 + \cdots + Y_{X(1)}, \tag{7.1}$$

we may apply the results from Section 6. An application of Theorem 6.1 yields

$$g_{X(2)}(t) = g_{X(1)}\big(g_{Y_1}(t)\big). \tag{7.2}$$

If we introduce the notations

$$g_n(t) = g_{X(n)}(t) \quad \text{for} \quad n = 1, 2, \ldots$$

and $g(t) = g_1(t) \ (= g_{X(1)}(t) = g_Y(t))$, (7.2) may be rewritten as

$$g_2(t) = g\big(g(t)\big). \tag{7.3}$$

Next, let $Y_1, Y_2, \ldots$ be the number of children obtained by the first, second, $\ldots$ individuals in generation $n - 1$. By arguing as before, we obtain

$$g_{X(n)}(t) = g_{X(n-1)}\big(g_{Y_1}(t)\big),$$

that is,

$$g_n(t) = g_{n-1}\big(g(t)\big). \tag{7.4}$$

This corresponds to the case $k = 1$ in the following result.

Theorem 7.1. *For a branching process as above we have*

$$g_n(t) = g_{n-k}\big(g_k(t)\big) \quad \text{for} \quad k = 1, 2, \ldots, n-1. \qquad \square$$

If, in addition, $E Y_1 < \infty$, it follows from Theorem 6.2(a) that

$$E X(2) = E X(1) \cdot E Y_1 = (E Y_1)^2,$$

which, after iteration, yields

$$E X(n) = (E Y_1)^n. \tag{7.5}$$

Since every individual is expected to produce $E Y_1$ children, this is, intuitively, a very reasonable relation.

An analogous, although slightly more complicated, formula for the variance can also be obtained.

Theorem 7.2. (a) *Suppose that* $m = E Y_1 < \infty$. *Then*

$$E X(n) = m^n.$$

(b) *Suppose, further, that* $\sigma^2 = \operatorname{Var} Y_1 < \infty$. *Then*

$$\operatorname{Var} X(n) = \sigma^2(m^{n-1} + m^n + \cdots + m^{2n-2}). \qquad \square$$

Exercise 7.1. Prove Theorems 7.1 and 7.2(b). $\qquad \square$

Remark 7.1. Theorem 7.2 may, of course, also be derived from Theorem 7.1 by differentiation (cf. Corollary 2.3.1). $\qquad \square$

Asymptotics

Suppose that $\sigma^2 = \operatorname{Var} Y_1 < \infty$. It follows from Theorem 7.2 that

$$E\,X(n) \to \begin{cases} 0, & \text{when } m < 1, \\ (=)1, & \text{when } m = 1, \\ +\infty, & \text{when } m > 1, \end{cases} \qquad (7.6)$$

and that

$$\operatorname{Var} X(n) \to \begin{cases} 0, & \text{when } m < 1, \\ +\infty, & \text{when } m \geq 1 \end{cases} \qquad (7.7)$$

as $n \to \infty$. It is easy to show that $P(X(n) > 0) \to 0$ as $n \to \infty$ when $m < 1$. Although we have not defined any concept of convergence yet (this will be done in Chapter 6), our intuition tells us that $X(n)$ should converge to zero as $n \to \infty$ in some sense in this case. Furthermore, (7.6) tells us that $X(n)$ increases indefinitely (on average) when $m > 1$. In this case, however, one might imagine that since the variance also grows the population may, by chance, die out at some finite time (in particular, at some early point in time). For the boundary case $m = 1$, it may be a little harder to guess what will happen in the long run. The following result puts our speculations into a stringent formulation.

Denote by η the probability of *(ultimate) extinction* of a branching process. For future reference we note that

$$\eta = P(\text{ultimate extinction}) = P(X(n) = 0 \text{ for some } n)$$
$$= P(\bigcup_{n=1}^{\infty} \{X(n) = 0\}). \qquad (7.8)$$

For obvious reasons we assume in the following that $P(X(1) = 0) > 0$.

Theorem 7.3. (a) η *satisfies the equation* $t = g(t)$.
(b) η *is the smallest nonnegative root of the equation* $t = g(t)$.
(c) $\eta = 1$ *for* $m \leq 1$ *and* $\eta < 1$ *for* $m > 1$.

Proof. (a) Let $A_k = \{\text{the founding member produces } k \text{ children}\}$, $k \geq 0$. By the law of total probability we have

$$\eta = \sum_{k=0}^{\infty} P(\text{extinction} \mid A_k) \cdot P(A_k). \qquad (7.9)$$

Now, $P(A_k) = p_k$, and by the independence assumptions we have

$$P(\text{extinction} \mid A_k) = \eta^k. \qquad (7.10)$$

These facts and (7.9) yield

$$\eta = \sum_{k=0}^{\infty} \eta^k p_k = g(\eta), \qquad (7.11)$$

which proves (a).

(b) Set $\eta_n = P(X(n) = 0)$ and suppose that η^* is *some* nonnegative root of the equation $t = g(t)$ (since $g(1) = 1$, such a root exists always). Since g is nondecreasing for $t \geq 0$, we have, by Theorem 7.1,

$$\eta_1 = g(0) \leq g(\eta^*) = \eta^*,$$
$$\eta_2 = g(\eta_1) \leq g(\eta^*) = \eta^*,$$

and, by induction,

$$\eta_{n+1} = g(\eta_n) \leq g(\eta^*) = \eta^*,$$

that is, $\eta_n \leq \eta^*$ for all n. Finally, in view of (7.8) and the fact that $\{X(n) = 0\} \subset \{X(n+1) = 0\}$ for all n, it follows that $\eta_n \nearrow \eta$ and hence that $\eta \leq \eta^*$, which was to be proved.

(c) Since g is an infinite series with nonnegative coefficients, it follows that $g'(t) \geq 0$ and $g''(t) \geq 0$ for $0 \leq t \leq 1$. This implies that g is convex and nondecreasing on $[0,1]$. Furthermore, $g(1) = 1$. By comparing the graphs of the functions $y = g(t)$ and $y = t$ in the three cases $m < 1$, $m = 1$, and $m > 1$, respectively, it follows that they intersect at $t = 1$ only when $m \leq 1$ (tangentially when $m = 1$) and at $t = \eta$ and $t = 1$ when $m > 1$ (see Figure 7.1).

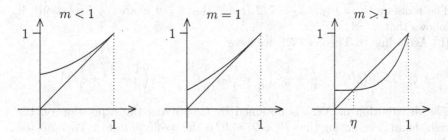

Figure 7.1

The proof of the theorem is complete.   □

We close this section with some computations to illustrate the theory. Given first is an example related to Example 7.2 as well as to a biological phenomenon called binary splitting.

Example 7.5. In this branching process, the neutrons or cells either split into two new "individuals" during their lifetime or die. Suppose that the probabilities for these alternatives are p and $q = 1 - p$, respectively.

Since $m = 0 \cdot q + 2 \cdot p = 2p$, it follows that the population becomes extinct with probability 1 when $p \leq 1/2$. For $p > 1/2$ we use Theorem 7.3. The equation $t = g(t)$ then becomes

$$t = q + p \cdot t^2,$$

the solutions of which are $t_1 = 1$ and $t_2 = q/p < 1$. Thus $\eta = q/p$ in this case.

Example 7.6. A branching process starts with one individual, who reproduces according to the following principle:

$$
\begin{array}{cccc}
\text{\# children} & 0 & 1 & 2 \\
\text{probability} & \frac{1}{6} & \frac{1}{2} & \frac{1}{3}
\end{array}
$$

The children reproduce according to the same rule, independently of each other, and so on.

(a) What is the probability of extinction?
(b) Determine the distribution of the number of grandchildren.

Solution. (a) We wish to apply Theorem 7.3. Since

$$
m = \frac{1}{6} \cdot 0 + \frac{1}{2} \cdot 1 + \frac{1}{3} \cdot 2 = \frac{7}{6} > 1,
$$

we solve the equation $t = g(t)$, that is,

$$
t = \frac{1}{6} + \frac{1}{2}t + \frac{1}{3}t^2.
$$

The roots are $t_1 = 1$ and $t_2 = 1/2$ (recall that $t = 1$ is always a solution). It follows that $\eta = 1/2$.
(b) According to Theorem 7.1, we have

$$
g_2(t) = g\big(g(t)\big) = \frac{1}{6} + \frac{1}{2} \cdot \left(\frac{1}{6} + \frac{1}{2}t + \frac{1}{3}t^2 \right) + \frac{1}{3} \cdot \left(\frac{1}{6} + \frac{1}{2}t + \frac{1}{3}t^2 \right)^2.
$$

The distribution of $X(2)$ is obtained by simplifying the expression on the right-hand side, noting that $P(X(2) = k)$ is the coefficient of t^k. We omit the details. □

Remark 7.2. The distribution may, of course, also be found by combinatorial methods (try it and check that the results are the same!). □

Finally, let us solve the problems posed in Example 7.4.

Regard failures of the first kind as children and failures of the second kind as grandchildren. Thus, $X(1) \in \text{Po}(\lambda)$ and $X(2) = Y_1 + Y_2 + \cdots + Y_{X(1)}$, where $Y_1, Y_2, \ldots \in \text{Bin}(n, p)$ are independent and independent of $X(1)$. We wish to find the expected value and the variance of $X(1) + X(2)$. Note, however, a discrepancy from the usual model in that the failures of the second kind do not have the same distribution as $X(1)$.
Since $E\,X(1) = \lambda$ and $E\,X(2) = E\,X(1) \cdot E\,Y_1 = \lambda np$, we obtain

$$
E\big(X(1) + X(2)\big) = \lambda + \lambda np.
$$

The computation of the variance is a little more tricky, since $X(1)$ and $X(2)$ are not independent. But

$$X(1) + X(2) = X(1) + Y_1 + \cdots + Y_{X(1)}$$
$$= (1 + Y_1) + (1 + Y_2) + \cdots + (1 + Y_{X(1)})$$
$$= \sum_{k=1}^{X(1)} (1 + Y_k),$$

and so

$$E\big(X(1) + X(2)\big) = E\,X(1)E(1 + Y_1) = \lambda(1 + np)$$

(as above) and

$$\mathrm{Var}\big(X(1) + X(2)\big) = E\,X(1)\mathrm{Var}(1 + Y_1) + \big(E(1 + Y_1)\big)^2 \mathrm{Var}X(1)$$
$$= \lambda npq + (1 + np)^2 \lambda = \lambda\big(npq + (1 + np)^2\big).$$

The same device can be used to find the generating function. Namely,

$$g_{X(1)+X(2)}(t) = g_{X(1)}\big(g_{1+Y_1}(t)\big),$$

which, together with the fact that

$$g_{1+Y_1}(t) = E\,t^{1+Y_1} = tE\,t^{Y_1} = tg_{Y_1}(t) = t(q + pt)^n,$$

yields

$$g_{X(1)+X(2)}(t) = e^{\lambda(t(q+pt)^n - 1)}. \qquad \square$$

8 Problems

1. The nonnegative, integer-valued, random variable X has generating function $g_X(t) = \log\big(1/(1 - qt)\big)$. Determine $P(X = k)$ for $k = 0, 1, 2, \ldots$, $E\,X$, and $\mathrm{Var}\,X$.
2. The random variable X has the property that all moments are equal, that is, $E\,X^n = c$ for all $n \geq 1$, for some constant c. Find the distribution of X (no proof of uniqueness is required).
3. The random variable X has the property that

$$E\,X^n = \frac{2^n}{n+1}, \quad n = 1, 2, \ldots.$$

Find some (in fact, the unique) distribution of X having these moments.
4. Suppose that Y is a random variable such that

$$E\,Y^k = \frac{1}{4} + 2^{k-1}, \quad k = 1, 2, \ldots.$$

Determine the distribution of Y.

5. Let $Y \in \beta(n,m)$ $(n, m$ integers).
   (a) Compute the moment generating function of $-\log Y$.
   (b) Show that $-\log Y$ has the same distribution as $\sum_{k=1}^{m} X_k$, where $X_1, X_2 \cdots$ are independent, exponentially distributed random variables.

   Remark. The formula $\Gamma(r+s)/\Gamma(r) = (r+s-1)\cdots(r+1)r$, which holds when s is an integer, might be useful.

6. Show, by using moment generating functions, that if $X \in L(1)$, then $X \stackrel{d}{=} Y_1 - Y_2$, where Y_1 and Y_2 are independent, exponentially distributed random variables.

7. In the previous problem we found that a standard Laplace-distributed random variable has the same distribution as the difference between two standard exponential random variables. It is therefore reasonable to believe that if Y_1 and Y_2 are independent $L(1)$-distributed, then

$$Y_1 + Y_2 \stackrel{d}{=} X_1 - X_2,$$

   where X_1 and X_2 are independent $\Gamma(2,1)$-distributed random variables. Prove, by checking moment generating functions, that this is in fact true.

8. Let $X \in \Gamma(p,a)$. Compute the (two-dimensional) moment generating function of $(X, \log X)$.

9. Let $X \in \mathrm{Bin}(n,p)$. Compute $E\, X^4$ with the aid of the moment generating function.

10. Let $X_1, X_2, \ldots, X_n$ be independent random variables with expectation 0 and finite third moments. Show, with the aid of characteristic functions, that
$$E(X_1 + X_2 + \cdots + X_n)^3 = E\,X_1^3 + E\,X_2^3 + \cdots + E\,X_n^3.$$

11. Let X and Y be independent random variables and suppose that Y is symmetric (around zero). Show that XY is symmetric.

12. The aim of the problem is to prove the double-angle formula

$$\sin 2t = 2 \sin t \cos t.$$

   Let X and Y be independent random variables, where $X \in U(-1,1)$ and Y assumes the values $+1$ and -1 with probabilities $1/2$.
   (a) Show that $Z = X + Y \in U(-2,2)$ by finding the distribution function of Z.
   (b) Translate this fact into a statement about the corresponding characteristic functions.
   (c) Rearrange.

13. Let $X_1, X_2 \ldots$ be independent $C(0,1)$-distributed random variables, and set $S_n = \sum_{k=1}^{n} X_k$, $n \geq 1$. Show that
   (a) $S_n/n \in C(0,1)$,
   (b) $(1/n)\sum_{k=1}^{n} S_k/k \in C(0,1)$.

   Remark. If $\{S_k/k,\ k \geq 1\}$ were independent, then (b) would follow immediately from (a).

14. For a positive, (absolutely) continuous random variable X we define the Laplace transform as

$$L_X(s) = E\,e^{-sX} = \int_0^\infty e^{-sx} f_X(x)\,dx, \quad s > 0.$$

Suppose that X is positive and stable with index $\alpha \in (0,1)$, which means that

$$L_X(s) = e^{-s^\alpha}, \quad s > 0.$$

Further, let $Y \in \text{Exp}(1)$ be independent of X. Show that

$$\left(\frac{Y}{X}\right)^\alpha \in \text{Exp}(1) \qquad \text{(which means that} \quad \left(\frac{Y}{X}\right)^\alpha \stackrel{d}{=} Y\text{)}.$$

15. Another transform: For a random variable X we define the *cumulant generating function*, $K_X(t) = \log \psi_X(t)$ as

$$K_X(t) = \sum_{n=1}^\infty \frac{1}{n!} k_n t^n,$$

where $k_n = k_n(X)$ is the so called nth *cumulant* or *semi-invariant* of X.
(a) Show that, if X and Y are independent random variables, then

$$k_n(X + Y) = k_n(X) + k_n(Y).$$

(b) Express k_1, k_2, and k_3 in terms of the moments $E\,X^k$, $k = 1,2,3$, of X.
16. Suppose that $X_1, X_2, \ldots$ are independent, identically Linnik(α)-distributed random variables, that $N \in \text{Fs}(p)$, and that N and $X_1, X_2, \ldots$ are independent. Show that $p^{1/\alpha}(X_1 + X_2 + \cdots + X_N)$ is, again, Linnik(α)-distributed.
Remark. The characteristic function of the Linnik(α)-distribution ($\alpha > 0$) is $\varphi(t) = (1 + |t|^\alpha)^{-1}$.
17. Suppose that the joint generating function of X and Y equals

$$g_{X,Y}(s,t) = E\,s^X t^Y = \exp\{\alpha(s-1) + \beta(t-1) + \gamma(st-1)\},$$

with $\alpha > 0$, $\beta > 0$, $\gamma \neq 0$.
(a) Show that X and Y both have a Poisson distribution, but that $X + Y$ does not.
(b) Are X and Y independent?
18. Let the random variables $Y, X_1, X_2, \ldots$ be independent, suppose that $Y \in \text{Fs}(p)$, where $0 < p < 1$, and suppose that $X_1, X_2, X_3, \ldots$ are all $\text{Exp}(1/a)$-distributed. Find the distribution of

$$Z = \sum_{j=1}^Y X_j.$$

19. Let $X_1, X_2, \ldots$ be $Ge(\alpha)$-distributed random variables, let $N \in Fs(p)$, suppose that all random variables are independent, and set

$$Y = X_1 + X_2 + \cdots + X_N.$$

(a) Show that $Y \in Ge(\beta)$, and determine β.

(b) Compute EY and $\operatorname{Var} Y$ with "the usual formulas", and check that the results agree with mean and variance of the distribution in (a).

20. Let $0 < p = 1 - q < 1$. Suppose that $X_1, X_2, \ldots$ are independent $Ge(q)$-distributed random variables and that $N \in Ge(p)$ is independent of $X_1, X_2, \ldots$.

(a) Find the distribution of $Z = X_1 + X_2 + \cdots + X_N$.

(b) Show that $Z \mid Z > 0 \in Fs(\alpha)$, and determine α.

21. Suppose that $X_1, X_2, \ldots$ are independent $L(a)$-distributed random variables, let $N_p \in Fs(p)$ be independent of $X_1, X_2, \ldots$, and set $Y_p = \sum_{k=1}^{N_p} X_k$. Show that

$$\sqrt{p} Y_p \in L(a).$$

22. Let $N, X_1, X_2, \ldots$ be independent random variables such that $N \in Po(1)$ and $X_k \in Po(2)$ for all k. Set $Z = \sum_{k=1}^{N} X_k$ (and $Z = 0$ when $N = 0$). Compute $E Z$, $\operatorname{Var} Z$, and $P(Z = 0)$.

23. Let $Y_1, Y_2, \ldots$ be i.i.d. random variables, and let N be a nonnegative, integer-valued random variable that is independent of $Y_1, Y_2, \ldots$. Compute $\operatorname{Cov} \left(\sum_{k=1}^{N} Y_k, N \right)$.

24. Let, for $m \neq 1$, $X_1, X_2, \ldots$ be independent random variables with $E X_n = m^n$, $n \geq 1$, let $N \in Po(\lambda)$ be independent of $X_1, X_2, \ldots$, and set

$$Z = X_1 + X_2 + \cdots + X_N.$$

Determine $E Z$.

Remark. Note that $X_1, X_2, \ldots$ are NOT identically distributed, that is, the usual "$E S_N = E N \cdot E X$" does NOT work; you have to modify the proof of that formula.

25. Let $N \in Bin(n, 1 - e^{-m})$, and let $X_1, X_2, \ldots$ have the same 0-truncated Poisson distribution, namely,

$$P(X_1 = x) = \frac{m^x / x!}{e^m - 1}, \quad x = 1, 2, 3, \ldots.$$

Further, assume that $N, X_1, X_2, \ldots$ are independent,

(a) Find the distribution of $Y = \sum_{k=1}^{N} X_k$ ($Y = 0$ when $N = 0$).

(b) Compute EY and $\operatorname{Var} Y$ without using (a).

26. The number of cars passing a road crossing during an hour is $Po(b)$-distributed. The number of passengers in each car is $Po(p)$-distributed. Find the generating function of the total number of passengers, Y, passing the road crossing during one hour, and compute EY and $\operatorname{Var} Y$.

27. A miner has been trapped in a mine with three doors. One takes him to freedom after one hour, one brings him back to the mine after 3 hours and the third one brings him back after 5 hours. Suppose that he wishes to get out of the mine and that he does so by choosing one of the three doors uniformly at random and continues to do so until he is free. Find the generating function, the mean and the variance for the time it takes him to reach freedom.

28. Lisa shoots at a target. The probability of a hit in each shot is 1/2. Given a hit, the probability of a bull's-eye is p. She shoots until she misses the target. Let X be the total number of bull's-eyes Lisa has obtained when she has finished shooting; find its distribution.

29. Karin has an unfair coin; the probability of heads is p $(0 < p < 1)$. She tosses the coin until she obtains heads. She then tosses a fair coin as many times as she tossed the unfair one. For every head she has obtained with the fair coin she finally throws a symmetric die. Determine the expected number and variance of the total number of dots Karin obtains by this procedure.

30. Philip throws a fair die until he obtains a four. Diane then tosses a coin as many times as Philip threw his die. Determine the expected value and variance of the number of

    (a) heads,

    (b) tails, and

    (c) heads and tails obtained by Diane.

31. Let p be the probability that the tip points downward after a person throws a drawing pin once. Miriam throws a drawing pin until it points downward for the first time. Let X be the number of throws for this to happen. She then throws the drawing pin another X times. Let Y be the number of times the drawing pin points downward in the latter series of throws. Find the distribution of Y (cf. Problem 2.6.38).

32. Let $X_1, X_2, \ldots$ be independent observations of a random variable X, whose density function is

$$f_X(x) = \tfrac{1}{2} e^{-|x|}, \quad -\infty < x < \infty.$$

Suppose we continue sampling until a negative observation appears. Let Y be the sum of the observations thus obtained (including the negative one). Show that the density function of Y is

$$f_Y(x) = \begin{cases} \tfrac{2}{3} e^x, & \text{for} \quad x < 0, \\ \tfrac{1}{6} e^{-x/2}, & \text{for} \quad x > 0. \end{cases}$$

33. At a certain black spot, the number of traffic accidents per year follows a Po(10,000)-distribution. The number of deaths per accident follows a Po(0.1)-distribution, and the number of casualties per accidents follows a Po(2)-distribution. The correlation coefficient between the number of

casualties and the number of deaths per accidents is 0.5. Compute the expectation and variance of the total number of deaths and casualties during a year.

34. Suppose that X is a nonnegative, integer-valued random variable, and let n and m be nonnegative integers. Show that

$$g_{nX+m}(t) = t^m \cdot g_X(t^n).$$

35. Suppose that the offspring distribution in a branching process is the $\mathrm{Ge}(p)$-distribution, and let $X(n)$ be the number of individuals in generation n, $n = 0, 1, 2, \ldots$.
    (a) What is the probability of extinction?
    (b) Find the probability that the population is extinct in the second generation.

36. Consider a branching process whose offspring distribution is $\mathrm{Bin}(n, p)$-distributed. Compute the expected value, the variance and the probability that there are 0 or 1 grandchild, that is, find, in the usual notation, $E\,X(2)$, $\mathrm{Var}\,X(2)$, $P(X(2) = 0)$, and $P(X(2) = 1)$.

37. Consider a branching process where the individuals reproduce according to the following pattern:

# of children	0	1	2
probability	$\frac{1}{6}$	$\frac{1}{3}$	$\frac{1}{2}$

    Individuals reproduce independently of each other and independently of the number of their sisters and brothers. Determine
    (a) the probability that the population becomes extinct;
    (b) the probability that the population has become extinct in the second generation;
    (c) the expected number of children given that there are no grandchildren.

38. One bacterium each of the two dangerous Alphomylia and Klaipeda tribes have escaped from a laboratory. They reproduce according to a standard branching process as follows:

# of children	0	1	2
probability Alphomylia	$\frac{1}{4}$	$\frac{1}{4}$	$\frac{1}{2}$
probability Klaipeda	$\frac{1}{6}$	$\frac{1}{6}$	$\frac{2}{3}$

    The two cultures reproduce independently of each other. Determine the probability that 0, 1, and 2 of the cultures, respectively, become extinct.

39. Suppose that the offspring distribution in a branching process is the $\mathrm{Ge}(p)$-distribution, and let $X(n)$ be the number of individuals in generation n, $n = 0, 1, 2, \ldots$.
    (a) What is the probability of extinction?
       Now suppose that $p = 1/2$, and set $g_n(t) = g_{X(n)}(t)$.

(b) Show that
$$g_n(t) = \frac{n - (n-1)t}{n + 1 - nt}, \quad n = 1, 2, \ldots .$$

(c) Show that
$$P(X(n) = k) = \begin{cases} \dfrac{n}{n+1}, & \text{for } k = 0, \\[2mm] \dfrac{n^{k-1}}{(n+1)^{k+1}}, & \text{for } k = 1, 2, \ldots . \end{cases}$$

(d) Show that
$$P(X(n) = k \mid X(n) > 0) = \frac{1}{n+1}\left(\frac{n}{n+1}\right)^{k-1}, \quad \text{for } k = 1, 2, \ldots ,$$

that is, show that the number of individuals in generation n, given that the population is not yet extinct, follows an $\mathrm{Fs}(1/(n+1))$-distribution. Finally, suppose that the population becomes extinct at generation number N.

(e) Show that
$$P(N = n) = g_{n-1}(\tfrac{1}{2}) - g_{n-1}(0), \quad n = 1, 2, \ldots .$$

(f) Show that $P(N = n) = 1/(n(n+1))$, $n = 1, 2, \ldots$ (and hence that $P(N < \infty) = 1$, i.e., $\eta = 1$).

(g) Compute $E\,N$. Why is this a reasonable answer?

40. The growth dynamics of pollen cells can be modeled by binary splitting as follows: After one unit of time, a cell either splits into two or dies. The new cells develop according to the same law independently of each other. The probabilities of dying and splitting are 0.46 and 0.54, respectively.

(a) Determine the maximal initial size of the population in order for the probability of extinction to be at least 0.3.

(b) What is the probability that the population is extinct after two generations if the initial population is the maximal number obtained in (a)?

41. Consider binary splitting, that is, the branching process where the distribution of $Y = $ the number of children is given by
$$P(Y = 2) = 1 - P(Y = 0) = p, \quad 0 < p < 1.$$

However, suppose that p is not known, that p is random, viz., consider the following setup: Assume that
$$P(Y = 2 \mid P = p) = p, \quad P(Y = 0 \mid P = p) = 1 - p, \quad \text{with}$$
$$f_P(x) = \begin{cases} 2x, & \text{for } 0 < x < 1, \\ 0, & \text{otherwise.} \end{cases}$$

(a) Find the distribution of Y.

(b) Determine the probability of extinction.

42. Consider the following modification of a branching process: A mature individual produces children according to the generating function $g(t)$. However, an individual becomes mature with probability α and dies before maturity with probability $1 - \alpha$. Throughout $X(0) = 1$, that is, we start with one immature individual.
    (a) Find the generating function of the number of individuals in the first two generations.
    (b) Suppose that the offspring distribution is geometric with parameter p. Determine the extinction probability.

43. Let $\{X(n), n \geq 0\}$ be the usual Galton–Watson process, starting with $X(0) = 1$. Suppose, in addition, that immigration is allowed in the sense that in addition to the children born in generation n there are Z_n individuals immigrating, where $\{Z_n, n \geq 1\}$ are i.i.d. random variables with the same distribution as $X(1)$.
    (a) What is the expected number of individuals in generation 1?
    (b) Find the generating function of the number of individuals in generations 1 and 2, respectively.
    (c) Determine/express the probability that the population is extinct after two generations.
    Remark. It may be helpful to let p_0 denote the probability that an individual does not have any children (which, in particular, means that $P(X(1) = 0) = p_0$).

44. Consider a branching process with reproduction mean $m < 1$. Suppose also, as before, that $X(0) = 1$.
    (a) What is the probability of extinction?
    (b) Determine the expected value of the total progeny.
    (c) Now suppose that $X(0) = k$, where k is an integer ≥ 2. What are the answers to the questions in (a) and (b) now?

45. The following model can be used to describe the number of women (mothers and daughters) in a given area. The number of mothers is a random variable $X \in \text{Po}(\lambda)$. Independently of the others, every mother gives birth to a $\text{Po}(\mu)$-distributed number of daughters. Let Y be the total number of daughters and hence $Z = X + Y$ be the total number of women in the area.
    (a) Find the generating function of Z.
    (b) Compute $E\,Z$ and $\text{Var}\,Z$.

46. Let $X(n)$ be the number of individuals in the nth generation of a branching process ($X(0) = 1$), and set $T_n = 1 + X(1) + \cdots + X(n)$, that is, T_n equals the total progeny up to and including generation number n. Let $g(t)$ and $G_n(t)$ be the generating functions of $X(1)$ and T_n, respectively. Prove the following formula:
$$G_n(t) = t \cdot g\big(G_{n-1}(t)\big).$$

47. Consider a branching process with a Po(m)-distributed offspring. Let $X(1)$ and $X(2)$ be the number of individuals in generations 1 and 2, respectively. Determine the generating function of
    (a) $X(1)$,
    (b) $X(2)$,
    (c) $X(1) + X(2)$,
    (d) Determine $\mathrm{Cov}(X(1), X(2))$.

48. Let X be the number of coin tosses until heads is obtained. Suppose that the probability of heads is unknown in the sense that we consider it to be a random variable $Y \in U(0,1)$. Find the distribution of X (cf. Problem 2.6.37).

4

Order Statistics

1 One-Dimensional Results

Let X_1, X_2, ... be a (random) sample from a distribution with distribution function F, and let X denote a generic random variable with this distribution. Very natural objects of interest are the largest observation, the smallest observation, the centermost observation (the median), among others. In this chapter we shall derive marginal as well as joint distributions of such objects.

Example 1.1. In a 100-meter Olympic race the running times can be considered to be $U(9.6, 10.0)$-distributed. Suppose that there are eight competitors in the finals. We wish to find the probability that the winner breaks the world record of 9.69 seconds. All units are seconds.

Example 1.2. One hundred numbers, uniformly distributed in the interval $(0, 1)$, are generated by a computer. What is the probability that the largest number is at most 0.9? What is the probability that the second smallest number is at least 0.002? □

Definition 1.1. *For $k = 1, 2, \ldots, n$, let*

$$X_{(k)} = the\ kth\ smallest\ of\ X_1, X_2, \ldots, X_n. \tag{1.1}$$

$(X_{(1)}, X_{(2)}, \ldots, X_{(n)})$ *is called the* order statistic *and $X_{(k)}$ the kth order variable, $k = 1, 2, \ldots, n$.* □

The order statistic is thus obtained from the original (unordered) sample through permutation; the observations are arranged in increasing order. It follows that

$$X_{(1)} \leq X_{(2)} \leq \cdots \leq X_{(n)}. \tag{1.2}$$

Remark 1.1. Actually, the order variables also depend on n; $X_{(k)}$ is the kth smallest of the n observations X_1, X_2, ..., X_n. To be completely descriptive, the notation should therefore also include an n, denoting the sample size. In the literature one sometimes finds the (more clumsy) notation $X_{1:n}$, $X_{2:n}$, ..., $X_{n:n}$. □

A. Gut, *An Intermediate course in Probabilty*, Springer Texts in Statistics, DOI: 10.1007/978-1-4419-0162-0_4, © Springer Science + Business Media, LLC 2009

Exercise 1.1. Suppose that F is continuous. Compute $P(X_k = X_{(k)},\ k = 1, 2, \ldots, n)$, that is, the probability that the original, unordered sample is in fact (already) ordered.

Exercise 1.2. Suppose that F is continuous and that we have a sample of size n. We now make one further observation. Compute, in the notation of Remark 1.1, $P(X_{k:n} = X_{k:n+1})$, that is, the probability that the kth smallest observation still is the kth smallest observation. $\qquad\square$

The extreme order variables are

$$X_{(1)} = \min\{X_1, X_2, \ldots, X_n\} \quad \text{and} \quad X_{(n)} = \max\{X_1, X_2, \ldots, X_n\},$$

whose distribution functions are obtained as follows:

$$F_{X_{(n)}}(x) = P(X_1 \leq x, X_2 \leq x, \ldots, X_n \leq x)$$
$$= \prod_{k=1}^{n} P(X_k \leq x) = \big(F(x)\big)^n$$

and

$$F_{X_{(1)}}(x) = 1 - P(X_{(1)} > x) = 1 - P(X_1 > x, X_2 > x, \ldots, X_n > x)$$
$$= 1 - \prod_{k=1}^{n} P(X_k > x) = 1 - \big(1 - F(x)\big)^n.$$

In the continuous case we additionally have the following expressions for the densities:

$$f_{X_{(n)}}(x) = n\big(F(x)\big)^{n-1} f(x) \quad \text{and} \quad f_{X_{(1)}}(x) = n\big(1 - F(x)\big)^{n-1} f(x).$$

Example 1.1 (continued). Let us solve the problem posed earlier. We wish to determine $P(X_{(1)} < 9.69)$.

Since $f_X(x) = 2.5$ for $9.6 < x < 10.0$ and zero otherwise, it follows that in the interval $9.8 < x < 10.2$ we have $F_X(x) = 2.5x - 24$ and hence

$$F_{X_{(1)}}(x) = 1 - (25 - 2.5x)^8$$

(since we assume that the running times are independent). Since the desired probability equals $F_{X_{(1)}}(9.69)$, the answer to our problem is $1 - (25 - 2.5 \cdot 9.69)^8 \approx 0.8699$. $\qquad\square$

Exercise 1.3. Solve the problem in Example 1.2. $\qquad\square$

These results are now generalized to arbitrary order variables.

Theorem 1.1. *For $k = 1, 2, \ldots, n$, we have*

$$F_{X_{(k)}}(x) = \frac{\Gamma(n+1)}{\Gamma(k)\Gamma(n+1-k)} \int_0^{F(x)} y^{k-1}(1-y)^{n-k}\, dy, \qquad (1.3)$$

that is,

$$F_{X_{(k)}}(x) = F_{\beta(k,n+1-k)}\big(F(x)\big).$$

In particular, if $X \in U(0,1)$, then

$$X_{(k)} \in \beta(k, n+1-k), \quad k = 1, 2, \ldots, n. \qquad (1.4)$$

Proof. For $i = 0, 1, 2, \ldots, n$, let

$$A_i(x) = \{\text{exactly } i \text{ of the variables } X_1, X_2, \ldots, X_n \le x\}.$$

Since these sets are disjoint and

$$\# \text{ observations } \le x \in \text{Bin}(n, F(x)), \qquad (1.5)$$

it follows that

$$\begin{aligned}
F_{X_{(k)}}(x) = P(X_{(k)} \le x) &= P\left(\bigcup_{i=k}^n A_i(x)\right) \\
&= \sum_{i=k}^n P(A_i(x)) = \sum_{i=k}^n \binom{n}{i} \big(F(x)\big)^i \big(1 - F(x)\big)^{n-i}.
\end{aligned}$$

A comparison with (1.3) shows that it remains to prove the following formula:

$$\sum_{i=k}^n \binom{n}{i} z^i (1-z)^{n-i} = \frac{\Gamma(n+1)}{\Gamma(k)\Gamma(n+1-k)} \int_0^z y^{k-1}(1-y)^{n-k}\, dy \qquad (1.6)$$

for $k = 1, 2, \ldots, n$ (and $0 \le z \le 1$). This is done by backward induction, that is, we begin with the case $k = n$ and move downward.

For $k = n$, both members in (1.6) equal z^n. Now suppose that relation (1.6) holds for $n, n-1, \ldots, k$.

Claim. Formula (1.6) holds for $k - 1$.

Proof. Set

$$a_i = \binom{n}{i} z^i (1-z)^{n-i}, \quad i = 1, 2, \ldots, n,$$

$$\Sigma_k = \sum_{i=k}^n a_i, \quad k = 1, 2, \ldots, n,$$

$$I_k = \frac{\Gamma(n+1)}{\Gamma(k)\Gamma(n+1-k)} \int_0^z y^{k-1}(1-y)^{n-k}\, dy, \quad k = 1, 2, \ldots, n.$$

We wish to show that $\Sigma_{k-1} = I_{k-1}$.

From the assumption and by partial integration, it follows that

$$\Sigma_k = I_k$$

$$= \left[\frac{\Gamma(n+1)}{\Gamma(k)\Gamma(n+1-k)} \cdot \left(-\frac{1}{n-k+1}\right) y^{k-1}(1-y)^{n-k+1} \right]_0^z$$

$$- \frac{\Gamma(n+1)}{\Gamma(k)\Gamma(n+1-k)} \left(-\frac{1}{n-k+1}\right)(k-1) \int_0^z y^{k-2}(1-y)^{n-k+1}\, dy$$

$$= -\frac{\Gamma(n+1)}{\Gamma(k)\Gamma(n+2-k)} z^{k-1}(1-z)^{n-k+1}$$

$$+ \frac{\Gamma(n+1)}{\Gamma(k-1)\Gamma(n+2-k)} \int_0^z y^{k-2}(1-y)^{n-k+1}\, dy \qquad (1.7)$$

$$= -\binom{n}{k-1} z^{k-1}(1-z)^{n-(k-1)}$$

$$+ \frac{\Gamma(n+1)}{\Gamma(k-1)\Gamma(n+1-(k-1))} \int_0^z y^{(k-1)-1}(1-y)^{n-(k-1)}\, dy$$

$$= -a_{k-1} + I_{k-1}.$$

The extreme members now tell us that $\Sigma_k = -a_{k-1} + I_{k-1}$, which, by moving a_{k-1} to the left-hand side, proves (1.3), from which the special case (1.4) follows immediately.

The proof of the theorem is thus complete.   □

Remark 1.2. Formula (1.6) will also appear in Chapter 8, where the members of the relation will be interpreted as a conditional probability for Poisson processes; see Remark 8.3.3.   □

In the continuous case, differentiation of $F_{X_{(k)}}(x)$ as given in (1.3) yields the density of $X_{(k)}$, $1 \le k \le n$.

Theorem 1.2. *Suppose that the distribution is continuous with density $f(x)$. For $k = 1, 2, \ldots, n$, the density of $X_{(k)}$ is given by*

$$f_{X_{(k)}}(x) = \frac{\Gamma(n+1)}{\Gamma(k)\Gamma(n+1-k)} \left(F(x)\right)^{k-1}\left(1-F(x)\right)^{n-k} f(x), \qquad (1.8)$$

that is,

$$f_{X_{(k)}}(x) = f_{\beta(k,n+1-k)}\left(F(x)\right) \cdot f(x).$$

□

Remark 1.3. For $k = 1$ and $k = n$, we rediscover, in both theorems, the familiar expressions for the distribution functions and density functions of the smallest and largest values.   □

Under the assumption that the density is (for instance) continuous, we can make the following heuristic derivation of (1.8): If h is "very small," then

$$F_{X_{(k)}}(x+h) - F_{X_{(k)}}(x) = P(x < X_{(k)} \leq x+h)$$
$$\approx P(k-1 \text{ obs } \leq x, \text{ 1 obs in } (x, x+h], n-k \text{ obs } > x+h),$$

because the probability that at least two observations fall into the interval $(x, x+h]$ is negligible.

Now, this is a multinomial probability, which equals

$$\frac{n!}{(k-1)!\,1!\,(n-k)!}\left(F(x)\right)^{k-1}\left(F(x+h) - F(x)\right)^1\left(1 - F(x+h)\right)^{n-k}$$
$$= \frac{\Gamma(n+1)}{\Gamma(k)\Gamma(n+1-k)}\left(F(x)\right)^{k-1}\left(F(x+h) - F(x)\right)\left(1 - F(x+h)\right)^{n-k}.$$

By the mean value theorem, $F(x+h) - F(x) = h \cdot f(\theta_{x,h})$, where $x \leq \theta_{x,h} \leq x+h$. Since h is small and f is (for instance) continuous, we further have $f(\theta_{x,h}) \approx f(x)$ and $F(x+h) \approx F(x)$, which yield

$$F_{X_{(k)}}(x+h) - F_{X_{(k)}}(x) \approx h \cdot \frac{\Gamma(n+1)}{\Gamma(k)\Gamma(n+1-k)}\left(F(x)\right)^{k-1}\left(1 - F(x)\right)^{n-k}f(x).$$

The conclusion now follows by dividing with h and letting $h \to 0$.

Remark 1.4. The probability we just computed is of the order of magnitude $O(h)$. With some additional work one can show that

$$P(\text{at least two observations in } (x, x+h]) = O(h^2) = o(h) \qquad (1.9)$$

as $h \to 0$, that is, what we considered negligible above is indeed negligible.

More formally, we thus have

$$F_{X_{(k)}}(x+h) - F_{X_{(k)}}(x)$$
$$= h\,\frac{\Gamma(n+1)}{\Gamma(k)\Gamma(n+1-k)}\left(F(x)\right)^{k-1}\left(1 - F(x)\right)^{n-k}f(x) + o(h)$$

as $h \to 0$ (where $o(h)$ also includes the other approximations). $\qquad \square$

2 The Joint Distribution of the Extremes

In the previous section we studied the distribution of a single order variable. Here we consider $X_{(1)}$ and $X_{(n)}$ jointly. The distribution is assumed to be continuous throughout.

Theorem 2.1. *The joint density of* $X_{(1)}$ *and* $X_{(n)}$ *is given by*

$$f_{X_{(1)}, X_{(n)}}(x, y) = \begin{cases} n(n-1)\left(F(y) - F(x)\right)^{n-2}f(y)f(x), & \text{for } x < y, \\ 0, & \text{otherwise.} \end{cases}$$

Proof. We modify the idea on which the derivation of the marginal distributions of $X_{(1)}$ and $X_{(n)}$ was based. The key observation is that

$$P(X_{(1)} > x, X_{(n)} \leq y) = P(x < X_k \leq y, \ k = 1, 2, \ldots, n)$$
$$= \prod_{k=1}^{n} P(x < X_k \leq y) = (F(y) - F(x))^n, \quad \text{for} \quad x < y. \tag{2.1}$$

For $x \geq y$ the probability is, of course, equal to zero.

Now, (2.1) and the fact that

$$P(X_{(1)} \leq x, X_{(n)} \leq y) + P(X_{(1)} > x, X_{(n)} \leq y) = P(X_{(n)} \leq y) \tag{2.2}$$

lead to

$$F_{X_{(1)}, X_{(n)}}(x, y) = F_{X_{(n)}}(y) - P(X_{(1)} > x, X_{(n)} \leq y)$$
$$= \begin{cases} (F(y))^n - (F(y) - F(x))^n, & \text{for} \quad x < y, \\ (F(y))^n, & \text{for} \quad x \geq y. \end{cases} \tag{2.3}$$

Differentiation with respect to x and y yields the desired result. $\square$

Exercise 2.1. Generalize the heuristic derivation of the density in Theorem 1.2 to the density in Theorem 2.1. $\square$

An important quantity related to $X_{(1)}$ and $X_{(n)}$ is the *range*

$$R_n = X_{(n)} - X_{(1)}, \tag{2.4}$$

which provides information of how spread out the underlying distribution might be. The distribution of R_n can be obtained by the methods of Chapter 1 by introducing the auxiliary random variable $U = X_{(1)}$. With the aid of Theorems 2.1 and 1.2.1 we then obtain an expression for $f_{R_n, U}(r, u)$. Integrating with respect to u yields the marginal density $f_{R_n}(r)$. The result is as follows:

Theorem 2.2. *The density of the range R_n, as defined in (2.4), is*

$$f_{R_n}(r) = n(n-1) \int_{-\infty}^{\infty} \left(F(u+r) - F(u)\right)^{n-2} f(u+r) f(u) \, du,$$

for $r > 0$. $\square$

Exercise 2.2. Give the details of the proof of Theorem 2.2. $\square$

Example 2.1. If $X \in U(0, 1)$, then

$$f_{R_n}(r) = n(n-1) \int_0^{1-r} (u + r - u)^{n-2} \cdot 1 \cdot 1 \, du$$
$$= n(n-1) r^{n-2} (1 - r), \quad 0 < r < 1,$$

that is, $R_n \in \beta(n-1, 2)$. Moreover,

$$E R_n = \int_0^1 r \cdot n(n-1)r^{n-2}(1-r) \, dr = n(n-1) \int_0^1 (r^{n-1} - r^n) \, dr$$

$$= n(n-1)\left(\frac{1}{n} - \frac{1}{n+1}\right) = \frac{n-1}{n+1}. \tag{2.5}$$

This may, alternatively, be read off from the $\beta(n-1, 2)$-distribution;

$$E R_n = \frac{n-1}{(n-1)+2} = \frac{n-1}{n+1}.$$

Furthermore, if one thinks intuitively about how n points in the unit interval are distributed on average, one realizes that the value

$$\frac{n}{n+1} - \frac{1}{n+1} \qquad \left(= \frac{n-1}{n+1}\right)$$

is to be expected for the range. □

Exercise 2.3. Find the probability that all runners in Example 1.1 finish within the time interval $(9.8, 9.9)$. □

Example 2.2. Let $X_1, X_2, \ldots, X_n$ be independent, Exp(1)-distributed random variables. Determine

(a) $f_{X_{(1)}, X_{(n)}}(x, y)$,
(b) $f_{R_n}(r)$.

Solution. (a) For $0 < x < y$,

$$f_{X_{(1)}, X_{(n)}}(x, y) = n(n-1)\big(1 - e^{-y} - (1 - e^{-x})\big)^{n-2} \cdot e^{-y} \cdot e^{-x}$$

$$= n(n-1)(e^{-x} - e^{-y})^{n-2} e^{-(x+y)},$$

(and zero otherwise).
(b) It follows from Theorem 2.2 that

$$f_{R_n}(r) = n(n-1) \int_0^\infty (e^{-u} - e^{-(u+r)})^{n-2} e^{-(2u+r)} \, du$$

$$= n(n-1) \int_0^\infty e^{-u(n-2)}(1 - e^{-r})^{n-2} e^{-(2u+r)} \, du$$

$$= n(n-1)(1 - e^{-r})^{n-2} e^{-r} \int_0^\infty e^{-nu} \, du$$

$$= (n-1)(1 - e^{-r})^{n-2} e^{-r}, \quad r > 0. \qquad □$$

Remark 2.1. We also observe that $F_{R_n}(r) = (1 - e^{-r})^{n-1} = (F(r))^{n-1}$. This can be explained by the fact that the exponential distribution has "no memory." A more thorough understanding of this fact is provided in Chapter 8, which is devoted to the study of the Poisson process. In that context the lack of memory amounts to the fact that $(X_{(2)} - X_{(1)}, X_{(3)} - X_{(1)}, \ldots, X_{(n)} - X_{(1)})$ can be interpreted as the order statistic corresponding to a sample of size $n - 1$ from an Exp(1)-distribution, which in turn implies that R_n can be interpreted as the largest of those $n - 1$ observations. In the language of Remark 1.1, we might say that $(X_{(2)} - X_{(1)}, X_{(3)} - X_{(1)}, \ldots, X_{(n)} - X_{(1)}) \stackrel{d}{=} (Y_{1:n-1}, Y_{2:n-1}, \ldots, Y_{n-1:n-1})$, in particular, $R_n \stackrel{d}{=} Y_{n-1:n-1}$, where $Y_1, Y_2, \ldots, Y_{n-1}$ is a sample (of size $n - 1$) from an Exp(1)-distribution. $\square$

Exercise 2.4. Consider Example 2.2 with $n = 2$. Then $R_n = R_2 = X_{(2)} - X_{(1)} \in \text{Exp}(1)$. In view of Remark 2.1 it is tempting to guess that $X_{(1)}$ and $X_{(2)} - X_{(1)}$ are independent. Show that this is indeed the case.

For an extension, see Problems 4.20(a) and 4.21(a).

Exercise 2.5. The geometric distribution is a discrete analog of the exponential distribution in the sense of lack of memory. More precisely, show that if X_1 and X_2 are independent, Ge(p)-distributed random variables, then $X_{(1)}$ and $X_{(2)} - X_{(1)}$ are independent. $\square$

Conditional distributions can also be obtained as is shown in the following example:

Example 2.3. Let X_1, X_2, and X_3 be independent, Exp(1)-distributed random variables. Compute

$$E(X_{(3)} \mid X_{(1)} = x).$$

Solution. By Theorem 2.1 (see also Example 2.2(a)) we have

$$f_{X_{(1)}, X_{(3)}}(x, y) = 3 \cdot 2(e^{-x} - e^{-y}) e^{-(x+y)} \quad \text{for} \quad 0 < x < y,$$

and hence

$$f_{X_{(3)} \mid X_{(1)} = x}(y) = \frac{f_{X_{(1)}, X_{(3)}}(x, y)}{f_{X_{(1)}}(x)} = \frac{6(e^{-x} - e^{-y})e^{-(x+y)}}{3e^{-3x}}$$

$$= 2(e^{-x} - e^{-y})e^{2x-y}, \quad \text{for} \quad 0 < x < y.$$

The conditional expectation thus becomes

$$E(X_{(3)} \mid X_{(1)} = x) = \int_x^\infty 2y(e^{-x} - e^{-y})e^{2x-y} \, dy$$

$$= \int_0^\infty 2(u + x)(e^{-x} - e^{-(u+x)})e^{2x-u-x} \, du$$

$$= 2 \int_0^\infty (u + x)(1 - e^{-u})e^{-u} \, du$$

$$= 2 \int_0^\infty u(e^{-u} - e^{-2u}) \, du + 2x \int_0^\infty (e^{-u} - e^{-2u}) \, du$$

$$= 2\left(1 - \frac{1}{2} \cdot \frac{1}{2}\right) + 2x\left(1 - \frac{1}{2}\right) = x + \frac{3}{2}. \qquad \square$$

Remark 2.2. As in the previous example, one can use properties of the Poisson process to justify the answer; see Problem 8.9.27.   □

Exercise 2.6. Suppose n points are chosen uniformly and independently of each other on the unit disc. Compute the expected value of the area of the annulus obtained by drawing circles through the extremes.   □

We conclude this section with a discrete version of Example 2.3.

Exercise 2.7. Independent repetitions of an experiment are performed. A is an event that occurs with probability p, $0 < p < 1$. Let T_k be the number of the performance at which A occurs the kth time, $k = 1, 2, \ldots$. Compute

(a) $E(T_3 \mid T_1 = 5)$,

(b) $E(T_1 \mid T_3 = 5)$.   □

3 The Joint Distribution of the Order Statistic

So far we have found the marginal distributions of the order variables and the joint distribution of the extremes. In general it might be of interest to know the distribution of an arbitrary collection of order variables. From Chapter 1 we know that once we are given a joint distribution we can always find marginal distributions by integrating the joint density with respect to the other variables. In this section we show how the joint distribution of the (whole) order statistic can be derived. The point of departure is that the joint density of the (unordered) sample is known and that the ordering, in fact, is a linear transformation to which Theorem 1.2.1 can be applied. However, it is not a 1-to-1 transformation, and so the arguments at the end of Section 1.2 must be used.

We are thus given the joint density of the unordered sample

$$f_{X_1,\ldots,X_n}(x_1,\ldots,x_n) = \prod_{i=1}^{n} f(x_i). \tag{3.1}$$

Consider the mapping $(X_1, X_2, \ldots, X_n) \to (X_{(1)}, X_{(2)}, \ldots, X_{(n)})$. We have already argued that it is a permutation; the observations are simply rearranged in increasing order. The transformation can thus be rewritten as

$$\begin{pmatrix} X_{(1)} \\ X_{(2)} \\ \vdots \\ X_{(n)} \end{pmatrix} = \mathbf{P} \begin{pmatrix} X_1 \\ X_2 \\ \vdots \\ X_n \end{pmatrix}, \tag{3.2}$$

where $\mathbf{P} = (P_{ij})$ is a permutation matrix, that is, a matrix with exactly one 1 in every row and every column and zeroes otherwise; $P_{ij} = 1$ means that

$X_{(i)} = X_j$. However, the mapping is not 1-to-1, since, by symmetry, there are $n!$ different outcomes that all generate the same order statistic. If, for example, $n = 3$ and the observations are 3,2,8, then the order statistic is (2,3,8). However, the results 2,3,8; 2,8,3; 3,8,2; 8,2,3; and 8,3,2 all would have yielded the same order statistic.

We therefore partition the space $\mathbf{R}^n$ into $n!$ equally shaped parts, departing from one "corner" each, so that the mapping from each part to $\mathbf{R}^n$ is 1-to-1 in the sense of the end of Section 1.2. By formula (1.2.2),

$$f_{X_{(1)},\ldots,X_{(n)}}(y_1,\ldots,y_n) = \sum_{i=1}^{n!} f_{X_1,\ldots,X_n}(x_{1i}(\mathbf{y}),\ldots,x_{ni}(\mathbf{y}))\cdot |\, \mathbf{J}_i\,|, \qquad (3.3)$$

where $\mathbf{J}_i$ is the Jacobian corresponding to the transformation from "domain" i to $\mathbf{R}^n$. Since a permutation matrix has determinant ± 1, it follows that $|\,\mathbf{J}_i\,| = 1$ for all i.

Now, by construction, each $x_{ki}(\mathbf{y})$ equals some y_j, so

$$f_{X_1,\ldots,X_n}(x_{1i}(\mathbf{y}),\ldots,x_{ni}(\mathbf{y})) = \prod_{k=1}^{n} f_{X_k}(x_{ki}(\mathbf{y})) = \prod_{k=1}^{n} f(y_k); \qquad (3.4)$$

namely, we multiply the original density f evaluated at the points $x_{ki}(\mathbf{y})$, $k = 1, 2, \ldots, n$, that is, at the points $y_1, y_2, \ldots, y_n$—however, in a different order. The density $f_{X_{(1)},\ldots,X_{(n)}}(y_1,\ldots,y_n)$ is therefore a sum of the $n!$ identical terms $\prod_{k=1}^{n} f(y_k) \cdot 1$. This leads to the following result:

Theorem 3.1. *The (joint) density of the order statistic is*

$$f_{X_{(1)},\ldots,X_{(n)}}(y_1,\ldots,y_n) = \begin{cases} n! \displaystyle\prod_{k=1}^{n} f(y_k), & \text{if } \ y_1 < y_2 < \cdots < y_n, \\ 0, & \text{otherwise.} \end{cases} \qquad \square$$

Exercise 3.1. Consider the event that we have exactly one observation in each of the intervals $(y_k, y_k + h_k]$, $k = 1, 2, \ldots, n$ (where $y_1, y_2, \ldots, y_n$ are given and $h_1, h_2, \ldots, h_n$ are so small that $y_1 < y_1 + h_1 \le y_2 < y_2 + h_2 \le \cdots \le y_{n-1} < y_{n-1} + h_{n-1} \le y_n < y_n + h_n$). The probability of this event equals

$$n! \cdot \prod_{k=1}^{n} \left(F(y_k + h_k) - F(y_k)\right), \qquad (3.5)$$

which, by the mean value theorem (and under the assumption that f is "nice" (cf. the end of Section 1)), implies that the expression in (3.5) is approximately equal to

$$n! \cdot \prod_{k=1}^{n} h_k \cdot f(y_k). \qquad (3.6)$$

Complete the heuristic derivation of $f_{X_{(1)},\ldots,X_{(n)}}(y_1,\ldots,y_n)$. $\qquad \square$

Next we observe that, given the joint density, we can obtain any desired marginal density by integration (recall Chapter 1). The $(n-1)$-dimensional integral

$$\int_{-\infty}^{\infty} \int_{-\infty}^{\infty} \cdots \int_{-\infty}^{\infty} f_{X_{(1)},\ldots,X_{(n)}}(y_1,\ldots,y_n)\, dy_1 \cdots dy_{k-1} dy_{k+1} \cdots dy_n, \quad (3.7)$$

for example, yields $f_{X_{(k)}}(y_k)$. This density was derived in Theorem 1.2 by one-dimensional arguments. The $(n-2)$-dimensional integral

$$\int_{-\infty}^{\infty} \int_{-\infty}^{\infty} \cdots \int_{-\infty}^{\infty} f_{X_{(1)},\ldots,X_{(n)}}(y_1,\ldots,y_n)\, dy_2 dy_3 \cdots dy_{n-1} \quad (3.8)$$

yields $f_{X_{(1)},X_{(n)}}(y_1, y_n)$, which was obtained earlier in Theorem 2.1. On the other hand, by integrating over all variables but X_j and X_k, $1 \le j < k \le n$, we obtain $f_{X_{(j)},X_{(k)}}(y_j, y_k)$, which has not been derived so far (for $j \ne 1$ or $k \ne n$).

As an illustration, let us derive $f_{X_{(k)}}(y_k)$ starting from the joint density as given in Theorem 3.1.

$$f_{X_{(k)}}(y_k)$$

$$= \int_{-\infty}^{y_k} \int_{-\infty}^{y_{k-1}} \cdots \int_{-\infty}^{y_2} \int_{y_k}^{\infty} \int_{y_{k+1}}^{\infty} \cdots \int_{y_{n-1}}^{\infty} n! \prod_{i=1}^{n} f(y_i)$$
$$\times\, dy_n \cdots dy_{k+1} dy_1 \cdots dy_{k-1}$$

$$= n! \int_{-\infty}^{y_k} \int_{-\infty}^{y_{k-1}} \cdots \int_{-\infty}^{y_2} \int_{y_k}^{\infty} \int_{y_{k+1}}^{\infty} \cdots \int_{y_{n-2}}^{\infty} \prod_{i=1}^{n-1} f(y_i) \big(1 - F(y_{n-1})\big)$$
$$\times\, dy_{n-1} \cdots dy_{k+1} dy_1 \cdots dy_{k-1}$$

$$= \cdots = \cdots$$

$$= n! \int_{-\infty}^{y_k} \int_{-\infty}^{y_{k-1}} \cdots \int_{-\infty}^{y_2} \prod_{i=1}^{k} f(y_i) \frac{\big(1 - F(y_k)\big)^{n-k}}{(n-k)!}\, dy_1 dy_2 \cdots dy_{k-1}$$

$$= \frac{\Gamma(n+1)}{\Gamma(n+1-k)} \big(1 - F(y_k)\big)^{n-k} \int_{-\infty}^{y_k} \int_{-\infty}^{y_{k-1}} \cdots \int_{-\infty}^{y_2} \prod_{i=1}^{k} f(y_i)$$
$$\times\, dy_1 dy_2 \cdots dy_{k-1}$$

$$= \frac{\Gamma(n+1)}{\Gamma(n+1-k)} \big(1 - F(y_k)\big)^{n-k} \int_{-\infty}^{y_k} \int_{-\infty}^{y_{k-1}} \cdots \int_{-\infty}^{y_3} \prod_{i=2}^{k} f(y_i) F(y_2)$$
$$\times\, dy_2 dy_3 \cdots dy_{k-1}$$

$$= \cdots = \frac{\Gamma(n+1)}{\Gamma(n+1-k)} \big(1 - F(y_k)\big)^{n-k} \cdot \frac{\big(F(y_k)\big)^{k-1}}{(k-1)!} \cdot f(y_k)$$

$$= \frac{\Gamma(n+1)}{\Gamma(k)\Gamma(n+1-k)} \big(F(y_k)\big)^{k-1} \big(1 - F(y_k)\big)^{n-k} \cdot f(y_k),$$

in accordance with Theorem 1.2.

We leave it to the reader to derive general two-dimensional densities. Let us consider, however, one example with $n = 3$.

Example 3.1. Let X_1, X_2, and X_3 be a sample from a $U(0,1)$-distribution. Compute the densities of $(X_{(1)}, X_{(2)})$, $(X_{(1)}, X_{(3)})$, and $(X_{(2)}, X_{(3)})$.

Solution. By Theorem 3.1 we have

$$f_{X_{(1)}, X_{(2)}, X_{(3)}}(y_1, y_2, y_3) = \begin{cases} 6, & \text{for} \quad 0 < y_1 < y_2 < y_3 < 1, \\ 0, & \text{otherwise.} \end{cases}$$

Consequently,

$$f_{X_{(1)}, X_{(2)}}(y_1, y_2) = \int_{y_2}^{1} 6 \, dy_3 = 6(1 - y_2), \quad 0 < y_1 < y_2 < 1, \tag{3.9}$$

$$f_{X_{(1)}, X_{(3)}}(y_1, y_3) = \int_{y_1}^{y_3} 6 \, dy_2 = 6(y_3 - y_1), \quad 0 < y_1 < y_3 < 1, \tag{3.10}$$

$$f_{X_{(2)}, X_{(3)}}(y_2, y_3) = \int_{0}^{y_2} 6 \, dy_1 = 6y_2, \quad 0 < y_2 < y_3 < 1, \tag{3.11}$$

and we are done. $\square$

Remark 3.1. From (3.9) we may further conclude that

$$f_{X_{(1)}}(y_1) = \int_{y_1}^{1} 6(1 - y_2) \, dy_2 = 3(1 - y_1)^2, \quad 0 < y_1 < 1,$$

and that

$$f_{X_{(2)}}(y_2) = \int_{0}^{y_2} 6(1 - y_2) \, dy_1 = 6y_2(1 - y_2), \quad 0 < y_2 < 1.$$

From (3.10) we similarly have

$$f_{X_{(1)}}(y_1) = \int_{y_1}^{1} 6(y_3 - y_1) \, dy_3 = 3(1 - y_1)^2, \quad 0 < y_1 < 1,$$

and

$$f_{X_{(3)}}(y_3) = \int_{0}^{y_3} 6(y_3 - y_1) \, dy_1 = 3y_3^2, \quad 0 < y_3 < 1.$$

Integration in (3.11) yields

$$f_{X_{(2)}}(y_2) = \int_{y_2}^{1} 6y_2 \, dy_3 = 6y_2(1 - y_2), \quad 0 < y_2 < 1,$$

and

$$f_{X_{(3)}}(y_3) = \int_{0}^{y_3} 6y_2 \, dy_2 = 3y_3^2, \quad 0 < y_3 < 1.$$

The densities of the extremes are, of course, the familiar ones, and the density of $X_{(2)}$ is easily identified as that of the $\beta(2,2)$-distribution (in accordance with Theorem 1.1 (and Remark 1.2)).   □

Exercise 3.2. Let X_1, X_2, X_3, and X_4 be a sample from a $U(0,1)$-distribution. Compute the marginal distributions of the order statistic. How many such marginal distributions are there?   □

4 Problems

1. Suppose that X, Y, and Z have a joint density function given by

$$f(x,y,z) = \begin{cases} e^{-(x+y+z)}, & \text{for } x,y,z > 0, \\ 0, & \text{otherwise.} \end{cases}$$

   Compute $P(X < Y < Z)$ and $P(X = Y < Z)$.

2. Two points are chosen uniformly and independently on the perimeter of a circle of radius 1. This divides the perimeter into two pieces. Determine the expected value of the length of the shorter piece.

3. Let X_1 and X_2 be independent, $U(0,1)$-distributed random variables, and let Y denote the point that is closest to an endpoint. Determine the distribution of Y.

4. The statistician Piggy has to wait an amount of time T_0 at the post office on an occasion when she is in a great hurry. In order to investigate whether or not chance makes her wait particularly long when she is in a hurry, she checks how many visits she makes to the post office until she has to wait longer than the first time. Formally, let T_1, T_2, ... be the successive waiting times and N be the number of times until some $T_k > T_0$, that is,

$$\{N = k\} = \{T_j \leq T_0,\ 1 \leq j < k,\ T_k > T_0\}.$$

   What is the distribution of N under the assumption that $\{T_n,\ n \geq 0\}$ are i.i.d. continuous random variables? What can be said about $E\,N$?

5. Let X_1, X_2, ..., X_n be independent, continuous random variables with common distribution function $F(x)$, and consider the order statistic $(X_{(1)}, X_{(2)}, \ldots, X_{(n)})$. Compute $E\big(F(X_{(n)}) - F(X_{(1)})\big)$.

6. Let X_1, X_2, X_3, and X_4 be independent, $U(0,1)$-distributed random variables. Compute
   (a) $P(X_{(3)} + X_{(4)} \leq 1)$,
   (b) $P(X_3 + X_4 \leq 1)$.

7. Let X_1, X_2, X_3 be independent $U(0,1)$-distributed random variables, and let $X_{(1)}$, $X_{(2)}$, $X_{(3)}$ be the corresponding order variables. Compute $P(X_{(1)} + X_{(3)} \leq 1)$.

8. Suppose that X_1, X_2, X_3, X_4 are independent $U(0,1)$-distributed random variables and let $(X_{(1)}, X_{(2)}, X_{(3)}, X_{(4)})$ be the corresponding order statistic. Compute
   (a) $P(X_{(2)} + X_{(3)} \leq 1)$,
   (b) $P(X_{(2)} \leq 3X_{(1)})$.

9. Suppose that X_1, X_2, X_3, X_4 are independent $U(0,1)$-distributed random variables and let $(X_{(1)}, X_{(2)}, X_{(3)}, X_{(4)})$ be the corresponding order statistic. Find the distribution of
   (a) $X_{(3)} - X_{(1)}$,
   (b) $X_{(4)} - X_{(2)}$.

10. Suppose that X_1, X_2, X_3 are independent $U(0,1)$-distributed random variables and let $(X_{(1)}, X_{(2)}, X_{(3)})$ be the corresponding order statistic. Compute $P(X_{(1)} + X_{(2)} > X_{(3)})$.
    Remark. A concrete example runs as follows: Take 3 sticks of length 1, break each of them uniformly at random, and pick one of the pieces from each stick. Find the probability that the 3 chosen pieces can be constructed into a triangle.

11. Suppose that X_1, X_2, X_3 are independent $U(0,1)$-distributed random variables and let $(X_{(1)}, X_{(2)}, X_{(3)})$ be the corresponding order statistic. It is of course a trivial observation that we always have $X_{(3)} \geq X_{(1)}$. However,
    (a) Compute $P(X_{(3)} > 2X_{(1)})$.
    (b) Determine a so that $P(X_{(3)} > aX_{(1)}) = 1/2$.

12. Let X_1, X_2, X_3 be independent $U(0,1)$-distributed random variables, and let $X_{(1)}, X_{(2)}, X_{(3)}$ be the ordered sample. Let $0 \leq a < b \leq 1$. Compute

$$E(X_{(2)} \mid X_{(1)} = a, X_{(3)} = b).$$

13. Let $X_1, X_2, \ldots, X_8$ be independent $\mathrm{Exp}(1)$-distributed random variables with order statistic $(X_{(1)}, X_{(2)}, \ldots, X_{(8)})$. Find

$$E(X_{(7)} \mid X_{(5)} = 10).$$

14. Let X_1, X_2, X_3 be independent $U(0,1)$-distributed random variables, and let $X_{(1)}, X_{(2)}, X_{(3)}$ be the order statistic. Prove the intuitively reasonable result that $X_{(1)}$ and $X_{(3)}$ are conditionally independent given $X_{(2)}$ and determine this (conditional) distribution.
    Remark. The problem thus is to determine the distribution of $(X_{(1)}, X_{(3)}) \mid X_{(2)} = x$.

15. The random variables X_1, X_2, and X_3 are independent and $\mathrm{Exp}(1)$-distributed. Compute the correlation coefficient $\rho_{X_{(1)}, X_{(3)}}$.

16. Let X_1 and X_2 be independent, $\mathrm{Exp}(a)$-distributed random variables.
    a) Show that $X_{(1)}$ and $X_{(2)} - X_{(1)}$ are independent, and determine their distributions.
    b) Compute $E(X_{(2)} \mid X_{(1)} = y)$ and $E(X_{(1)} \mid X_{(2)} = x)$.

17. Let X_1, X_2, and X_3 be independent, $U(0,1)$-distributed random variables. Compute $P(X_{(3)} > 1/2 \mid X_{(1)} = x)$.

18. Suppose that $X \in U(0,1)$. Let $X_{(1)}, X_{(2)}, \ldots, X_{(n)}$ be the order variables corresponding to a sample of n independent observations of X, and set

$$V_i = \frac{X_{(i)}}{X_{(i+1)}}, \quad i = 1, 2, \cdots, n-1, \quad \text{and} \quad V_n = X_{(n)}.$$

Show that
(a) $V_1, V_2, \ldots, V_n$ are independent,
(b) $V_i^i \in U(0,1)$ for $i = 1, 2, \ldots, n$.

19. The random variables $X_1, X_2, \ldots, X_n, Y_1, Y_2, \ldots, Y_n$ are independent and $U(0,a)$-distributed. Determine the distribution of

$$Z_n = n \cdot \log\left(\frac{\max\{X_{(n)}, Y_{(n)}\}}{\min\{X_{(n)}, Y_{(n)}\}}\right).$$

20. Let $X_1, X_2, \ldots, X_n$ be independent, $\text{Exp}(a)$-distributed random variables, and set

$$Y_1 = X_{(1)} \quad \text{and} \quad Y_k = X_{(k)} - X_{(k-1)}, \quad \text{for} \quad 2 \le k \le n.$$

(a) Show that $Y_1, Y_2, \ldots, Y_n$ are independent, and determine their distributions.
(b) Determine $E X_{(n)}$ and $\text{Var}\, X_{(n)}$.

21. The purpose of this problem is to provide a probabilistic proof of the relation

$$\int_0^\infty nx(1 - e^{-x})^{n-1} e^{-x}\, dx = 1 + \frac{1}{2} + \frac{1}{3} + \cdots + \frac{1}{n}.$$

Let $X_1, X_2, \ldots, X_n$ be independent, $\text{Exp}(1)$-distributed random variables. Consider the usual order variables $X_{(1)}, X_{(2)}, \ldots, X_{(n)}$, and set

$$Y_1 = X_{(1)} \quad \text{and} \quad Y_k = X_{(k)} - X_{(k-1)}, \quad k = 2, 3, \ldots, n.$$

(a) Show that $Y_1, Y_2, \ldots, Y_n$ are independent, and determine their distributions.
(b) Use (a) and the fact that $X_{(n)} = Y_1 + Y_2 + \cdots + Y_n$ to prove the desired formula.

Remark 1. The independence of $Y_1, Y_2, \ldots, Y_n$ is not needed for the proof of the formula.

Remark 2. For a proof using properties of the Poisson process, see Subsection 8.5.4.

22. Let $X_1, X_2, \ldots, X_n$ be independent, $\text{Exp}(1)$-distributed random variables, and set

$$Z_n = nX_{(1)} + (n-1)X_{(2)} + \cdots + 2X_{(n-1)} + X_{(n)}.$$

Compute $E Z_n$ and $\text{Var}\, Z_n$.

23. The random variables $X_1, X_2, \ldots, X_n$ are independent and Exp(1)-distributed. Set

$$V_n = X_{(n)} \quad \text{and} \quad W_n = X_1 + \frac{1}{2}X_2 + \frac{1}{3}X_3 + \cdots + \frac{1}{n}X_n.$$

Show that $V_n \overset{d}{=} W_n$.

24. Let $X_1, X_2, \ldots, X_n$ be independent, Exp(a)-distributed random variables. Determine the distribution of $\sum_{k=1}^{n} X_{(k)}$.

25. Let $X_1, X_2, \ldots, X_n$ be i.i.d. random variables and let $X_{(1)}, X_{(2)}, \ldots, X_{(n)}$ be the order variables. Determine

$$E(X_1 \mid X_{(1)}, X_{(2)}, \ldots, X_{(n)}).$$

26. The number of individuals N in a tribe is Fs(p)-distributed. The lifetimes of the individuals in the tribe are independent, Exp($1/a$)-distributed random variables, which, further, are independent of N. Determine the distribution of the shortest lifetime.

27. Let $X_1, X_2, \ldots$ be independent, $U(0,1)$-distributed random variables, and let $N \in \text{Po}(\lambda)$ be independent of $X_1, X_2, \ldots$. Set

$$V = \max\{X_1, X_2, \ldots, X_N\}$$

($V = 0$ when $N = 0$). Determine the distribution of V, and compute $E\,V$.

28. Let $X_1, X_2, \ldots$ be Exp(θ)-distributed random variables, let $N \in \text{Po}(\lambda)$, and suppose that all random variables are independent. Set

$$Y = \max\{X_1, X_2, \ldots, X_N\} \quad \text{with} \quad Y = 0 \text{ for } N = 0.$$

Show that $Y \overset{d}{=} \max\{0, V\}$, where V has a Gumbel type distribution.
Remark. The distribution function of the standard Gumbel distribution equals

$$\Lambda(x) = e^{-e^{-x}}, \quad -\infty < x < \infty.$$

29. Suppose that the random variables $X_1, X_2, \ldots$ are independent with common distribution function $F(x)$. Suppose, further, that N is a positive, integer-valued random variable with generating function $g(t)$. Finally, suppose that N and $X_1, X_2, \ldots$ are independent. Set

$$Y = \max\{X_1, X_2, \ldots, X_N\}.$$

Show that

$$F_Y(y) = g\big(F(y)\big).$$

5

The Multivariate Normal Distribution

1 Preliminaries from Linear Algebra

In Chapter 1 we studied how to handle (linear transformations of) random vectors, that is, vectors whose components are random variables. Since the normal distribution is (one of) the most important distribution(s) and since there are special properties, methods, and devices pertaining to this distribution, we devote this chapter to the study of the *multivariate normal distribution*, or, equivalently, to the study of *normal random vectors*. We show, for example, that the sample mean and the sample variance in a (one-dimensional) sample are independent, a property that, in fact, characterizes this distribution and is essential, for example, in the so called *t*-test, which is used to test hypotheses about the mean in the (univariate) normal distribution when the variance is unknown. In fact, along the way we will encounter three different ways to show this independence. Another interesting fact that will be established is that if the components of a normal random vector are uncorrelated, then they are in fact independent. One section is devoted to quadratic forms of normal random vectors, which are of great importance in many branches of statistics. The main result, Cohran's theorem, states that, under certain conditions, one can split the sum of the squares of the observations into a number of quadratic forms, each of them pertaining to some cause of variation in an experiment in such a way that these quadratic forms are independent, and (essentially) χ^2-distributed random variables. This can be used to test whether or not a certain cause of variation influences the outcome of the experiment. For more on the statistical aspects, we refer to the literature cited in Appendix A.

We begin, however, by recalling some basic facts from linear algebra. Vectors are always column vectors (recall Remark 1.1.2). For convenience, however, we sometimes write $\mathbf{x} = (x_1, x_2, \ldots, x_n)'$. A square matrix $\mathbf{A} = \{a_{ij}, \ i,j = 1, 2, \ldots, n\}$ is *symmetric* if $a_{ij} = a_{ji}$ and all elements are real. All *eigenvalues* of a real, symmetric matrix are real. In this chapter all matrices are real.

A. Gut, *An Intermediate course in Probabilty*, Springer Texts in Statistics,
DOI: 10.1007/978-1-4419-0162-0_5,
© Springer Science + Business Media, LLC 2009

A square matrix $\mathbf{C}$ is orthogonal if $\mathbf{C}'\mathbf{C} = \mathbf{I}$, where $\mathbf{I}$ is the identity matrix. Note that since, trivially, $\mathbf{C}^{-1}\mathbf{C} = \mathbf{C}\mathbf{C}^{-1} = \mathbf{I}$, it follows that

$$\mathbf{C}^{-1} = \mathbf{C}'. \tag{1.1}$$

Moreover, $\det \mathbf{C} = \pm 1$.

Remark 1.1. Orthogonality means that the rows (and columns) of an orthogonal matrix, considered as vectors, are orthonormal, that is, they have length 1 and are orthogonal; the scalar products between them are zero.   □

Let $\mathbf{x}$ be an n-vector, let $\mathbf{C}$ be an orthogonal $n \times n$ matrix, and set $\mathbf{y} = \mathbf{C}\mathbf{x}$; $\mathbf{y}$ is also an n-vector. A consequence of the orthogonality is that $\mathbf{x}$ and $\mathbf{y}$ have the same length. Indeed,

$$\mathbf{y}'\mathbf{y} = (\mathbf{C}\mathbf{x})'\mathbf{C}\mathbf{x} = \mathbf{x}'\mathbf{C}'\mathbf{C}\mathbf{x} = \mathbf{x}'\mathbf{x}. \tag{1.2}$$

Now, let $\mathbf{A}$ be a symmetric matrix. A fundamental result is that there exists an orthogonal matrix $\mathbf{C}$ such that

$$\mathbf{C}'\mathbf{A}\mathbf{C} = \mathbf{D}, \tag{1.3}$$

where $\mathbf{D}$ is a diagonal matrix, the elements of the diagonal being the eigenvalues, $\lambda_1, \lambda_2, \ldots, \lambda_n$, of $\mathbf{A}$. It also follows that

$$\det \mathbf{A} = \det \mathbf{D} = \prod_{k=1}^{n} \lambda_k. \tag{1.4}$$

A *quadratic form* $Q = Q(\mathbf{x})$ based on the symmetric matrix $\mathbf{A}$ is defined by

$$Q(\mathbf{x}) = \mathbf{x}'\mathbf{A}\mathbf{x} \quad \left(= \sum_{i=1}^{n}\sum_{j=1}^{n} a_{ij}x_ix_j \right), \quad \mathbf{x} \in \mathbf{R}^n. \tag{1.5}$$

Q is *positive-definite* if $Q(\mathbf{x}) > 0$ for all $\mathbf{x} \neq \mathbf{0}$ and *nonnegative-definite* (positive-semidefinite) if $Q(\mathbf{x}) \geq 0$ for all $\mathbf{x}$.

One can show that Q is positive- (nonnegative-)definite iff all eigenvalues are positive (nonnegative). Another useful criterion is to check all subdeterminants of $\mathbf{A}$, that is, $\det \mathbf{A}_k$, where $\mathbf{A}_k = \{a_{ij}, i, j = 1, 2, \ldots, k\}$ and $k = 1, 2, \ldots, n$. Then Q is positive- (nonnegative-)definite iff $\det \mathbf{A}_k > 0$ (≥ 0) for all $k = 1, 2, \ldots, n$.

A matrix is positive- (nonnegative-)definite iff the corresponding quadratic form is positive- (nonnegative-)definite.

Now, let $\mathbf{A}$ be a square matrix whose inverse exists. The *algebraic complement* $\mathbf{A}_{ij}$ of the element a_{ij} is defined as the matrix that remains after deleting the ith row and the jth column of $\mathbf{A}$. For the element a_{ij}^{-1} of the inverse $\mathbf{A}^{-1}$ of $\mathbf{A}$, we have

$$a_{ij}^{-1} = (-1)^{i+j} \frac{\det \mathbf{A}_{ji}}{\det \mathbf{A}}. \tag{1.6}$$

In particular, if $\mathbf{A}$ is symmetric, it follows that $\mathbf{A}_{ij} = \mathbf{A}_{ji}'$, from which we conclude that $\det \mathbf{A}_{ij} = \det \mathbf{A}_{ji}$ and hence that $a_{ij}^{-1} = a_{ji}^{-1}$ and that $\mathbf{A}^{-1}$ is symmetric.

Finally, we need to define the square root of a nonnegative-definite symmetric matrix. For a diagonal matrix $\mathbf{D}$ it is easy to see that the diagonal matrix whose diagonal elements are the square roots of those of $\mathbf{D}$ has the property that the square equals $\mathbf{D}$. For the general case we know, from (1.3), that there exists an orthogonal matrix $\mathbf{C}$ such that $\mathbf{C}'\mathbf{A}\mathbf{C} = \mathbf{D}$, that is, such that

$$\mathbf{A} = \mathbf{C}\mathbf{D}\mathbf{C}', \tag{1.7}$$

where $\mathbf{D}$ is the diagonal matrix whose diagonal elements are the eigenvalues of $\mathbf{A}$; $d_{ii} = \lambda_i$, $i = 1, 2, \ldots, n$.

Let us denote the square root of $\mathbf{D}$, as described above, by $\tilde{\mathbf{D}}$. We thus have $\tilde{d}_{ii} = \sqrt{\lambda_i}$, $i = 1, 2, \ldots, n$ and $\tilde{\mathbf{D}}^2 = \mathbf{D}$. Set $\mathbf{B} = \mathbf{C}\tilde{\mathbf{D}}\mathbf{C}'$. Then

$$\mathbf{B}^2 = \mathbf{B}\mathbf{B} = \mathbf{C}\tilde{\mathbf{D}}\mathbf{C}'\mathbf{C}\tilde{\mathbf{D}}\mathbf{C}' = \mathbf{C}\tilde{\mathbf{D}}\tilde{\mathbf{D}}\mathbf{C}' = \mathbf{C}\mathbf{D}\mathbf{C}' = \mathbf{A}, \tag{1.8}$$

that is, $\mathbf{B}$ is a square root of $\mathbf{A}$. A common notation is $\mathbf{A}^{1/2}$.

Now, this holds true for any of the 2^n choices of square roots. However, in order to ensure that the square root is nonnegative-definite we tacitly assume in the following that the nonnegative square root of the eigenvalues has been chosen, viz., that throughout $\tilde{d}_{ii} = +\sqrt{\lambda_i}$.

If, in addition, $\mathbf{A}$ has an inverse, one can show that

$$(\mathbf{A}^{-1})^{1/2} = (\mathbf{A}^{1/2})^{-1}, \tag{1.9}$$

which is denoted by $\mathbf{A}^{-1/2}$.

Exercise 1.1. Verify formula (1.9).

Exercise 1.2. Show that $\det \mathbf{A}^{-1/2} = (\det \mathbf{A})^{-1/2}$.   □

Remark 1.2. The reader who is less used to working with vectors and matrices might like to spell out certain formulas explicitly as sums or double sums, and so forth.   □

2 The Covariance Matrix

Let $\mathbf{X}$ be a random n-vector whose components have finite variance.

Definition 2.1. *The* mean vector *of* $\mathbf{X}$ *is* $\boldsymbol{\mu} = E\mathbf{X}$, *the components of which are* $\mu_i = EX_i$, $i = 1, 2, \ldots, n$.

The covariance matrix *of* $\mathbf{X}$ *is* $\boldsymbol{\Lambda} = E(\mathbf{X} - \boldsymbol{\mu})(\mathbf{X} - \boldsymbol{\mu})'$, *whose elements are* $\lambda_{ij} = E(X_i - \mu_i)(X_j - \mu_j)$, $i, j = 1, 2, \ldots, n$.   □

Thus, $\lambda_{ii} = \operatorname{Var} X_i$, $i = 1, 2, \ldots, n$, and $\lambda_{ij} = \operatorname{Cov}(X_i, X_j) = \lambda_{ji}$, $i, j = 1, 2, \ldots, n$ (and $i \neq j$, or else $\operatorname{Cov}(X_i, X_i) = \operatorname{Var} X_i$). In particular, every covariance matrix is symmetric.

Theorem 2.1. *Every covariance matrix is nonnegative-definite.*

Proof. The proof is immediate from the fact that, for any $\mathbf{y} \in \mathbf{R}^n$,

$$Q(\mathbf{y}) = \mathbf{y}'\Lambda\mathbf{y} = \mathbf{y}'E(\mathbf{X} - \boldsymbol{\mu})(\mathbf{X} - \boldsymbol{\mu})'\mathbf{y} = \operatorname{Var}(\mathbf{y}'(\mathbf{X} - \boldsymbol{\mu})) \geq 0. \qquad \square$$

Remark 2.1. If $\det \Lambda > 0$, the probability distribution of $\mathbf{X}$ is truly n-dimensional in the sense that it cannot be concentrated on a subspace of lower dimension. If $\det \Lambda = 0$ it can be concentrated on such a subspace; we call it the *singular* case (as opposed to the nonsingular case). $\qquad \square$

Next we consider linear transformations.

Theorem 2.2. *Let $\mathbf{X}$ be a random n-vector with mean vector $\boldsymbol{\mu}$ and covariance matrix Λ. Further, let $\mathbf{B}$ be an $m \times n$ matrix, let $\mathbf{b}$ be a constant m-vector, and set $\mathbf{Y} = \mathbf{BX} + \mathbf{b}$. Then*

$$E\,\mathbf{Y} = \mathbf{B}\boldsymbol{\mu} + \mathbf{b} \quad and \quad \operatorname{Cov}\mathbf{Y} = \mathbf{B}\Lambda\mathbf{B}'.$$

Proof. We have

$$E\,\mathbf{Y} = \mathbf{B}E\,\mathbf{X} + \mathbf{b} = \mathbf{B}\boldsymbol{\mu} + \mathbf{b}$$

and

$$\operatorname{Cov}\mathbf{Y} = E(\mathbf{Y} - E\,\mathbf{Y})(\mathbf{Y} - E\,\mathbf{Y})' = E\,\mathbf{B}(\mathbf{X} - \boldsymbol{\mu})(\mathbf{X} - \boldsymbol{\mu})'\mathbf{B}'$$
$$= \mathbf{B}E\{(\mathbf{X} - \boldsymbol{\mu})(\mathbf{X} - \boldsymbol{\mu})'\}\mathbf{B}' = \mathbf{B}\Lambda\mathbf{B}'. \qquad \square$$

Remark 2.2. Note that for $n = 1$ the theorem reduces to the well-known facts $E\,Y = aE\,X + b$ and $\operatorname{Var} Y = a^2\operatorname{Var} X$ (where $Y = aX + b$).

Remark 2.3. We will permit ourselves, at times, to be somewhat careless about specifying dimensions of matrices and vectors. It will always be tacitly understood that the dimensions are compatible with the arithmetic of the situation at hand. $\qquad \square$

3 A First Definition

We will provide three definitions of the multivariate normal distribution. In this section we present the first one, which states that a random vector is normal iff every linear combination of its components is normal. In Section 4 we provide a definition based on the characteristic function, and in Section 5 we give a definition based on the density function. We also prove that the first two definitions are always equivalent (i.e., when the covariance matrix is nonnegative-definite) and that the three of them are equivalent in the nonsingular case (i.e., when the covariance matrix is positive-definite). A fourth definition is given in Problem 10.1.

Definition I. *The random n-vector* **X** *is normal iff, for every n-vector* **a**, *the (one-dimensional) random variable* **a′X** *is normal. The notation* $\mathbf{X} \in N(\boldsymbol{\mu}, \boldsymbol{\Lambda})$ *is used to denote that* **X** *has a (multivariate) normal distribution with mean vector* $\boldsymbol{\mu}$ *and covariance matrix* $\boldsymbol{\Lambda}$. □

Remark 3.1. The actual distribution of **a′X** depends, of course, on **a**. The degenerate normal distribution (meaning variance equal to zero) is also included as a possible distribution of **a′X**.

Remark 3.2. Note that no assumption whatsoever is made about independence between the components of **X**. □

Surprisingly enough, this somewhat abstract definition is extremely applicable and useful. Moreover, several proofs, which otherwise become complicated, become very "simple" (and beautiful). For example, the following three properties are immediate consequences of this definition:

(a) Every component of **X** is normal.
(b) $X_1 + X_2 + \cdots + X_n$ is normal.
(c) Every marginal distribution is normal.

Indeed, to see that X_k is normal for $k = 1, 2, \ldots, n$, we choose **a** such that $a_k = 1$ and $a_j = 0$ otherwise.

To see that the sum of all components is normal, we simply choose $a_k = 1$ for all k.

As for (c) we argue as follows: To show that $(X_{i_1}, X_{i_2}, \ldots, X_{i_k})'$ is normal for some $k = (1,) 2, \ldots, n - 1$, amounts to checking that all linear combinations of these components are normal. However, since we know that **X** is normal, we know that **a′X** is normal for *every* **a**, in particular for all **a**, such that $a_j = 0$ for $j \neq i_1, i_2, \ldots, i_k$, which establishes the desired conclusion.

We also observe that, from a first course in probability theory, we know that any linear combination of *independent* normal random variables is normal (via the convolution formula and/or the moment generating function—recall Theorem 3.3.2), that is, the condition in Definition I is satisfied. It follows, in particular, that

(d) if **X** has independent normal components, then **X** is normal.

Another important result is as follows:

Theorem 3.1. *Suppose that* $\mathbf{X} \in N(\boldsymbol{\mu}, \boldsymbol{\Lambda})$ *and set* $\mathbf{Y} = \mathbf{BX} + \mathbf{b}$. *Then* $\mathbf{Y} \in N(\mathbf{B}\boldsymbol{\mu} + \mathbf{b}, \mathbf{B}\boldsymbol{\Lambda}\mathbf{B}')$.

Proof. The first part of the proof merely amounts to establishing the fact that a linear combination of the components of **Y** is a (some other) linear combination of the components of **X**. Namely, we wish to show that **a′Y** is normal for every **a**. However,

$$\mathbf{a}'\mathbf{Y} = \mathbf{a}'\mathbf{BX} + \mathbf{a}'\mathbf{b} = (\mathbf{B}'\mathbf{a})'\mathbf{X} + \mathbf{a}'\mathbf{b} = \mathbf{c}'\mathbf{X} + d, \tag{3.1}$$

where $\mathbf{c} = \mathbf{B}'\mathbf{a}$ and $d = \mathbf{a}'\mathbf{b}$. Since $\mathbf{c}'\mathbf{X}$ is normal according to Definition I (and d is a constant), it follows that $\mathbf{a}'\mathbf{Y}$ is normal. The correctness of the parameters follows from Theorem 2.2.   □

Exercise 3.1. Let X_1, X_2, X_3, and X_4 be independent, $N(0,1)$-distributed random variables. Set $Y_1 = X_1 + 2X_2 + 3X_3 + 4X_4$ and $Y_2 = 4X_1 + 3X_2 + 2X_3 + X_4$. Determine the distribution of $\mathbf{Y}$.

Exercise 3.2. Let $\mathbf{X} \in N + (\begin{pmatrix} 1 \\ 2 \end{pmatrix}, \begin{pmatrix} 1 & -2 \\ -2 & 7 \end{pmatrix})$. Set

$$Y_1 = X_1 + X_2 \quad \text{and} \quad Y_2 = 2X_1 - 3X_2.$$

Determine the distribution of $\mathbf{Y}$.   □

A word of caution is appropriate at this point. We noted above that all marginal distributions of a normal random vector $\mathbf{X}$ are normal. The *joint normality* of all components of $\mathbf{X}$ was essential here. In the following example we define two random variables that are normal but not jointly normal. This shows that a general converse does not hold; there exist normal random variables that are not jointly normal.

Example 3.1. Let $X \in N(0,1)$ and let Z be independent of X and such that $P(Z = 1) = P(Z = -1) = 1/2$. Set $Y = Z \cdot X$. Then

$$P(Y \leq x) = \frac{1}{2}P(X \leq x) + \frac{1}{2}P(-X \leq x) = \frac{1}{2}\Phi(x) + \frac{1}{2}(1 - \Phi(-x)) = \Phi(x),$$

that is, $Y \in N(0,1)$. Thus, X and Y are both (standard) normal. However, since

$$P(X + Y = 0) = P(Z = -1) = \frac{1}{2},$$

it follows from Definition I that $X + Y$ cannot be normal and, hence, that $(X, Y)'$ is not normal.   □

For a further example, see Problem 10.7.

Another kind of converse one might consider is the following. An obvious consequence of Theorem 3.1 is that if $\mathbf{X} \in N(\boldsymbol{\mu}, \boldsymbol{\Lambda})$, and if the matrices $\mathbf{A}$ and $\mathbf{B}$ are such that $\mathbf{A} = \mathbf{B}$, then $\mathbf{AX} \stackrel{d}{=} \mathbf{BX}$. A natural question is whether or not the converse holds, viz., if $\mathbf{AX} \stackrel{d}{=} \mathbf{BX}$, does it then follow that $\mathbf{A} = \mathbf{B}$?

Exercise 3.3. Let X_1 and X_2 be independent standard normal random variables and put

$$Y_1 = X_1 + X_2, \quad Y_2 = 2X_1 + X_2 \quad \text{and} \quad Z_1 = X_1\sqrt{2}, \quad Z_2 = \frac{3}{\sqrt{2}}X_1 + \frac{1}{\sqrt{2}}X_2.$$

(a) Determine the corresponding matrices $\mathbf{A}$ and $\mathbf{B}$?
(b) Check that $\mathbf{A} \neq \mathbf{B}$.
(c) Show that (nevertheless) $\mathbf{Y}$ and $\mathbf{Z}$ are have the same normal distribution (which one?).   □

4 The Characteristic Function: Another Definition

The characteristic function of a random vector $\mathbf{X}$ is (recall Definition 3.4.2)

$$\varphi_{\mathbf{X}}(\mathbf{t}) = E\, e^{i\mathbf{t}'\mathbf{X}}. \tag{4.1}$$

Now, suppose that $\mathbf{X} \in N(\boldsymbol{\mu}, \boldsymbol{\Lambda})$. We observe that $Z = \mathbf{t}'\mathbf{X}$ in (4.1) has a one-dimensional normal distribution by Definition I. The parameters are $m = E\,Z = \mathbf{t}'\boldsymbol{\mu}$ and $\sigma^2 = \operatorname{Var} Z = \mathbf{t}'\boldsymbol{\Lambda}\mathbf{t}$. Since

$$\varphi_{\mathbf{X}}(\mathbf{t}) = \varphi_Z(1) = \exp\{im - \tfrac{1}{2}\sigma^2\}, \tag{4.2}$$

we have established the following result:

Theorem 4.1. *For* $\mathbf{X} \in N(\boldsymbol{\mu}, \boldsymbol{\Lambda})$*, we have*

$$\varphi_{\mathbf{X}}(\mathbf{t}) = \exp\{i\mathbf{t}'\boldsymbol{\mu} - \tfrac{1}{2}\mathbf{t}'\boldsymbol{\Lambda}\mathbf{t}\}. \qquad \square$$

It turns out that we can, in fact, establish a converse to this result and thereby obtain another, equivalent, definition of the multivariate normal distribution. We therefore temporarily "forget" the above and begin by proving the following fact:

Lemma 4.1. *For any nonnegative-definite symmetric matrix* $\boldsymbol{\Lambda}$*, the function*

$$\varphi^*(\mathbf{t}) = \exp\{i\mathbf{t}'\boldsymbol{\mu} - \tfrac{1}{2}\mathbf{t}'\boldsymbol{\Lambda}\mathbf{t}\}$$

is the characteristic function of a random vector $\mathbf{X}$ *with* $E\,\mathbf{X} = \boldsymbol{\mu}$ *and* $\operatorname{Cov}\mathbf{X} = \boldsymbol{\Lambda}$*.*

Proof. Let $\mathbf{Y}$ be a random vector whose components $Y_1, Y_2, \ldots, Y_n$ are independent, $N(0,1)$-distributed random variables, and set

$$\mathbf{X} = \boldsymbol{\Lambda}^{1/2}\mathbf{Y} + \boldsymbol{\mu}. \tag{4.3}$$

Since $\operatorname{Cov}\mathbf{Y} = \mathbf{I}$, it follows from Theorem 2.2 that

$$E\,\mathbf{X} = \boldsymbol{\mu} \quad \text{and} \quad \operatorname{Cov}\mathbf{X} = \boldsymbol{\Lambda}. \tag{4.4}$$

Furthermore, an easy computation shows that

$$\varphi_{\mathbf{Y}}(\mathbf{t}) = E \exp\{i\mathbf{t}'\mathbf{Y}\} = \exp\{-\tfrac{1}{2}\mathbf{t}'\mathbf{t}\}. \tag{4.5}$$

It finally follows that

$$\begin{aligned}
\varphi_{\mathbf{X}}(\mathbf{t}) &= E \exp\{i\mathbf{t}'\mathbf{X}\} = E \exp\{i\mathbf{t}'(\boldsymbol{\Lambda}^{1/2}\mathbf{Y} + \boldsymbol{\mu})\} \\
&= \exp\{i\mathbf{t}'\boldsymbol{\mu}\} \cdot E \exp\{i\mathbf{t}'\boldsymbol{\Lambda}^{1/2}\mathbf{Y}\} \\
&= \exp\{i\mathbf{t}'\boldsymbol{\mu}\} \cdot E \exp\{i(\boldsymbol{\Lambda}^{1/2}\mathbf{t})'\mathbf{Y}\} \\
&= \exp\{i\mathbf{t}'\boldsymbol{\mu}\} \cdot \varphi_{\mathbf{Y}}(\boldsymbol{\Lambda}^{1/2}\mathbf{t}) \\
&= \exp\{i\mathbf{t}'\boldsymbol{\mu}\} \cdot \exp\{-\tfrac{1}{2}(\boldsymbol{\Lambda}^{1/2}\mathbf{t})'(\boldsymbol{\Lambda}^{1/2}\mathbf{t})\} \\
&= \exp\{i\mathbf{t}'\boldsymbol{\mu} - \tfrac{1}{2}\mathbf{t}'\boldsymbol{\Lambda}\mathbf{t}\},
\end{aligned}$$

as desired. $\square$

Note that at this point we do not (yet) know that $\mathbf{X}$ is normal.

The next step is to show that if $\mathbf{X}$ has a characteristic function given as in the lemma, then $\mathbf{X}$ is normal in the sense of Definition I. Thus, let $\mathbf{X}$ be given as described and let $\mathbf{a}$ be an arbitrary n-vector. Then

$$\varphi_{\mathbf{a}'\mathbf{X}}(u) = E \exp\{iu\,\mathbf{a}'\mathbf{X}\} = \varphi_{\mathbf{X}}(u\mathbf{a})$$
$$= \exp\{i(u\mathbf{a})'\boldsymbol{\mu} - \tfrac{1}{2}(u\mathbf{a})'\boldsymbol{\Lambda}(u\mathbf{a})\}$$
$$= \exp\{ium - \tfrac{1}{2}u^2\sigma^2\},$$

where $m = \mathbf{a}'\boldsymbol{\mu}$ and $\sigma^2 = \mathbf{a}'\boldsymbol{\Lambda}\mathbf{a} \geq 0$, which proves that $\mathbf{a}'\mathbf{X} \in N(m,\sigma^2)$ and hence that $\mathbf{X}$ is normal in the sense of Definition I.

Alternatively, we may argue as in the proof of Theorem 3.1:

$$\mathbf{a}'\mathbf{X} = \mathbf{a}'\left(\boldsymbol{\Lambda}^{1/2}\mathbf{Y} + \boldsymbol{\mu}\right) = \mathbf{a}'\boldsymbol{\Lambda}^{1/2}\mathbf{Y} + \mathbf{a}'\boldsymbol{\mu} = \left(\boldsymbol{\Lambda}^{1/2}\mathbf{a}\right)'\mathbf{Y} + \mathbf{a}'\boldsymbol{\mu},$$

which shows that a linear combination of the components of $\mathbf{X}$ is equal to (another) linear combination of the components of $\mathbf{Y}$, which, in turn, we know is normal, since $\mathbf{Y}$ has *independent components*.

We have thus shown that the function defined in Lemma 4.1 is, indeed, a characteristic function and that the linear combinations of the components of the corresponding random vector are normal. This motivates the following alternative definition of the multivariate normal distribution.

Definition II. *A random vector* $\mathbf{X}$ *is normal iff its characteristic function is of the form*

$$\varphi_{\mathbf{X}}(\mathbf{t}) = \exp\{i\mathbf{t}'\boldsymbol{\mu} - \tfrac{1}{2}\mathbf{t}'\boldsymbol{\Lambda}\mathbf{t}\},$$

for some vector $\boldsymbol{\mu}$ *and nonnegative-definite matrix* $\boldsymbol{\Lambda}$. □

We have also established the following fact:

Theorem 4.2. *Definitions I and II are equivalent.* □

Remark 4.1. The definition and expression for the moment generating function are the obvious ones:

$$\psi_{\mathbf{X}}(\mathbf{t}) = E\,e^{\mathbf{t}'\mathbf{X}} = \exp\{\mathbf{t}'\boldsymbol{\mu} + \tfrac{1}{2}\mathbf{t}'\boldsymbol{\Lambda}\mathbf{t}\}.$$
□

Exercise 4.1. Suppose that $\mathbf{X} = (X_1, X_2)'$ has characteristic function

$$\varphi_{\mathbf{X}}(\mathbf{t}) = \exp\{it_1 + 2it_2 - \tfrac{1}{2}t_1^2 + 2t_1t_2 - 6t_2^2\}.$$

Determine the distribution of $\mathbf{X}$.

Exercise 4.2. Suppose that $\mathbf{X} = (X_1, X_2)'$ has characteristic function

$$\varphi(t,u) = \exp\{it - 2t^2 - u^2 - tu\}.$$

Find the distribution of $X_1 + X_2$.

Exercise 4.3. Suppose that X and Y have a (joint) moment generating function given by

$$\psi(t,u) = \exp\{t^2 + 2tu + 4u^2\}.$$

Compute $P(2X < Y + 2)$. □

5 The Density: A Third Definition

Let $\mathbf{X} \in N(\boldsymbol{\mu}, \boldsymbol{\Lambda})$. If $\det \boldsymbol{\Lambda} = 0$, the distribution is singular, as mentioned before, and no density exists. If, however, $\det \boldsymbol{\Lambda} > 0$, then there exists a density function that, moreover, is uniquely determined by the parameters $\boldsymbol{\mu}$ and $\boldsymbol{\Lambda}$.

In order to determine the density, it is therefore sufficient to find it for a normal distribution constructed in some convenient way. To this end, let $\mathbf{Y}$ and $\mathbf{X}$ be defined as in the proof of Lemma 4.1, that is, $\mathbf{Y}$ has independent, standard normal components and $\mathbf{X} = \boldsymbol{\Lambda}^{1/2} \mathbf{Y} + \boldsymbol{\mu}$. Then $\mathbf{X} \in N(\boldsymbol{\mu}, \boldsymbol{\Lambda})$ by Theorem 3.1, as desired.

Now, since the density of $\mathbf{Y}$ is known, it is easy to compute the density of $\mathbf{X}$ with the aid of the transformation theorem. Namely,

$$f_{\mathbf{Y}}(\mathbf{y}) = \prod_{k=1}^{n} f_{Y_k}(y_k) = \prod_{k=1}^{n} \frac{1}{\sqrt{2\pi}} e^{-y_k^2/2}$$
$$= \left(\frac{1}{2\pi}\right)^{n/2} e^{-\frac{1}{2}\sum_{k=1}^{n} y_k^2} = \left(\frac{1}{2\pi}\right)^{n/2} e^{-\frac{1}{2}\mathbf{y}'\mathbf{y}}, \quad \mathbf{y} \in \mathbf{R}^n.$$

Further, since $\det \boldsymbol{\Lambda} > 0$, we know that the inverse $\boldsymbol{\Lambda}^{-1}$ exists, that

$$\mathbf{Y} = \boldsymbol{\Lambda}^{-1/2}(\mathbf{X} - \boldsymbol{\mu}), \tag{5.1}$$

and hence that the Jacobian is $\det \boldsymbol{\Lambda}^{-1/2} = (\det \boldsymbol{\Lambda})^{-1/2}$ (Exercise 1.2). The following result emerges.

Theorem 5.1. *For* $\mathbf{X} \in N(\boldsymbol{\mu}, \boldsymbol{\Lambda})$ *with* $\det \boldsymbol{\Lambda} > 0$, *we have*

$$f_{\mathbf{X}}(\mathbf{x}) = \left(\frac{1}{2\pi}\right)^{n/2} \frac{1}{\sqrt{\det \boldsymbol{\Lambda}}} \exp\left\{-\tfrac{1}{2}(\mathbf{x} - \boldsymbol{\mu})' \boldsymbol{\Lambda}^{-1}(\mathbf{x} - \boldsymbol{\mu})\right\}. \qquad \square$$

Exercise 5.1. We have tacitly used the fact that if $\mathbf{X}$ is a random vector and $\mathbf{Y} = \mathbf{B}\mathbf{X}$ then

$$\left|\frac{d(\mathbf{y})}{d(\mathbf{x})}\right| = \det \mathbf{B}.$$

Prove that this is correct. $\qquad \square$

We are now ready for our third definition.

Definition III. *A random vector* $\mathbf{X}$ *with* $E\mathbf{X} = \boldsymbol{\mu}$ *and* $\mathrm{Cov}\,\mathbf{X} = \boldsymbol{\Lambda}$, *such that* $\det \boldsymbol{\Lambda} > 0$, *is* $N(\boldsymbol{\mu}, \boldsymbol{\Lambda})$-*distributed iff the density equals*

$$f_{\mathbf{X}}(\mathbf{x}) = \left(\frac{1}{2\pi}\right)^{n/2} \frac{1}{\sqrt{\det \boldsymbol{\Lambda}}} \exp\left\{-\tfrac{1}{2}(\mathbf{x} - \boldsymbol{\mu})' \boldsymbol{\Lambda}^{-1}(\mathbf{x} - \boldsymbol{\mu})\right\}, \quad \mathbf{x} \in \mathbf{R}^n. \quad \square$$

Theorem 5.2. *Definitions I, II, and III are equivalent (in the nonsingular case).*

Proof. The equivalence of Definitions I and II was established in Section 4. The equivalence of Definitions II and III (in the nonsingular case) is a consequence of the uniqueness theorem for characteristic functions.   □

Now let us see how the density function can be computed explicitly. Let Λ_{ij} be the algebraic complement of $\lambda_{ij} = \mathrm{Cov}(X_i, X_j)$ and set $\triangle_{ij} = (-1)^{i+j} \det \Lambda_{ij}$ $(= \triangle_{ji}$, since Λ is symmetric). Since the elements of Λ^{-1} are $\triangle_{ij}/\triangle$, $i, j = 1, 2, \ldots, n$, where $\triangle = \det \Lambda$, it follows that

$$f_{\mathbf{X}}(\mathbf{x}) = \left(\frac{1}{2\pi}\right)^{n/2} \frac{1}{\sqrt{\triangle}} \exp\left\{-\frac{1}{2}\sum_{i=1}^{n}\sum_{j=1}^{n}\frac{\triangle_{ij}}{\triangle}(x_i - \mu_i)(x_j - \mu_j)\right\}. \qquad (5.2)$$

In particular, the following holds for the case $n = 2$: Set $\mu_k = E\,X_k$ and $\sigma_k^2 = \mathrm{Var}\,X_k$, $k = 1, 2$, and $\sigma_{12} = \mathrm{Cov}(X_1, X_2)$, and let $\rho = \sigma_{12}/\sigma_1\sigma_2$ be the correlation coefficient, where $|\rho| < 1$ (since $\det \Lambda > 0$). Then $\triangle = \sigma_1^2\sigma_2^2(1-\rho^2)$, $\triangle_{11} = \sigma_2^2$, $\triangle_{22} = \sigma_1^2$, $\triangle_{12} = \triangle_{21} = -\rho\sigma_1\sigma_2$, and hence

$$\Lambda = \begin{pmatrix} \sigma_1^2 & \rho\sigma_1\sigma_2 \\ \rho\sigma_1\sigma_2 & \sigma_2^2 \end{pmatrix} \quad \text{and} \quad \Lambda^{-1} = \frac{1}{1-\rho^2}\begin{pmatrix} \dfrac{1}{\sigma_1^2} & -\dfrac{\rho}{\sigma_1\sigma_2} \\ -\dfrac{\rho}{\sigma_1\sigma_2} & \dfrac{1}{\sigma_2^2} \end{pmatrix}.$$

It follows that

$$f_{X_1, X_2}(x_1, x_2) = \frac{1}{2\pi\sigma_1\sigma_2\sqrt{1-\rho^2}}$$
$$\times \exp\left\{-\frac{1}{2(1-\rho^2)}\left(\left(\frac{x_1 - \mu_1}{\sigma_1}\right)^2 - 2\rho\frac{(x_1 - \mu_1)(x_2 - \mu_2)}{\sigma_1\sigma_2} + \left(\frac{x_2 - \mu_2}{\sigma_2}\right)^2\right)\right\}.$$

Exercise 5.2. Let the (joint) moment generating function of $\mathbf{X}$ be

$$\psi(t, u) = \exp\{t^2 + 3tu + 4u^2\}.$$

Determine the density function of $\mathbf{X}$.

Exercise 5.3. Suppose that $\mathbf{X} \in N(\mathbf{0}, \Lambda)$, where

$$\Lambda = \begin{pmatrix} \frac{7}{2} & \frac{1}{2} & -1 \\ \frac{1}{2} & \frac{1}{2} & 0 \\ -1 & 0 & \frac{1}{2} \end{pmatrix}.$$

Put $Y_1 = X_2 + X_3$, $Y_2 = X_1 + X_3$, and $Y_3 = X_1 + X_2$. Determine the density function of $\mathbf{Y}$.   □

6 Conditional Distributions

Let $\mathbf{X} \in N(\boldsymbol{\mu}, \boldsymbol{\Lambda})$, and suppose that $\det \boldsymbol{\Lambda} > 0$. The density thus exists as given in Section 5. Conditional densities are defined (Chapter 2) as the ratio of the relevant joint and marginal densities. One can show that all marginal distributions of a nonsingular normal distribution are nonsingular and hence possess densities.

Let us consider the case $n = 2$ in some detail. Suppose that $(X, Y)' \in N(\boldsymbol{\mu}, \boldsymbol{\Lambda})$, where $EX = \mu_x, EY = \mu_y$, $\operatorname{Var} X = \sigma_x^2$, $\operatorname{Var} Y = \sigma_y^2$, and $\rho_{X,Y} = \rho$, where $|\rho| < 1$. Then

$$
\begin{aligned}
f_{Y|X=x}(y) &= \frac{f_{X,Y}(x,y)}{f_X(x)} \\
&= \frac{\frac{1}{2\pi\sigma_x\sigma_y\sqrt{1-\rho^2}} \exp\{-\frac{1}{2(1-\rho^2)}((\frac{x-\mu_x}{\sigma_x})^2 - 2\rho\frac{(x-\mu_x)(y-\mu_y)}{\sigma_x\sigma_y} + (\frac{y-\mu_y}{\sigma_y})^2)\}}{\frac{1}{\sqrt{2\pi}\sigma_x} \exp\{-\frac{1}{2}(\frac{x-\mu_x}{\sigma_x})^2\}} \\
&= \frac{1}{\sqrt{2\pi}\sigma_y\sqrt{1-\rho^2}} \exp\{-\frac{1}{2(1-\rho^2)}((\frac{x-\mu_x}{\sigma_x})^2\rho^2 - 2\rho\frac{(x-\mu_x)(y-\mu_y)}{\sigma_x\sigma_y} + (\frac{y-\mu_y}{\sigma_y})^2)\} \\
&= \frac{1}{\sqrt{2\pi}\sigma_y\sqrt{1-\rho^2}} \exp\left\{-\frac{1}{2\sigma_y^2(1-\rho^2)}\left(y - \mu_y - \rho\frac{\sigma_y}{\sigma_x}(x - \mu_x)\right)^2\right\}. \quad (6.1)
\end{aligned}
$$

This density is easily recognized as the density of a normal distribution with mean $\mu_y + \rho\frac{\sigma_y}{\sigma_x}(x - \mu_x)$ and variance $\sigma_y^2(1 - \rho^2)$. It follows, in particular, that

$$
\begin{aligned}
E(Y \mid X = x) &= \mu_y + \rho\frac{\sigma_y}{\sigma_x}(x - \mu_x), \\
\operatorname{Var}(Y \mid X = x) &= \sigma_y^2(1 - \rho^2).
\end{aligned} \quad (6.2)
$$

As a special feature we observe that the regression function is linear (and coinciding with the regression line) and that the conditional variance equals the residual variance. For the former statement we refer back to Remark 2.5.4 and for the latter to Theorem 2.5.3. Further, recall that the residual variance is independent of x.

Example 6.1. Suppose the density of $(X, Y)'$ is given by

$$
f(x, y) = \frac{1}{2\pi} \exp\{-\tfrac{1}{2}(x^2 - 2xy + 2y^2)\}.
$$

Determine the conditional distributions, particularly the conditional expectations and the conditional variances.

Solution. The function $x^2 - 2xy + 2y^2 = (x - y)^2 + y^2$ is positive-definite. We thus identify the joint distribution as normal. An inspection of the density shows that

$$EX = EY = 0 \quad \text{and} \quad \Lambda^{-1} = \begin{pmatrix} 1 & -1 \\ -1 & 2 \end{pmatrix}, \tag{6.3}$$

which implies that

$$\begin{pmatrix} X \\ Y \end{pmatrix} \in N(\mathbf{0}, \Lambda), \quad \text{where} \quad \Lambda = \begin{pmatrix} 2 & 1 \\ 1 & 1 \end{pmatrix}. \tag{6.4}$$

It follows that $\text{Var}\,X = 2$, $\text{Var}\,Y = 1$, and $\text{Cov}(X, Y) = 1$, and hence that $\rho_{X,Y} = 1/\sqrt{2}$.

A comparison with (6.2) shows that

$$E(Y \mid X = x) = \frac{x}{2} \quad \text{and} \quad \text{Var}\,(Y \mid X = x) = \frac{1}{2},$$

$$E(X \mid Y = y) = y \quad \text{and} \quad \text{Var}(X \mid Y = y) = 1.$$

The conditional distributions are the normal distributions with corresponding parameters. □

Remark 6.1. Instead of having to remember formula (6.2), it is often as simple to perform the computations leading to (6.1) directly in each case. Indeed, in higher dimensions this is necessary. As an illustration, let us compute $f_{Y|X=x}(y)$.

Following (6.4) or by using the fact that $f_X(x) = \int_{-\infty}^{\infty} f_{X,Y}(x, y)\, dy$, we have

$$f_{Y|X=x}(y) = \frac{\frac{1}{2\pi} \exp\{-\frac{1}{2}(x^2 - 2xy + 2y^2)\}}{\frac{1}{\sqrt{2\pi}\sqrt{2}} \exp\{-\frac{1}{2} \cdot \frac{x^2}{2}\}}$$

$$= \frac{1}{\sqrt{2\pi}\sqrt{1/2}} \exp\{-\frac{1}{2}(\frac{x^2}{2} - 2xy + 2y^2)\}$$

$$= \frac{1}{\sqrt{2\pi}\sqrt{1/2}} \exp\{-\frac{1}{2}\frac{(y - x/2)^2}{1/2}\},$$

which is the density of the $N(x/2, 1/2)$-distribution. □

Exercise 6.1. Compute $f_{X|Y=y}(x)$ similarly. □

Example 6.2. Suppose that $\mathbf{X} \in N(\boldsymbol{\mu}, \Lambda)$, where $\boldsymbol{\mu} = \mathbf{1}$ and

$$\Lambda = \begin{pmatrix} 3 & 1 \\ 1 & 2 \end{pmatrix}.$$

Find the conditional distribution of $X_1 + X_2$ given that $X_1 - X_2 = 0$.

Solution. We introduce the random variables $Y_1 = X_1 + X_2$ and $Y_2 = X_1 - X_2$ to reduce the problem to the standard case; we are then faced with the problem of finding the conditional distribution of Y_1 given that $Y_2 = 0$.

Since we can write $\mathbf{Y} = \mathbf{BX}$, where

$$\mathbf{B} = \begin{pmatrix} 1 & 1 \\ 1 & -1 \end{pmatrix},$$

it follows that $\mathbf{Y} \in N(\mathbf{B}\boldsymbol{\mu}, \mathbf{B}\boldsymbol{\Lambda}\mathbf{B}')$, that is, that

$$\mathbf{Y} \in N\left(\begin{pmatrix} 2 \\ 0 \end{pmatrix}, \begin{pmatrix} 7 & 1 \\ 1 & 3 \end{pmatrix} \right),$$

and hence that

$$f_{\mathbf{Y}}(\mathbf{y}) = \frac{1}{2\pi\sqrt{20}} \exp\left\{ -\frac{1}{2}\left(\frac{3(y_1-2)^2}{20} - \frac{(y_1-2)y_2}{10} + \frac{7y_2^2}{20} \right) \right\}.$$

Further, since $Y_2 \in N(0,3)$, we have

$$f_{Y_2}(y_2) = \frac{1}{\sqrt{2\pi\sqrt{3}}} \exp\left\{ -\frac{1}{2} \cdot \frac{y_2^2}{3} \right\}.$$

Finally,

$$f_{Y_1|Y_2=0}(y_1) = \frac{f_{Y_1,Y_2}(y_1,0)}{f_{Y_2}(0)} = \frac{\frac{1}{2\pi\sqrt{20}} \exp\{-\frac{1}{2}\cdot\frac{3(y_1-2)^2}{20}\}}{\frac{1}{\sqrt{2\pi\sqrt{3}}} \exp\{-\frac{1}{2}\cdot 0\}}$$

$$= \frac{1}{\sqrt{2\pi}\sqrt{20/3}} \exp\left\{ -\frac{1}{2}\frac{(y_1-2)^2}{20/3} \right\},$$

which we identify as the density of the $N(2,20/3)$-distribution.   □

Remark 6.2. It follows from the general formula (6.1) that the final exponent must be a square. This provides an extra check of one's computations. Also, the variance appears twice (in the last example it is $20/3$) and must be the same in both places.   □

Let us conclude by briefly considering the general case $n \geq 2$. Thus, $\mathbf{X} \in N(\boldsymbol{\mu}, \boldsymbol{\Lambda})$ with $\det \boldsymbol{\Lambda} > 0$. Let $\widetilde{\mathbf{X}}_1 = (X_{i_1}, X_{i_2}, ..., X_{i_k})'$ and $\widetilde{\mathbf{X}}_2 = (X_{j_1}, X_{j_2}, ..., X_{j_m})'$ be subvectors of $\mathbf{X}$, that is, vectors whose components consist of k and m of the components of $\mathbf{X}$, respectively, where $1 \leq k < n$ and $1 \leq m < n$. The components of $\widetilde{\mathbf{X}}_1$ and $\widetilde{\mathbf{X}}_2$ are assumed to be different. By definition we then have

$$f_{\widetilde{\mathbf{X}}_2|\widetilde{\mathbf{X}}_1=\widetilde{\mathbf{x}}_1}(\widetilde{\mathbf{x}}_2) = \frac{f_{\widetilde{\mathbf{X}}_1,\widetilde{\mathbf{X}}_2}(\widetilde{\mathbf{x}}_1,\widetilde{\mathbf{x}}_2)}{f_{\widetilde{\mathbf{X}}_1}(\widetilde{\mathbf{x}}_1)}. \tag{6.5}$$

Given the formula for normal densities (Theorem 5.1) and the fact that the coordinates of $\widetilde{\mathbf{x}}_1$ are constants, the ratio in (6.5) must be the density of some normal distribution. The conclusion is that *conditional distributions of multivariate normal distributions are normal.*

Exercise 6.2. Let $\mathbf{X} \in N(\mathbf{0}, \Lambda)$, where

$$\Lambda = \begin{pmatrix} 1 & 2 & -1 \\ 2 & 6 & 0 \\ -1 & 0 & 4 \end{pmatrix}.$$

Set $Y_1 = X_1 + X_3$, $Y_2 = 2X_1 - X_2$, and $Y_3 = 2X_3 - X_2$. Find the conditional distribution of Y_3 given that $Y_1 = 0$ and $Y_2 = 1$.

7 Independence

A very special property of the multivariate normal distribution is the following:

Theorem 7.1. *Let* $\mathbf{X}$ *be a normal random vector. The components of* $\mathbf{X}$ *are independent iff they are uncorrelated.*

Proof. We only need to show that uncorrelated components are independent, the converse always being true.

Thus, by assumption, $\text{Cov}(X_i, X_j) = 0$, $i \neq j$. This implies that the covariance matrix is diagonal, the diagonal elements being $\sigma_1^2, \sigma_2^2, \ldots, \sigma_n^2$. If some $\sigma_k^2 = 0$, then that component is degenerate and hence independent of the others. We therefore may assume that all variances are positive in the following. It then follows that the inverse Λ^{-1} of the covariance matrix exists; it is a diagonal matrix with diagonal elements $1/\sigma_1^2, 1/\sigma_2^2, \ldots, 1/\sigma_n^2$. The corresponding density function therefore equals

$$f_{\mathbf{X}}(\mathbf{x}) = \left(\frac{1}{2\pi}\right)^{n/2} \frac{1}{\prod_{k=1}^{n} \sigma_k} \cdot \exp\left\{-\frac{1}{2} \sum_{k=1}^{n} \frac{(x_k - \mu_k)^2}{\sigma_k^2}\right\}$$

$$= \prod_{k=1}^{n} \frac{1}{\sqrt{2\pi}\sigma_k} \cdot \exp\left\{-\frac{(x_k - \mu_k)^2}{2\sigma_k^2}\right\},$$

which proves the desired independence.   □

Example 7.1. Let X_1 and X_2 be independent, $N(0,1)$-distributed random variables. Show that $X_1 + X_2$ and $X_1 - X_2$ are independent.

Solution. It is easily checked that $\text{Cov}(X_1 + X_2, X_1 - X_2) = 0$, which implies that $X_1 + X_2$ and $X_1 - X_2$ are uncorrelated. By Theorem 7.1 they are also independent.   □

Remark 7.1. We have already encountered Example 7.1 in Chapter 1; see Example 1.2.4. There independence was proved with the aid of transformation (Theorem 1.2.1) and factorization. The solution here illustrates the power of Theorem 7.1.   □

Exercise 7.1. Let X and Y be jointly normal with correlation coefficient ρ and suppose that $\operatorname{Var} X = \operatorname{Var} Y$. Show that X and $Y - \rho X$ are independent.

Exercise 7.2. Let X and Y be jointly normal with $EX = EY = 0$, $\operatorname{Var} X = \operatorname{Var} Y = 1$, and correlation coefficient ρ. Find θ such that $X \cos\theta + Y \sin\theta$ and $X \cos\theta - Y \sin\theta$ are independent.

Exercise 7.3. Generalize the results of Example 7.1 and Exercise 7.1 to the case of nonequal variances. $\qquad\square$

Remark 7.2. In Example 3.1 we stressed the importance of the assumption that the distribution was *jointly* normal. The example is also suited to illustrate the importance of that assumption with respect to Theorem 7.1. Namely, since $EX = EY = 0$ and $EXY = EX^2 Z = EX^2 \cdot EZ = 0$, it follows that X and Y are *uncorrelated*. However, since $|X| = |Y|$, it is clear that X and Y are *not independent*. $\qquad\square$

We conclude by stating the following generalization of Theorem 7.1, the proof of which we leave as an exercise:

Theorem 7.2. *Suppose that* $\mathbf{X} \in N(\boldsymbol{\mu}, \boldsymbol{\Lambda})$, *where* $\boldsymbol{\Lambda}$ *can be partitioned as follows:*

$$\boldsymbol{\Lambda} = \begin{pmatrix} \boldsymbol{\Lambda}_1 & \mathbf{0} & \mathbf{0} & \mathbf{0} \\ \mathbf{0} & \boldsymbol{\Lambda}_2 & \mathbf{0} & \mathbf{0} \\ \mathbf{0} & \mathbf{0} & \ddots & \mathbf{0} \\ \mathbf{0} & \mathbf{0} & \mathbf{0} & \boldsymbol{\Lambda}_k \end{pmatrix}$$

(possibly after reordering the components), where $\boldsymbol{\Lambda}_1, \boldsymbol{\Lambda}_2, \ldots, \boldsymbol{\Lambda}_k$ *are matrices along the diagonal of* $\boldsymbol{\Lambda}$. *Then* $\mathbf{X}$ *can be partitioned into vectors* $\mathbf{X}^{(1)}, \mathbf{X}^{(2)}, \ldots, \mathbf{X}^{(k)}$ *with* $\operatorname{Cov}(\mathbf{X}^{(i)}) = \boldsymbol{\Lambda}_i$, $i = 1, 2, \ldots, k$, *in such a way that these random vectors are independent.* $\qquad\square$

Example 7.2. Suppose that $\mathbf{X} \in N(\mathbf{0}, \boldsymbol{\Lambda})$, where

$$\boldsymbol{\Lambda} = \begin{pmatrix} 1 & 0 & 0 \\ 0 & 2 & 4 \\ 0 & 4 & 9 \end{pmatrix}.$$

Then X_1 and $(X_2, X_3)'$ are independent. $\qquad\square$

8 Linear Transformations

A major consequence of Theorem 7.1 is that it is possible to make linear transformations of normal vectors in such a way that the new vector has independent components. In particular, any orthogonal transformation of a normal vector whose components are independent and have common variance produces a new normal random vector with independent components. As a major application, we show in Example 8.3 how these relatively simple facts can be used to prove the rather delicate result that states that the sample mean and the sample variance in a normal sample are independent. For further details concerning applications in statistics we refer to Appendix A, where some references are given.

We first recall from Section 3 that a linear transformation of a normal random vector is normal. Now suppose that $\mathbf{X} \in N(\boldsymbol{\mu}, \boldsymbol{\Lambda})$. Since $\boldsymbol{\Lambda}$ is nonnegative-definite, there exists (formula (1.3)) an orthogonal matrix $\mathbf{C}$, such that $\mathbf{C}'\boldsymbol{\Lambda}\mathbf{C} = \mathbf{D}$, where $\mathbf{D}$ is a diagonal matrix whose diagonal elements are the eigenvalues $\lambda_1, \lambda_2, \ldots, \lambda_n$ of $\boldsymbol{\Lambda}$.

Set $\mathbf{Y} = \mathbf{C}'\mathbf{X}$. It follows from Theorem 3.1 that $\mathbf{Y} \in N(\mathbf{C}'\boldsymbol{\mu}, \mathbf{D})$. The components of $\mathbf{Y}$ are thus uncorrelated and, in view of Theorem 7.1, *independent*, which establishes the following result:

Theorem 8.1. *Let* $\mathbf{X} \in N(\boldsymbol{\mu}, \boldsymbol{\Lambda})$, *and set* $\mathbf{Y} = \mathbf{C}'\mathbf{X}$, *where the orthogonal matrix* $\mathbf{C}$ *is such that* $\mathbf{C}'\boldsymbol{\Lambda}\mathbf{C} = \mathbf{D}$. *Then* $\mathbf{Y} \in N(\mathbf{C}'\boldsymbol{\mu}, \mathbf{D})$. *Moreover, the components of* $\mathbf{Y}$ *are independent and* $\operatorname{Var} Y_k = \lambda_k$, $k = 1, 2, \ldots, n$, *where* $\lambda_1, \lambda_2, \ldots, \lambda_n$ *are the eigenvalues of* $\boldsymbol{\Lambda}$. $\square$

Remark 8.1. In particular, it may occur that some eigenvalues are equal to zero, in which case the corresponding component is degenerate.

Remark 8.2. As a special corollary it follows that the statement "$\mathbf{X} \in N(\mathbf{0}, \mathbf{I})$" is equivalent to the statement "$X_1, X_2, \ldots, X_n$ are independent, standard normal random variables."

Remark 8.3. The primary use of Theorem 8.1 is in proofs and for theoretical arguments. In practice it may be cumbersome to apply the theorem when n is large, since the computation of the eigenvalues of $\boldsymbol{\Lambda}$ amounts to solving an algebraic equation of degree n. $\square$

Another situation of considerable importance in statistics is orthogonal transformations of independent, normal random variables with the same variance, the point being that the transformed random variables also are independent. That this is indeed the case may easily be proved with the aid of Theorem 8.1. Namely, let $\mathbf{X} \in N(\boldsymbol{\mu}, \sigma^2 \mathbf{I})$, where $\sigma^2 > 0$, and set $\mathbf{Y} = \mathbf{C}\mathbf{X}$, where $\mathbf{C}$ is an orthogonal matrix. Then $\operatorname{Cov} \mathbf{Y} = \mathbf{C}\sigma^2 \mathbf{I}\mathbf{C}' = \sigma^2 \mathbf{I}$, which, in view of Theorem 7.1, yields the following result:

Theorem 8.2. *Let* $\mathbf{X} \in N(\boldsymbol{\mu}, \sigma^2 \mathbf{I})$, *where* $\sigma^2 > 0$, *let* $\mathbf{C}$ *be an arbitrary orthogonal matrix, and set* $\mathbf{Y} = \mathbf{C}\mathbf{X}$. *Then* $\mathbf{Y} \in N(\mathbf{C}\boldsymbol{\mu}, \sigma^2 \mathbf{I})$; *in particular,* $Y_1, Y_2, \ldots, Y_n$ *are independent normal random variables with the same variance,* σ^2. $\square$

As a first application we reexamine Example 7.1.

Example 8.1. Thus, X and Y are independent, $N(0,1)$-distributed random variables, and we wish to show that $X + Y$ and $X - Y$ are independent.

It is clearly equivalent to prove that $U = (X+Y)/\sqrt{2}$ and $V = (X-Y)/\sqrt{2}$ are independent. Now, $(X,Y)' \in N(\mathbf{0}, \mathbf{I})$ and

$$\begin{pmatrix} U \\ V \end{pmatrix} = \mathbf{B} \begin{pmatrix} X \\ Y \end{pmatrix}, \quad \text{where} \quad \mathbf{B} = \begin{pmatrix} \frac{1}{\sqrt{2}} & \frac{1}{\sqrt{2}} \\ \frac{1}{\sqrt{2}} & -\frac{1}{\sqrt{2}} \end{pmatrix},$$

that is, $\mathbf{B}$ is orthogonal. The conclusion follows immediately from Theorem 8.2.

Example 8.2. Let $X_1, X_2, \ldots, X_n$ be independent, $N(0,1)$-distributed random variables, and let $a_1, a_2, \ldots, a_n$ be reals, such that $\sum_{k=1}^n a_k^2 \neq 0$. Find the conditional distribution of $\sum_{k=1}^n X_k^2$ given that $\sum_{k=1}^n a_k X_k = 0$.

Solution. We first observe that $\sum_{k=1}^n X_k^2 \in \chi^2(n)$ (recall Exercise 3.3.6 for the case $n = 2$). In order to determine the desired conditional distribution, we define an orthogonal matrix $\mathbf{C}$, whose first row consists of the elements $a_1/a, a_2/a, \ldots, a_n/a$, where $a = \sqrt{\sum_{k=1}^n a_k^2}$; note that $\sum_{k=1}^n (a_k/a)^2 = 1$. From linear algebra we know that the matrix $\mathbf{C}$ can be completed in such a way that it becomes an orthogonal matrix. Next we set $\mathbf{Y} = \mathbf{C}\mathbf{X}$, note that $\mathbf{Y} \in N(\mathbf{0}, \mathbf{I})$ by Theorem 8.2, and observe that, in particular, $aY_1 = \sum_{k=1}^n a_k X_k$. Moreover, since $\mathbf{C}$ is orthogonal, we have $\sum_{k=1}^n Y_k^2 = \sum_{k=1}^n X_k^2$ (formula (1.2)). It follows that the desired conditional distribution is the same as the conditional distribution of $\sum_{k=1}^n Y_k^2$ given that $Y_1 = 0$, that is, as the distribution of $\sum_{k=2}^n Y_k^2$, which is $\chi^2(n-1)$. $\qquad \square$

Exercise 8.1. Study the case $n = 2$ and $a_1 = a_2 = 1$ in detail. Try also to reach the conclusion via the random variables U and V in Example 8.1. $\quad \square$

Example 8.3. There exists a famous characterization of the normal distribution to the effect that it is the only distribution such that the arithmetic mean and the sample variance are independent. This independence is, for example, exploited in order to verify that the t-statistic, which is used for testing the mean in a normal population when the variance is unknown, actually follows a t-distribution.

Here we prove the "if" part; the other one is much harder. Thus, let $X_1, X_2, \ldots, X_n$ be independent, $N(0,1)$-distributed random variables, set $\bar{X}_n = \frac{1}{n} \sum_{k=1}^n X_k$ and $s_n^2 = \frac{1}{n-1} \sum_{k=1}^n (X_k - \bar{X}_n)^2$.

The first step is to determine the distribution of

$$(\bar{X}_n, X_1 - \bar{X}_n, X_2 - \bar{X}_n, \ldots, X_n - \bar{X}_n)'.$$

Since the vector can be written as $\mathbf{B}\mathbf{X}$, where

$$\mathbf{B} = \begin{pmatrix} \frac{1}{n} & \frac{1}{n} & \frac{1}{n} & \cdots & \frac{1}{n} \\ 1-\frac{1}{n} & -\frac{1}{n} & -\frac{1}{n} & \cdots & -\frac{1}{n} \\ -\frac{1}{n} & 1-\frac{1}{n} & -\frac{1}{n} & \cdots & -\frac{1}{n} \\ \vdots & \vdots & & \ddots & \vdots \\ -\frac{1}{n} & -\frac{1}{n} & -\frac{1}{n} & \cdots & 1-\frac{1}{n} \end{pmatrix},$$

we know that the vector is normal with mean $\mathbf{0}$ and covariance matrix

$$\mathbf{BB'} = \begin{pmatrix} \frac{1}{n} & \mathbf{0} \\ \mathbf{0} & \mathbf{A} \end{pmatrix},$$

where $\mathbf{A}$ is some matrix the exact expression of which is of no importance here. Namely, the point is that we may apply Theorem 7.2 in order to conclude that $\bar{X}_n$ and $(X_1 - \bar{X}_n, X_2 - \bar{X}_n, \ldots, X_n - \bar{X}_n)$ are independent, and since s_n^2 is simply a function of $(X_1 - \bar{X}_n, X_2 - \bar{X}_n, \ldots, X_n - \bar{X}_n)$ it follows that $\bar{X}_n$ and s_n^2 are independent random variables.   □

Exercise 8.2. Suppose that $\mathbf{X} \in N(\boldsymbol{\mu}, \sigma^2 \mathbf{I})$, where $\sigma^2 > 0$. Show that if $\mathbf{B}$ is any matrix such that $\mathbf{BB'} = \mathbf{D}$, a diagonal matrix, then the components of $\mathbf{Y} = \mathbf{BX}$ are independent, normal random variables; this generalizes Theorem 8.2. As an application, reconsider Example 8.1.   □

Theorem 8.3. *(Daly's theorem) Let $\mathbf{X} \in N(\boldsymbol{\mu}, \sigma^2 \mathbf{I})$ and set $\bar{X}_n = \frac{1}{n}\sum_{k=1}^{n} X_k$. Suppose that $g(\mathbf{x})$ is translation invariant, that is, for all $\mathbf{x} \in \mathbf{R}^n$, we have $g(\mathbf{x} + a \cdot \mathbf{1}) = g(\mathbf{x})$ for all a. Then $\bar{X}_n$ and $g(\mathbf{X})$ are independent.*

Proof. Throughout the proof we assume, without restriction, that $\boldsymbol{\mu} = \mathbf{0}$ and that $\sigma^2 = 1$. The translation invariance of g implies that g is, in fact, living in the $(n-1)$-dimensional hyperplane $x_1 + x_2 + \cdots + x_n =$ constant, on which $\bar{X}_n$ is constant. We therefore make a change of variable similar to that of Example 8.2. Namely, define an orthogonal matrix $\mathbf{C}$ such that the first row has all elements equal to $1/\sqrt{n}$, and set $\mathbf{Y} = \mathbf{CX}$. Then, by construction, we have $Y_1 = \sqrt{n} \cdot \bar{X}_n$ and, by Theorem 8.2, that $\mathbf{Y} \in N(\mathbf{0}, \mathbf{I})$. The translation invariance implies, in view of the above, that g depends only on $Y_2, Y_3, \ldots, Y_n$ and hence, by Theorem 7.2, is independent of Y_1.   □

Example 8.4. Since the sample variance s_n^2 as defined in Example 8.3 is translation invariant, the conclusion of that example follows, alternatively, from Daly's theorem. Note, however, that Daly's theorem can be viewed as an extension of that very example.

Example 8.5. The range $R_n = X_{(n)} - X_{(1)}$ (which was defined in Section 4.2) is obviously translation invariant. It follows that $\bar{X}_n$ and R_n are independent (in normal samples).   □

There also exist useful linear transformations that are not orthogonal. One important example, in the two-dimensional case, is the following, a special case of which was considered in Exercise 7.1.

Suppose that $\mathbf{X} \in N(\boldsymbol{\mu}, \boldsymbol{\Lambda})$, where

$$\boldsymbol{\mu} = \begin{pmatrix} \mu_1 \\ \mu_2 \end{pmatrix} \quad \text{and} \quad \boldsymbol{\Lambda} = \begin{pmatrix} \sigma_1^2 & \rho\sigma_1\sigma_2 \\ \rho\sigma_1\sigma_2 & \sigma_2^2 \end{pmatrix}$$

with $|\rho| < 1$. Define $\mathbf{Y}$ through the relations

$$\begin{aligned} X_1 &= \mu_1 + \sigma_1 Y_1, \\ X_2 &= \mu_2 + \rho\sigma_2 Y_1 + \sigma_2\sqrt{1 - \rho^2}Y_2. \end{aligned} \tag{8.1}$$

This means that $\mathbf{X}$ and $\mathbf{Y}$ are connected via $\mathbf{X} = \boldsymbol{\mu} + \mathbf{B}\mathbf{Y}$, where

$$\mathbf{B} = \begin{pmatrix} \sigma_1 & 0 \\ \rho\sigma_2 & \sigma_2\sqrt{1 - \rho^2} \end{pmatrix},$$

which is not orthogonal. However, a simple computation shows that $\mathbf{Y} \in N(\mathbf{0}, \mathbf{I})$, that is, Y_1 and Y_2 are independent, standard normal random variables.

Example 8.6. If X_1 and X_2 are independent and $N(0,1)$-distributed, then X_1^2 and X_2^2 are independent, $\chi^2(1)$-distributed random variables, from which it follows that $X_1^2 + X_2^2 \in \chi^2(2)$ (Exercise 3.3.6(b)). Now, assume that $\mathbf{X}$ is normal with $E X_1 = E X_2 = 0$, $\mathrm{Var}\, X_1 = \mathrm{Var}\, X_2 = 1$, and $\rho_{X_1, X_2} = \rho$ with $|\rho| < 1$. Find the distribution of $X_1^2 - 2\rho X_1 X_2 + X_2^2$.

To solve this problem, we first observe that for $\rho = 0$ it reduces to Exercise 3.3.6(b) (why?). In the general case,

$$X_1^2 - 2\rho X_1 X_2 + X_2^2 = (X_1 - \rho X_2)^2 + (1 - \rho^2)X_2^2. \tag{8.2}$$

From above (or Exercise 7.1) we know that $X_1 - \rho X_2$ and X_2 are independent, in fact,

$$\begin{pmatrix} X_1 - \rho X_2 \\ X_2 \end{pmatrix} = \begin{pmatrix} 1 & -\rho \\ 0 & 1 \end{pmatrix} \cdot \begin{pmatrix} X_1 \\ X_2 \end{pmatrix} \in N\left(\mathbf{0}, \begin{pmatrix} 1 - \rho^2 & 0 \\ 0 & 1 \end{pmatrix}\right).$$

It follows that

$$X_1^2 - 2\rho X_1 X_2 + X_2^2 = (1 - \rho^2)\left\{ \left(\frac{X_1 - \rho X_2}{\sqrt{1 - \rho^2}}\right)^2 + X_2^2 \right\} \in (1 - \rho^2) \cdot \chi^2(2),$$

and since $\chi^2(2) = \mathrm{Exp}(2)$ we conclude, from the scaling property of the exponential distribution, that $X_1^2 - 2\rho X_1 X_2 + X_2^2 \in \mathrm{Exp}(2(1 - \rho^2))$.

We shall return to this example in a more general setting in Section 9; see also Problem 10.37. □

9 Quadratic Forms and Cochran's Theorem

Quadratic forms of normal random vectors are of great importance in many branches of statistics, such as least-squares methods, the analysis of variance, regression analysis, and experimental design. The general idea is to split the sum of the squares of the observations into a number of quadratic forms, each corresponding to some cause of variation. In an agricultural experiment, for example, the yield of crop varies. The reason for this may be differences in fertilization, watering, climate, and other factors in the various areas where the experiment is performed. For future purposes one would like to investigate, if possible, how much (or if at all) the various treatments influence the variability of the result. The splitting of the sum of squares mentioned above separates the causes of variability in such a way that each quadratic form corresponds to one cause, with a final form—the residual form—that measures the random errors involved in the experiment. The conclusion of Cochran's theorem (Theorem 9.2) is that, under the assumption of normality, the various quadratic forms are independent and χ^2-distributed (except for a constant factor). This can then be used for testing hypotheses concerning the influence of the different treatments. Once again, we remind the reader that some books on statistics for further study are mentioned in Appendix A.

We begin by investigating a particular quadratic form, after which we prove the important Cochran's theorem.

Let $\mathbf{X} \in N(\boldsymbol{\mu}, \boldsymbol{\Lambda})$, where $\boldsymbol{\Lambda}$ is nonsingular, and consider the quadratic form $(\mathbf{X} - \boldsymbol{\mu})'\boldsymbol{\Lambda}^{-1}(\mathbf{X} - \boldsymbol{\mu})$, which appears in the exponent of the normal density. In the special case $\boldsymbol{\mu} = \mathbf{0}$ and $\boldsymbol{\Lambda} = \mathbf{I}$ it reduces to $\mathbf{X}'\mathbf{X}$, which is $\chi^2(n)$-distributed (n is the dimension of $\mathbf{X}$). The following result shows that this is also true in the general case.

Theorem 9.1. *Suppose that* $\mathbf{X} \in N(\boldsymbol{\mu}, \boldsymbol{\Lambda})$ *with* $\det \boldsymbol{\Lambda} > 0$. *Then*

$$(\mathbf{X} - \boldsymbol{\mu})'\boldsymbol{\Lambda}^{-1}(\mathbf{X} - \boldsymbol{\mu}) \in \chi^2(n),$$

where n *is the dimension of* $\mathbf{X}$.

Proof. Set $\mathbf{Y} = \boldsymbol{\Lambda}^{-1/2}(\mathbf{X} - \boldsymbol{\mu})$. Then

$$E\mathbf{Y} = \mathbf{0} \quad \text{and} \quad \text{Cov}\,\mathbf{Y} = \boldsymbol{\Lambda}^{-1/2}\boldsymbol{\Lambda}\boldsymbol{\Lambda}^{-1/2} = \mathbf{I},$$

that is, $\mathbf{Y} \in N(\mathbf{0}, \mathbf{I})$, and it follows that

$$(\mathbf{X} - \boldsymbol{\mu})'\boldsymbol{\Lambda}^{-1}(\mathbf{X} - \boldsymbol{\mu}) = (\boldsymbol{\Lambda}^{-1/2}(\mathbf{X} - \boldsymbol{\mu}))'(\boldsymbol{\Lambda}^{-1/2}(\mathbf{X} - \boldsymbol{\mu})) = \mathbf{Y}'\mathbf{Y} \in \chi^2(n),$$

as was shown above.   □

Remark 9.1. Let $n = 2$. With the usual notation the theorem amounts to the fact that

$$\frac{1}{1 - \rho^2}\left\{\frac{(X_1 - \mu_1)^2}{\sigma_1^2} - 2\rho\frac{(X_1 - \mu_1)(X_2 - \mu_2)}{\sigma_1\sigma_2} + \frac{(X_2 - \mu_2)^2}{\sigma_2^2}\right\} \in \chi^2(2).   □$$

As an introduction to Cochran's theorem, we study the following situation. Suppose that $X_1, X_2, \ldots, X_n$ is a sample of $X \in N(0, \sigma^2)$. Set $\bar{X}_n = \frac{1}{n} \sum_{k=1}^n X_k$, and consider the following identity:

$$\sum_{k=1}^n X_k^2 = \sum_{k=1}^n (X_k - \bar{X}_n)^2 + n \cdot \bar{X}_n^2. \tag{9.1}$$

The first term on the right-hand side equals $(n-1)s_n^2$, where s_n^2 is the sample variance. It is a $\sigma^2 \cdot \chi^2(n-1)$-distributed quadratic form. The second term is $\sigma^2 \cdot \chi^2(1)$-distributed. The terms are independent. The left-hand side is $\sigma^2 \cdot \chi^2(n)$-distributed. We have thus split the sum of the squares of the observations into a sum of two independent quadratic forms that both follow some χ^2-distribution (except for the factor σ^2).

The statistical significance of this is that the splitting of the sum of the squares $\sum_{k=1}^n X_k^2$ is the following. Namely, the first term on the right-hand side of (9.1) is large if the sample is very much spread out, and the second term is large if the mean is not "close" to zero. Thus, if the sum of squares is large we may, via the decomposition (9.1) find out the cause; is the variance large or is it not true that the mean is zero (or both)?

In Example 8.3 we found that the terms on the right-hand side of (9.1) were independent. This leads to the t-test, which is used for testing whether or not the mean equals zero. More generally, representations of the sum of squares as a sum of nonnegative-definite quadratic forms play a fundamental role in statistics, as pointed out before. The problem is to assert that the various terms on the right-hand side of such representations are independent and χ^2-distributed. Cochran's theorem provides a solution to this problem.

As a preliminary we need the following lemma:

Lemma 9.1. *Let* $x_1, x_2, \ldots, x_n$ *be real numbers. Suppose that* $\sum_{i=1}^n x_i^2$ *can be split into a sum of nonnegative-definite quadratic forms, that is, suppose that*

$$\sum_{i=1}^n x_i^2 = Q_1 + Q_2 + \cdots + Q_k,$$

where $Q_i = \mathbf{x}' \mathbf{A}_i \mathbf{x}$ *and* $(\mathrm{Rank}\, Q_i =) \; \mathrm{Rank}\, \mathbf{A}_i = r_i$ *for* $i = 1, 2, \ldots, k$. *If* $\sum_{i=1}^k r_i = n$, *then there exists an orthogonal matrix* $\mathbf{C}$ *such that, with* $\mathbf{x} = \mathbf{C}\mathbf{y}$, *we have*

$$Q_1 = y_1^2 + y_2^2 + \cdots + y_{r_1}^2,$$
$$Q_2 = y_{r_1+1}^2 + y_{r_1+2}^2 + \cdots + y_{r_1+r_2}^2,$$
$$Q_3 = y_{r_1+r_2+1}^2 + y_{r_1+r_2+2}^2 + \cdots + y_{r_1+r_2+r_3}^2,$$
$$\vdots$$
$$Q_k = y_{n-r_k+1}^2 + y_{n-r_k+2}^2 + \cdots + y_n^2. \qquad \square$$

Remark 9.2. Note that different quadratic forms contain different y variables and that the number of terms in each Q_i equals the rank r_i of Q_i.   □

We confine ourselves to proving the lemma for the case $k = 2$. The general case is obtained by induction.

Proof. Recall the assumption that $k = 2$. We thus have

$$Q = \sum_{i=1}^{n} x_i^2 = \mathbf{x}'\mathbf{A}_1\mathbf{x} + \mathbf{x}'\mathbf{A}_2\mathbf{x} \quad (= Q_1 + Q_2), \qquad (9.2)$$

where $\mathbf{A}_1$ and $\mathbf{A}_2$ are nonnegative-definite matrices with ranks r_1 and r_2, respectively, and $r_1 + r_2 = n$. Since A_1 is nonnegative-definite, there exists an orthogonal matrix $\mathbf{C}$ such that

$$\mathbf{C}'\mathbf{A}_1\mathbf{C} = \mathbf{D},$$

where $\mathbf{D}$ is a diagonal matrix, the diagonal elements $\lambda_1, \lambda_2, \ldots, \lambda_n$ of which are the eigenvalues of $\mathbf{A}_1$. Since Rank $\mathbf{A}_1 = r_1$, r_1 λ-values are positive and $n - r_1$ λ-values equal zero. Suppose, without restriction, that $\lambda_i > 0$ for $i = 1, 2, \ldots, r_1$ and that $\lambda_{r_1+1} = \lambda_{r_1+2} = \cdots = \lambda_n = 0$, and set $\mathbf{x} = \mathbf{Cy}$. Then (recall (1.2) for the first equality)

$$Q = \sum_{i=1}^{n} y_i^2 = \sum_{i=1}^{r_1} \lambda_i \cdot y_i^2 + \mathbf{y}'\mathbf{C}'\mathbf{A}_2\mathbf{Cy},$$

or, equivalently,

$$\sum_{i=1}^{r_1}(1 - \lambda_i) \cdot y_i^2 + \sum_{i=r_1+1}^{n} y_i^2 = \mathbf{y}'\mathbf{C}'\mathbf{A}_2\mathbf{Cy}. \qquad (9.3)$$

Since the rank of the right-hand side of (9.3) equals $r_2 (= n - r_1)$, it follows that $\lambda_1 = \lambda_2 = \cdots = \lambda_{r_1} = 1$, which shows that

$$Q_1 = \sum_{i=1}^{r_1} y_i^2 \quad \text{and} \quad Q_2 = \sum_{i=r_1+1}^{n} y_i^2. \qquad (9.4)$$

This proves the lemma for the case $k = 2$.   □

Theorem 9.2. *(Cochran's theorem) Suppose that $X_1, X_2, \ldots, X_n$ are independent, $N(0, \sigma^2)$-distributed random variables, and that*

$$\sum_{i=1}^{n} X_i^2 = Q_1 + Q_2 + \cdots + Q_k,$$

where $Q_1, Q_2, \ldots, Q_k$ are nonnegative-definite quadratic forms in the random variables $X_1, X_2, \ldots, X_n$, that is,

$$Q_i = \mathbf{X}'\mathbf{A}_i\mathbf{X}, \quad i = 1, 2, \ldots, k.$$

Set $\text{Rank}\,\mathbf{A}_i = r_i$, $i = 1, 2, \ldots, k$. *If*

$$r_1 + r_2 + \cdots + r_k = n,$$

then

(a) $Q_1, Q_2, \ldots, Q_k$ *are independent;*

(b) $Q_i \in \sigma^2 \chi^2(r_i)$, $i = 1, 2, \ldots, k$.

Proof. It follows from Lemma 9.1 that there exists an orthogonal matrix $\mathbf{C}$ such that the transformation $\mathbf{X} = \mathbf{CY}$ yields

$$Q_1 = Y_1^2 + Y_2^2 + \cdots + Y_{r_1}^2,$$
$$Q_2 = Y_{r_1+1}^2 + Y_{r_1+2}^2 + \cdots + Y_{r_1+r_2}^2,$$
$$\vdots$$
$$Q_k = Y_{n-r_k+1}^2 + Y_{n-r_k+2}^2 + \cdots + Y_n^2.$$

Since, by Theorem 8.2, $Y_1, Y_2, \ldots, Y_n$ are independent, $N(0, \sigma^2)$-distributed random variables, and since every Y^2 occurs in exactly one Q_j, the conclusion follows. $\square$

Remark 9.3. It suffices to assume that $\text{Rank}\,\mathbf{A}_i \leq r_i$ for $i = 1, 2, \ldots, k$, with $r_1 + r_2 + \cdots + r_k = n$, in order for Theorem 9.2 to hold. This follows from a result in linear algebra, namely that if $\mathbf{A}, \mathbf{B}$, and $\mathbf{C}$ are matrices such that $\mathbf{A} + \mathbf{B} = \mathbf{C}$, then $\text{Rank}\,\mathbf{C} \leq \text{Rank}\,\mathbf{A} + \text{Rank}\,\mathbf{B}$. An application of this result yields

$$n \leq \sum_{i=1}^{k} \text{Rank}\,\mathbf{A}_i \leq \sum_{i=1}^{k} r_i = n, \tag{9.5}$$

which, in view of the assumption, forces $\text{Rank}\,\mathbf{A}_i$ to be equal to r_i for all i. $\square$

Example 9.1. We have already proved (twice) in Section 8 that the sample mean and the sample variance are independent in a normal sample. By using the partition in formula (9.1) and Cochran's theorem (and Remark 9.2) we may obtain a third proof of that fact. $\square$

In applications the quadratic forms can frequently be written as

$$Q = L_1^2 + L_2^2 + \cdots + L_p^2, \tag{9.6}$$

where $L_1, L_2, \ldots, L_p$ are linear forms in $X_1, X_2, \ldots, X_n$. It may therefore be useful to know some method for determining the rank of a quadratic form of this kind.

Theorem 9.3. *Suppose that the nonnegative-definite form $Q = Q(\mathbf{x})$ is of the form (9.6), where*

$$L_i = \mathbf{a}_i'\mathbf{x}, \quad i = 1, 2, \ldots, p,$$

and set $\mathbf{L} = (L_1, L_2, ..., L_p)'$. If there exist exactly m linear relations $\mathbf{d}_j'\mathbf{L} = 0$, $j = 1, 2, \ldots, m$, then $\text{Rank}\,Q = p - m$.

Proof. Put $\mathbf{L} = \mathbf{A}\mathbf{x}$, where $\mathbf{A}$ is a $p \times n$ matrix. Then $\mathrm{Rank}\,\mathbf{A} = p - m$. However, since

$$Q = \mathbf{L}'\mathbf{L} = \mathbf{x}'\mathbf{A}'\mathbf{A}\mathbf{x},$$

it follows (from linear algebra) that $\mathrm{Rank}\,\mathbf{A}'\mathbf{A} = \mathrm{Rank}\,\mathbf{A}$.   □

Example 9.1 (continued). Thus, let $\mathbf{X} \in N(\mathbf{0}, \sigma^2\mathbf{I})$, and consider the partition (9.1). Then $Q_1 = \sum_{k=1}^{n}(X_k - \bar{X}_n)^2$ is of the kind described in Theorem 9.3, since $\sum_{k=1}^{n}(X_k - \bar{X}_n) = 0$.   □

10 Problems

1. In this chapter we have (so far) met three equivalent definitions of a multivariate normal distribution. Here is a fourth one: $\mathbf{X}$ is normal if and only if there exists an orthogonal transformation $\mathbf{C}$ such that the random vector $\mathbf{C}\mathbf{X}$ has independent, normal components. Show that this definition is indeed equivalent to the usual ones (e.g., by showing that it is equivalent to the first one).

2. Suppose that X and Y have a two-dimensional normal distribution with means 0, variances 1, and correlation coefficient ρ, $|\rho| < 1$. Let (R, Θ) be the polar coordinates. Determine the distribution of Θ.

3. The random variables X_1 and X_2 are independent and $N(0,1)$-distributed. Set
$$Y_1 = \frac{X_1^2 - X_2^2}{\sqrt{X_1^2 + X_2^2}} \quad \text{and} \quad Y_2 = \frac{2X_1 \cdot X_2}{\sqrt{X_1^2 + X_2^2}}.$$
Show that Y_1 and Y_2 are independent, $N(0,1)$-distributed random variables.

4. The random vector $(X, Y)'$ has a two-dimensional normal distribution with $\mathrm{Var}\,X = \mathrm{Var}\,Y$. Show that $X + Y$ and $X - Y$ are independent random variables.

5. Suppose that X and Y have a joint normal distribution with $E\,X = E\,Y = 0$, $\mathrm{Var}\,X = \sigma_x^2$, $\mathrm{Var}\,Y = \sigma_y^2$, and correlation coefficient ρ. Compute $E\,XY$ and $\mathrm{Var}\,XY$.
Remark. One may use the fact that X and a suitable linear combination of X and Y are independent.

6. The random variables X and Y are independent and $N(0,1)$-distributed. Determine
   (a) $E(X \mid X > Y)$,
   (b) $E(X + Y \mid X > Y)$.

7. We know from Section 7 that if X and Y are jointly normally distributed then they are independent iff they are uncorrelated. Now, let $X \in N(0,1)$ and $c \geq 0$. Define Y as follows:
$$Y = \begin{cases} X, & \text{for } |X| \leq c, \\ -X, & \text{for } |X| > c. \end{cases}$$

(a) Show that $Y \in N(0,1)$.
(b) Show that X and Y are not jointly normal.
    Next, let $g(c) = \mathrm{Cov}\,(X,Y)$.
(c) Show that $g(0) = -1$ and that $g(c) \to 1$ as $c \to \infty$. Show that there exists c_0 such that $g(c_0) = 0$ (i.e., such that X and Y are uncorrelated).
(d) Show that X and Y are not independent (when $c = c_0$).

8. In Section 6 we found that conditional distributions of normal vectors are normal. The converse is, however, not true. Namely, consider the bivariate density

$$f_{X,Y}(x,y) = C \cdot \exp\{-(1+x^2)(1+y^2)\}, \quad -\infty < x, y < \infty,$$

where C is a normalizing constant. This is *not* a bivariate normal density. Show that in spite of this the conditional distributions are normal, that is, compute the conditional densities $f_{Y|X=x}(y)$ and $f_{X|Y=y}(x)$ and show that they are normal densities.

9. Suppose that the random variables X and Y are independent and $N(0,\sigma^2)$-distributed.
(a) Show that $X/Y \in C(0,1)$.
(b) Show that $X + Y$ and $X - Y$ are independent.
(c) Determine the distribution of $(X - Y)/(X + Y)$ (see also Problem 1.43(b)).

10. Suppose that the moment generating function of $(X,Y)'$ is

$$\psi_{X,Y}(t,u) = \exp\{2t + 3u + t^2 + atu + 2u^2\}.$$

Determine a so that $X + 2Y$ and $2X - Y$ become independent.

11. Let $\mathbf{X}$ have a three-dimensional normal distribution. Show that if X_1 and $X_2 + X_3$ are independent, X_2 and $X_1 + X_3$ are independent, and X_3 and $X_1 + X_2$ are independent, then X_1, X_2, and X_3 are independent.

12. Let X_1 and X_2 be independent, $N(0,1)$-distributed random variables. Set $Y_1 = X_1 - 3X_2 + 2$ and $Y_2 = 2X_1 - X_2 - 1$. Determine the distribution of
(a) $\mathbf{Y}$, and
(b) $Y_1 \mid Y_2 = y$.

13. Let X_1, X_2, and X_3 be independent, $N(1,1)$-distributed random variables. Set $U = 2X_1 - X_2 + X_3$ and $V = X_1 + 2X_2 + 3X_3$. Determine the conditional distribution of V given that $U = 3$.

14. Let X_1, X_2, X_3 be independent $N(2,1)$-distributed random variables. Determine the distribution of $X_1 + 3X_2 - 2X_3$ given that $2X_1 - X_2 = 1$.

15. Let Y_1, Y_2, and Y_3 be independent, $N(0,1)$-distributed random variables, and set

$$X_1 = Y_1 - Y_3,$$
$$X_2 = 2Y_1 + Y_2 - 2Y_3,$$
$$X_3 = -2Y_1 + 3Y_3.$$

Determine the conditional distribution of X_2 given that $X_1 + X_3 = x$.

16. The random variables X_1, X_2, and X_3 are independent and $N(0,1)$-distributed. Consider the random variables

$$Y_1 = X_2 + X_3,$$
$$Y_2 = X_1 + X_3,$$
$$Y_3 = X_1 + X_2.$$

Determine the conditional density of Y_1 given that $Y_2 = Y_3 = 0$.

17. The random vector $\mathbf{X}$ has a three-dimensional normal distribution with mean vector $\mathbf{0}$ and covariance matrix Λ given by

$$\Lambda = \begin{pmatrix} 2 & 0 & -1 \\ 0 & 3 & 1 \\ -1 & 1 & 5 \end{pmatrix}.$$

Find the distribution of X_2 given that $X_1 - X_3 = 1$ and that $X_2 + X_3 = 0$.

18. The random vector $\mathbf{X}$ has a three-dimensional normal distribution with expectation $\mathbf{0}$ and covariance matrix Λ given by

$$\Lambda = \begin{pmatrix} 1 & 2 & -1 \\ 2 & 4 & 0 \\ -1 & 0 & 7 \end{pmatrix}.$$

Find the distribution of X_3 given that $X_1 = 1$.

19. The random vector $\mathbf{X}$ has a three-dimensional normal distribution with expectation $\mathbf{0}$ and covariance matrix Λ given by

$$\Lambda = \begin{pmatrix} 2 & 1 & -1 \\ 1 & 3 & 0 \\ -1 & 0 & 5 \end{pmatrix}.$$

Find the distribution of X_2 given that $X_1 + X_3 = 1$.

20. The random vector $\mathbf{X}$ has a three-dimensional normal distribution with mean vector $\mu = \mathbf{0}$ and covariance matrix

$$\Lambda = \begin{pmatrix} 3 & -2 & 1 \\ -2 & 2 & 0 \\ 1 & 0 & 1 \end{pmatrix}.$$

Find the distribution of $X_1 + X_3$ given that
(a) $X_2 = 0$,
(b) $X_2 = 2$.

21. Let $\mathbf{X} \in N(\mu, \Lambda)$, where

$$\mu = \begin{pmatrix} 2 \\ 0 \\ 1 \end{pmatrix} \quad \text{and} \quad \Lambda = \begin{pmatrix} 3 & -2 & 1 \\ -2 & 2 & 0 \\ 1 & 0 & 2 \end{pmatrix}.$$

Determine the conditional distribution of $X_1 - X_3$ given that $X_2 = -1$.

22. Let $\mathbf{X} \in N(\boldsymbol{\mu}, \boldsymbol{\Lambda})$, where

$$\boldsymbol{\mu} = \begin{pmatrix} 2 \\ 0 \\ 1 \end{pmatrix} \quad \text{and} \quad \boldsymbol{\Lambda} = \begin{pmatrix} 3 & -2 & 1 \\ -2 & 2 & 0 \\ 1 & 0 & 3 \end{pmatrix}.$$

Determine the conditional distribution of $X_1 + X_2$ given that $X_3 = 1$.

23. The random vector $\mathbf{X}$ has a three-dimensional normal distribution with expectation $\boldsymbol{\mu}$ and covariance matrix $\boldsymbol{\Lambda}$ given by

$$\boldsymbol{\mu} = \begin{pmatrix} 1 \\ 1 \\ 0 \end{pmatrix} \quad \text{and} \quad \boldsymbol{\Lambda} = \begin{pmatrix} 4 & -2 & 1 \\ -2 & 3 & 0 \\ 1 & 0 & 1 \end{pmatrix}.$$

Find the conditional distribution of $X_1 + 2X_2$ given that
(a) $X_2 - X_3 = 1$.
(b) $X_2 + X_3 = 1$.

24. The random vector $\mathbf{X}$ has a three-dimensional normal distribution with mean vector $\boldsymbol{\mu}$ and covariance matrix $\boldsymbol{\Lambda}$ given by

$$\boldsymbol{\mu} = \begin{pmatrix} 1 \\ 0 \\ -2 \end{pmatrix} \quad \text{and} \quad \boldsymbol{\Lambda} = \begin{pmatrix} 3 & -2 & 1 \\ -2 & 4 & -1 \\ 1 & -1 & 2 \end{pmatrix}.$$

Find the conditional distribution of X_1 given that $X_1 = -X_2$.

25. Let $\mathbf{X}$ have a three-dimensional normal distribution with mean vector and covariance matrix

$$\boldsymbol{\mu} = \begin{pmatrix} 1 \\ 1 \\ 1 \end{pmatrix} \quad \text{and} \quad \boldsymbol{\Lambda} = \begin{pmatrix} 2 & 1 & 1 \\ 1 & 3 & -1 \\ 1 & -1 & 2 \end{pmatrix},$$

respectively. Set $Y_1 = X_1 + X_2 + X_3$ and $Y_2 = X_1 + X_3$. Determine the conditional distribution of Y_1 given that $Y_2 = 0$.

26. Let $\mathbf{X} \in N(\mathbf{0}, \boldsymbol{\Lambda})$, where

$$\boldsymbol{\Lambda} = \begin{pmatrix} 2 & 1 & -1 \\ 1 & 3 & 0 \\ -1 & 0 & 5 \end{pmatrix}.$$

Find the conditional distribution of X_1 given that $X_1 = X_2$ and $X_1 + X_2 + X_3 = 0$.

27. The random vector $\mathbf{X}$ has a three-dimensional normal distribution with expectation $\mathbf{0}$ and covariance matrix $\boldsymbol{\Lambda}$ given by

$$\boldsymbol{\Lambda} = \begin{pmatrix} 2 & 1 & 0 \\ 1 & 2 & 1 \\ 0 & 1 & 2 \end{pmatrix}.$$

Find the distribution of X_2 given that $X_1 = X_2 = X_3$.

28. Let $\mathbf{X} \in N(\mathbf{0}, \mathbf{\Lambda})$, where

$$\mathbf{\Lambda} = \begin{pmatrix} 1 & -\frac{1}{2} & \frac{3}{2} \\ -\frac{1}{2} & 2 & -1 \\ \frac{3}{2} & -1 & 4 \end{pmatrix}.$$

Determine the conditional distribution of $(X_1, X_1 + X_2)'$ given that $X_1 + X_2 + X_3 = 0$.

29. Suppose that the characteristic function of $(X, Y, Z)'$ is

$$\varphi(s, t, u) = \exp\{2is - s^2 - 2t^2 - 4u^2 - 2st + 2su\}.$$

Compute the conditional distribution of $X + Z$ given that $X + Y = 0$.

30. Let X_1, X_2, and X_3 have a joint moment generating function as follows:

$$\psi(t_1, t_2, t_3) = \exp\{2t_1 - t_3 + t_1^2 + 2t_2^2 + 3t_3^2 + 2t_1t_2 - 2t_1t_3\}.$$

Determine the conditional distribution of $X_1 + X_3$ given that $X_1 + X_2 = 1$.

31. The moment generating function of $(X, Y, Z)'$ is

$$\psi(s, t, u) = \exp\left\{\frac{s^2}{2} + t^2 + 2u^2 - \frac{st}{2} + \frac{3su}{2} - \frac{tu}{2}\right\}.$$

Determine the conditional distribution of X given that $X + Z = 0$ and $Y + Z = 1$.

32. Suppose $(X, Y, Z)'$ is normal with density

$$C \cdot \exp\left\{-\frac{1}{2}(4x^2 + 3y^2 + 5z^2 + 2xy + 6xz + 4zy)\right\},$$

where C is a normalizing constant. Determine the conditional distribution of X given that $X + Z = 1$ and $Y + Z = 0$.

33. Let X and Y be random variables, such that

$$Y \mid X = x \in N(x, \tau^2) \quad \text{with} \quad X \in N(\mu, \sigma^2).$$

(a) Compute EY, $\operatorname{Var} Y$ and $\operatorname{Cov}(X, Y)$.
(b) Determine the distribution of the vector $(X, Y)'$.
(c) Determine the (posterior) distribution of $X \mid Y = y$.

34. Let X and Y be jointly normal with means 0, variances 1, and correlation coefficient ρ. Compute the moment generating function of $X \cdot Y$ for
(a) $\rho = 0$, and
(b) general ρ.

35. Suppose X_1, X_2, and X_3 are independent and $N(0, 1)$-distributed. Compute the moment generating function of $Y = X_1X_2 + X_1X_3 + X_2X_3$.

36. If X and Y are independent, $N(0, 1)$-distributed random variables, then $X^2 + Y^2 \in \chi^2(2)$ (recall Exercise 3.3.6). Now, let X and Y be jointly normal with means 0, variances 1, and correlation coefficient ρ. In this case $X^2 + Y^2$ has a *noncentral* $\chi^2(2)$-distribution. Determine the moment generating function of that distribution.

37. Let $(X, Y)'$ have a two-dimensional normal distribution with means 0, variances 1, and correlation coefficient ρ, $|\rho| < 1$. Determine the distribution of $(X^2 - 2\rho XY + Y^2)/(1 - \rho^2)$ by computing its moment generating function.

    Remark. Recall Example 8.6 and Remark 9.1.

38. Let $X_1, X_2, \ldots, X_n$ be independent, $N(0, 1)$-distributed random variables, and set $\bar{X}_k = \frac{1}{k-1} \sum_{i=1}^{k-1} X_i$, $2 \le k \le n$. Show that

$$Q = \sum_{k=2}^{n} \frac{k-1}{k}(X_k - \bar{X}_k)^2$$

is χ^2-distributed. What is the number of degrees of freedom?

39. Let X_1, X_2, and X_3 be independent, $N(1, 1)$-distributed random variables. Set $U = X_1 + X_2 + X_3$ and $V = X_1 + 2X_2 + 3X_3$. Determine the constants a and b so that $E(U - a - bV)^2$ is minimized.

40. Let X and Y be independent, $N(0, 1)$-distributed random variables. Then $X + Y$ and $X - Y$ are independent; see Example 7.1. The purpose of this problem is to point out a (partial) converse. Suppose that X and Y are independent random variables with common distribution function F. Suppose, further, that F is symmetric and that $\sigma^2 = E X^2 < \infty$. Let φ be the characteristic function of X (and Y). Show that if $X + Y$ and $X - Y$ are independent then we have

$$\varphi(t) = \big(\varphi(t/2)\big)^4.$$

Use this relation to show that $\varphi(t) = e^{-\sigma^2 t^2/2}$. Finally, conclude that F is the distribution function of a normal distribution $(N(0, \sigma^2))$.

Remark 1. The assumptions that the distribution is symmetric and the variance is finite are not necessary. However, without them the problem becomes much more difficult.

Remark 2. Results of this kind are called *characterization theorems*. Another characterization of the normal distribution is provided by the following famous theorem due to the Swedish probabilist and statistician Harald Cramér (1893–1985): If X and Y are independent random variables such that $X + Y$ has a normal distribution, then X and Y are both normal.

6

Convergence

1 Definitions

There are several convergence concepts in probability theory. We shall discuss four of them here.

Let X_1, X_2, ... be random variables.

Definition 1.1. X_n *converges* almost surely (a.s.) *to the random variable X as $n \to \infty$ iff*

$$P(\{\omega : X_n(\omega) \to X(\omega) \text{ as } n \to \infty\}) = 1.$$

Notation: $X_n \xrightarrow{a.s.} X$ as $n \to \infty$.

Definition 1.2. X_n *converges in* probability *to the random variable X as $n \to \infty$ iff, $\forall \varepsilon > 0$,*

$$P(|X_n - X| > \varepsilon) \to 0 \quad \text{as} \quad n \to \infty.$$

Notation: $X_n \xrightarrow{p} X$ as $n \to \infty$.

Definition 1.3. X_n *converges in r-mean to the random variable X as $n \to \infty$ iff*

$$E|X_n - X|^r \to 0 \quad \text{as} \quad n \to \infty.$$

Notation: $X_n \xrightarrow{r} X$ as $n \to \infty$.

Definition 1.4. X_n *converges in* distribution *to the random variable X as $n \to \infty$ iff*

$$F_{X_n}(x) \to F_X(x) \quad \text{as} \quad n \to \infty \quad \text{for all} \quad x \in C(F_X),$$

where $C(F_X) = \{x : F_X(x) \text{ is continuous at } x\} = $ *the continuity set of F_X.*
Notation: $X_n \xrightarrow{d} X$ as $n \to \infty$. □

A. Gut, *An Intermediate course in Probabilty*, Springer Texts in Statistics,
DOI: 10.1007/978-1-4419-0162-0_6,
© Springer Science + Business Media, LLC 2009

Remark 1.1. When dealing with almost-sure convergence, we consider every $\omega \in \Omega$ and check whether or not the real numbers $X_n(\omega)$ converge to the real number $X(\omega)$ as $n \to \infty$. We have almost-sure convergence if the ω-set for which there is convergence has probability 1 or, equivalently, if the ω-set for which we do not have convergence has probability 0. Almost-sure convergence is also called convergence with probability 1 (w.p.1).

Remark 1.2. Convergence in 2-mean ($r = 2$ in Definition 1.3) is usually called convergence in square mean (or mean-square convergence).

Remark 1.3. Note that in Definition 1.4 the random variables are present only in terms of their distribution functions. Thus, they need not be defined on the same probability space.

Remark 1.4. We will permit ourselves the convenient abuse of notation such as $X_n \xrightarrow{d} N(0,1)$ or $X_n \xrightarrow{d} Po(\lambda)$ as $n \to \infty$ instead of the formally more correct, but lengthier $X_n \xrightarrow{d} X$ as $n \to \infty$, where $X \in N(0,1)$, and $X_n \xrightarrow{d} X$ as $n \to \infty$, where $X \in Po(\lambda)$, respectively.

Remark 1.5. As mentioned in Section 4 of the Introduction, one can show that a distribution function has at most only a countable number of discontinuities. As a consequence, $C(F_X)$ equals the whole real line except, possibly, for at most a countable number of points.   □

Before proceeding with the theory, we present some examples.

Example 1.1. Let $X_n \in \Gamma(n, 1/n)$. Show that $X_n \xrightarrow{p} 1$ as $n \to \infty$.

We first note that $E\,X_n = 1$ and that $\text{Var}\,X_n = 1/n$. An application of Chebyshev's inequality now shows that, for all $\varepsilon > 0$,

$$P(|X_n - 1| > \varepsilon) \leq \frac{1}{n\varepsilon^2} \to 0 \quad \text{as} \quad n \to \infty.$$

Example 1.2. Let $X_1, X_2, \ldots$ be independent random variables with common density

$$f(x) = \begin{cases} \alpha x^{-\alpha-1}, & \text{for } x > 1, \quad alpha > 0, \\ 0, & \text{otherwise,} \end{cases}$$

and set $Y_n = n^{-1/\alpha} \cdot \max_{1 \leq k \leq n} X_k$, $n \geq 1$. Show that Y_n converges in distribution as $n \to \infty$, and determine the limit distribution.

In order to solve this problem we first compute the common distribution function:

$$F(x) = \begin{cases} \int_1^x \alpha y^{-\alpha-1}\,dy = 1 - x^{-\alpha}, & \text{for } x > 1, \\ 0, & \text{otherwise,} \end{cases}$$

from which it follows that, for any $x > 0$,

$$F_{Y_n}(x) = P\left(\max_{1 \le k \le n} X_k \le xn^{1/\alpha}\right) = \left(F(xn^{1/\alpha})\right)^n$$

$$= \left(1 - \frac{1}{nx^\alpha}\right)^n \to e^{-x^{-\alpha}} \quad \text{as} \quad n \to \infty.$$

Example 1.3. The law of large numbers. This important result will be proved in full generality in Section 5 ahead. However, in Section 8 of the Introduction it was mentioned that a weaker version assuming finite variance usually is proved in a first course in probability. More precisely, let $X_1, X_2, \ldots$ be a sequence of i.i.d. random variables with mean μ and finite variance σ^2 and set $S_n = X_1 + X_2 + \cdots + X_n$, $n \ge 1$. The law of large numbers states that

$$P\left(|\frac{S_n}{n} - \mu| > \varepsilon\right) \to 0 \quad \text{as} \quad n \to \infty \quad \text{for all} \quad \varepsilon > 0,$$

that is,

$$\frac{S_n}{n} \xrightarrow{p} \mu \quad \text{as} \quad n \to \infty.$$

The proof of this statement under the above assumptions follows from Chebyshev's inequality:

$$P\left(|\frac{S_n}{n} - \mu| > \varepsilon\right) \le \frac{\sigma^2}{n\varepsilon^2} \to 0 \quad \text{as} \quad n \to \infty. \qquad \square$$

The following example, which involves convergence in distribution, deals with a special case of the Poisson approximation of the binomial distribution. The general result states that if X_n is binomial with n "large" and p "small" we may approximate X_n with a suitable Poisson distribution.

Example 1.4. Suppose that $X_n \in \text{Bin}(n, \lambda/n)$. Then

$$X_n \xrightarrow{d} \text{Po}(\lambda) \quad \text{as} \quad n \to \infty.$$

The elementary proof involves showing that, for fixed k,

$$\binom{n}{k}\left(\frac{\lambda}{n}\right)^k\left(1 - \frac{\lambda}{n}\right)^{n-k} \to e^{-\lambda}\frac{\lambda^k}{k!} \quad \text{as} \quad n \to \infty.$$

We omit the details. Another solution, involving transforms, will be given in Section 4. $\qquad \square$

We close this section with two exercises.

Exercise 1.1. Let $X_1, X_2, \ldots$ be a sample from the distribution whose density is

$$f(x) = \begin{cases} \frac{1}{2}(1+x)e^{-x}, & \text{for} \quad x > 0, \\ 0, & \text{otherwise.} \end{cases}$$

Set $Y_n = \min\{X_1, X_2, \ldots, X_n\}$. Show that $n \cdot Y_n$ converges in distribution as $n \to \infty$, and find the limit distribution.

Exercise 1.2. Let $X_1, X_2, \ldots$ be random variables defined by the relations

$$P(X_n = 0) = 1 - \frac{1}{n}, \quad P(X_n = 1) = \frac{1}{2n}, \quad \text{and} \quad P(X_n = -1) = \frac{1}{2n}, \quad n \geq 1.$$

Show that

(a) $X_n \xrightarrow{p} 0$ as $n \to \infty$,
(b) $X_n \xrightarrow{r} 0$ as $n \to \infty$, for any $r > 0$.   □

2 Uniqueness

We begin by proving that convergence is unique—in other words, that the limiting random variable is uniquely defined in the following sense: If $X_n \to X$ and $X_n \to Y$ almost surely, in probability, or in r-mean, then $X = Y$ almost surely, that is, $P(X = Y) = 1$ (or, equivalently, $P(\{\omega : X(\omega) \neq Y(\omega)\}) = 0$). For distributional convergence, uniqueness means $F_X(x) = F_Y(x)$ for all x, that is, $X \stackrel{d}{=} Y$.

As a preparation, we recall how uniqueness is proved in analysis. Let $a_1, a_2, \ldots$ be a convergent sequence of real numbers. We claim that the limit is unique. In order to prove this, one shows that if there are reals a and b such that

$$a_n \to a \quad \text{and} \quad a_n \to b \quad \text{as} \quad n \to \infty, \tag{2.1}$$

then, necessarily, $a = b$.

The conclusion follows from the triangle inequality:

$$|a - b| \leq |a - a_n| + |a_n - b| \to 0 + 0 = 0 \quad \text{as} \quad n \to \infty.$$

Since $a - b$ does not depend on n, it follows that $|a - b| = 0$, that is, $a = b$.

A proof using reductio ad absurdum runs as follows. Suppose that $a \neq b$. This implies that $|a - b| > \varepsilon$ for some $\varepsilon > 0$. Let such an $\varepsilon > 0$ be given. For every n, we must have either

$$|a_n - a| > \varepsilon/2 \quad \text{or} \quad |a_n - b| > \varepsilon/2, \tag{2.2}$$

that is, there must exist infinitely many n such that (at least) one of the inequalities in (2.2) holds. Therefore, (at least) one of the statements $a_n \to a$ as $n \to \infty$ or $a_n \to b$ as $n \to \infty$ cannot hold, which contradicts the assumption and hence shows that indeed $a = b$.

This is, of course, a rather inelegant proof. We present it only because the proof for convergence in probability is closely related.

To prove uniqueness for our new convergence concepts, we proceed analogously.

Theorem 2.1. *Let $X_1, X_2, \ldots$ be a sequence of random variables. If X_n converges almost surely, in probability, in r-mean, or in distribution as $n \to \infty$, then the limiting random variable (distribution) is unique.*

Proof. (i)   Suppose first that $X_n \xrightarrow{a.s.} X$ and $X_n \xrightarrow{a.s.} Y$ as $n \to \infty$. Let

$$N_X = \{\omega : X_n(\omega) \not\to X(\omega) \text{ as } n \to \infty\}$$

and

$$N_Y = \{\omega : X_n(\omega) \not\to Y(\omega) \text{ as } n \to \infty\}.$$

Clearly, $P(N_X) = P(N_Y) = 0$.

Now let $\omega \in (N_X \cup N_Y)^c$. By the triangle inequality it follows that

$$|X(\omega) - Y(\omega)| \leq |X(\omega) - X_n(\omega)| + |X_n(\omega) - Y(\omega)| \to 0 \qquad (2.3)$$

as $n \to \infty$ and hence that

$$X(\omega) = Y(\omega) \quad \text{whenever} \quad \omega \notin N_X \cup N_Y.$$

Consequently,

$$P(X \neq Y) \leq P(N_X \cup N_Y) \leq P(N_X) + P(N_Y) = 0,$$

which proves uniqueness in this case.

(ii)   Next suppose that $X_n \xrightarrow{p} X$ and $X_n \xrightarrow{p} Y$ as $n \to \infty$, and let $\varepsilon > 0$ be arbitrary. Since

$$|X - Y| \leq |X - X_n| + |X_n - Y|, \qquad (2.4)$$

it follows that if $|X - Y| > \varepsilon$ for some $\omega \in \Omega$, then either $|X - X_n| > \varepsilon/2$ or $|X_n - Y| > \varepsilon/2$ (cf. (2.2)). More formally,

$$\{\omega : |X - Y| > \varepsilon\} \subset \left\{\omega : |X - X_n| > \frac{\varepsilon}{2}\right\} \cup \left\{\omega : |X_n - Y| > \frac{\varepsilon}{2}\right\}. \qquad (2.5)$$

Thus,

$$P(|X - Y| > \varepsilon) \leq P\left(|X - X_n| > \frac{\varepsilon}{2}\right) + P\left(|X_n - Y| > \frac{\varepsilon}{2}\right) \to 0 \qquad (2.6)$$

as $n \to \infty$, that is,

$$P(|X - Y| > \varepsilon) = 0 \quad \text{for all} \quad \varepsilon > 0,$$

which implies that

$$P(|X - Y| > 0) = 0, \quad \text{that is,} \quad P(X = Y) = 1.$$

(iii)   Now suppose that $X_n \xrightarrow{r} X$ and $X_n \xrightarrow{r} Y$ as $n \to \infty$. For this case we need a replacement for the triangle inequality when $r \neq 1$.

Lemma 2.1. *Let $r > 0$. Suppose that U and V are random variables such that $E|U|^r < \infty$ and $E|V|^r < \infty$. Then*

$$E|U + V|^r \leq 2^r (E|U|^r + E|V|^r).$$

Proof. Let a and b be reals. Then

$$|a + b|^r \leq (|a| + |b|)^r \leq (2 \cdot \max\{|a|, |b|\})^r$$
$$= 2^r \cdot \max\{|a|^r, |b|^r\} \leq 2^r \cdot (|a|^r + |b|^r).$$

For every $\omega \in \Omega$, we thus have

$$|U(\omega) + V(\omega)|^r \leq 2^r (|U(\omega)|^r + |V(\omega)|^r).$$

Taking expectations in both members yields

$$E|U + V|^r \leq 2^r (E|U|^r + E|V|^r). \qquad \qquad \square$$

Remark 2.1. The constant 2^r can be improved to $\max\{1, 2^{r-1}\}$. $\qquad \square$

In order to prove (iii), we now note that by Lemma 2.1

$$E|X - Y|^r \leq 2^r (E|X - X_n|^r + E|X_n - Y|^r) \to 0 \quad \text{as} \quad n \to \infty. \qquad (2.7)$$

This implies that $E|X - Y|^r = 0$, which yields $P(|X - Y| = 0) = 1$ (i.e., $P(X = Y) = 1$).
(iv) Finally, suppose that

$$X_n \xrightarrow{d} X \quad \text{and} \quad X_n \xrightarrow{d} Y \quad \text{as} \quad n \to \infty,$$

and let $x \in C(F_X) \cap C(F_Y)$ (note that $(C(F_X) \cap C(F_Y))^c$ contains at most a countable number of points). Then

$$|F_X(x) - F_Y(x)| \leq |F_X(x) - F_{X_n}(x)| + |F_{X_n}(x) - F_Y(x)| \to 0 \qquad (2.8)$$

as $n \to \infty$, which shows that $F_X(x) = F_Y(x)$ for all $x \in C(F_X) \cap C(F_Y)$. As a last step we would have to show that in fact $F_X(x) = F_Y(x)$ for *all* x. We confine ourselves to claiming that this is a consequence of the right continuity of distribution functions. $\qquad \square$

3 Relations Between the Convergence Concepts

The obvious first question is whether or not the convergence concepts we have introduced really are different and if they are, whether or not they can be ordered in some sense. These problems are the topic of the present section. Instead of beginning with a big theorem, we prefer to proceed step by step and state the result at the end.

One can show that $X_n \xrightarrow{a.s.} X$ as $n \to \infty$ iff, $\forall \varepsilon > 0$ and δ, $0 < \delta < 1$, $\exists n_0$ such that, $\forall n > n_0$,

$$P\left(\bigcap_{m>n} \{|X_m - X| < \varepsilon\}\right) > 1 - \delta, \tag{3.1}$$

or, equivalently,

$$P\left(\bigcup_{m>n} \{|X_m - X| > \varepsilon\}\right) < \delta. \tag{3.2}$$

Since, for $m > n$,

$$\{|X_m - X| > \varepsilon\} \subset \bigcup_{k>n} \{|X_k - X| > \varepsilon\},$$

we have made plausible the fact that

I. $X_n \xrightarrow{a.s.} X$ as $n \to \infty$ $\implies$ $X_n \xrightarrow{p} X$ as $n \to \infty$. $\qquad\square$

Remark 3.1. An approximate way of verbalizing the conclusion is that, for convergence in probability, the set where X_m and X are not close is small for m large. *But,* we may have different sets of discrepancy for different (large) values of m. For a.s. convergence, however, the discrepancy set is *fixed*, common, for all large m. $\qquad\square$

The following example shows that the two convergence concepts are not equivalent:

Example 3.1. Let $X_2, X_3, \ldots$ be independent random variables such that

$$P(X_n = 1) = 1 - \frac{1}{n} \quad \text{and} \quad P(X_n = n) = \frac{1}{n}, \quad n \geq 2.$$

Clearly,

$$P(|X_n - 1| > \varepsilon) = P(X_n = n) = \frac{1}{n} \to 0 \quad \text{as} \quad n \to \infty,$$

for every $\varepsilon > 0$, that is,

$$X_n \xrightarrow{p} 1 \quad \text{as} \quad n \to \infty. \tag{3.3}$$

We now show that X_n does not converge a.s. to 1 as $n \to \infty$. Namely, for every $\varepsilon > 0$, $\delta \in (0, 1)$, and $N > n$, we have

$$P\left(\bigcap_{m>n} \{|X_m - 1| < \varepsilon\}\right) \leq P\left(\bigcap_{m=n+1}^{N} \{|X_m - 1| < \varepsilon\}\right)$$

$$= \prod_{m=n+1}^{N} P(|X_m - 1| < \varepsilon) = \prod_{m=n+1}^{N} P(X_m = 1) = \prod_{m=n+1}^{N} \left(1 - \frac{1}{m}\right)$$

$$= \prod_{m=n+1}^{N} \frac{m-1}{m} = \frac{n}{N} < 1 - \delta, \tag{3.4}$$

no matter how large n is chosen, provided we then choose N such that $N > n/(1 - \delta)$. This shows that there exists no n_0 for which (3.1) can hold, and hence that X_n does not converge a.s. to 1 as $n \to \infty$. Moreover, it follows from Theorem 2.1 (uniqueness) that we cannot have a.s. convergence to any other random variable either, since we then would also have convergence in probability to that random variable, which in turn would contradict (3.3). $\square$

It is actually possible to compute the left-hand side of (3.4):

$$P\Big(\bigcap_{m>n} \{|X_m - 1| < \varepsilon\} \Big) = P\Big(\lim_{N \to \infty} \bigcap_{m=n+1}^{N} \{|X_m - 1| < \varepsilon\} \Big)$$

$$= \lim_{N \to \infty} P\Big(\bigcap_{m=n+1}^{N} \{|X_m - 1| < \varepsilon\} \Big) = \cdots = \lim_{N \to \infty} \frac{n}{N} = 0 < 1 - \delta$$

for every δ, $0 < \delta < 1$. However, in order to do this properly, we would need the following lemma for the second equality sign.

Lemma 3.1. *Suppose that B and $\{B_n, n \geq 1\}$ are subsets of Ω, such that $B_n \uparrow B$ as $n \to \infty$. Then $P(B_n) \to P(B)$ as $n \to \infty$.* $\square$

Exercise 3.1. Prove Lemma 3.1. $\square$

Remark 3.2. The lemma amounts to showing that

$$P(\lim_{n \to \infty} B_n) = \lim_{n \to \infty} P(B_n),$$

that is, we must verify the interchange of taking limits and computing probabilities (i.e., summation or integration).

Remark 3.3. Please note that our proofs for a.s. convergence have not been carried out in detail. In fact, complete proofs would lead beyond the scope of this book. Our hope was to "make plausible" the results. $\square$

The next step is to show that

II. $X_n \xrightarrow{r} X$ as $n \to \infty$ $\implies$ $X_n \xrightarrow{p} X$ as $n \to \infty$.

This part is easy since, by Markov's inequality (recall formula (8.2) of the Introduction),

$$P(|X_n - X| > \varepsilon) \leq \frac{E|X_n - X|^r}{\varepsilon^r} \to 0 \quad \text{as} \quad n \to \infty, \tag{3.5}$$

which proves the conclusion. $\square$

That the converse need not hold follows trivially from the fact that $E|X_n - X|^r$ might not even exist. There are, however, cases when $X_n \xrightarrow{p} X$ as $n \to \infty$, whereas $E|X_n - X|^r \not\to 0$ as $n \to \infty$. For $r = 1$ we may use Example 3.1. We prefer, however, to modify the example in order to make it more general.

Example 3.2. Let $\alpha > 0$ and let $X_2, X_3, \ldots$ be random variables such that

$$P(X_n = 1) = 1 - \frac{1}{n^\alpha} \quad \text{and} \quad P(X_n = n) = \frac{1}{n^\alpha}, \quad n \geq 2.$$

Since $P(|X_n - 1| > \varepsilon) = P(X_n = n) = 1/n^\alpha \to 0$ as $n \to \infty$, it follows that

$$X_n \xrightarrow{p} 1 \quad \text{as} \quad n \to \infty. \tag{3.6}$$

Furthermore,

$$E|X_n - 1|^r = 0^r \cdot \left(1 - \frac{1}{n^\alpha}\right) + |n - 1|^r \cdot \frac{1}{n^\alpha} = \frac{(n-1)^r}{n^\alpha},$$

from which it follows that

$$E|X_n - 1|^r \to \begin{cases} 0, & \text{for } r < \alpha, \\ 1, & \text{for } r = \alpha, \\ +\infty, & \text{for } r > \alpha. \end{cases} \tag{3.7}$$

This shows that $X_n \xrightarrow{r} 1$ as $n \to \infty$ when $r < \alpha$ but that X_n does not converge in r-mean as $n \to \infty$ when $r \geq \alpha$. Convergence in r-mean thus is a strictly stronger concept than convergence in probability. $\qquad\square$

Remark 3.4. If $\alpha = 1$ and if, in addition, $X_2, X_3, \ldots$ are independent, then

$$X_n \xrightarrow{p} 1 \quad \text{as} \quad n \to \infty,$$

$$X_n \overset{\text{a.s.}}{\not\longrightarrow} \quad \text{as} \quad n \to \infty,$$

$$E X_n \to 2 \quad \text{as} \quad n \to \infty,$$

$$X_n \xrightarrow{r} 1 \quad \text{as} \quad n \to \infty \quad \text{for } 0 < r < 1,$$

$$X_n \overset{r}{\not\longrightarrow} \quad \text{as} \quad n \to \infty \quad \text{for } r \geq 1.$$

Remark 3.5. If $\alpha = 2$ and in addition $X_2, X_3, \ldots$ are independent, then

$$X_n \xrightarrow{p} 1 \quad \text{as} \quad n \to \infty,$$

$$X_n \xrightarrow{\text{a.s.}} 1 \quad \text{as} \quad n \to \infty \quad \text{(try to prove that!)},$$

$$E X_n \to 1 \quad \text{and} \quad \operatorname{Var} X_n \to 1 \quad \text{as} \quad n \to \infty,$$

$$X_n \xrightarrow{r} 1 \quad \text{as} \quad n \to \infty \quad \text{for } 0 < r < 2,$$

$$X_n \overset{r}{\not\longrightarrow} \quad \text{as} \quad n \to \infty \quad \text{for } r \geq 2. \qquad\square$$

III. The concepts a.s. convergence and convergence in r-mean cannot be ordered; neither implies the other.

To see this, we inspect Remarks 3.4 and 3.5. In the former, X_n converges in r-mean for $0 < r < 1$, but not almost surely, and in the latter X_n converges almost surely, but not in r-mean, if $r \geq 2$. $\qquad\square$

Note also that if $r \geq 1$ in Remark 3.4, then X_n converges in probability, but neither almost surely nor in r-mean; whereas if $0 < r < 2$ in Remark 3.5, then X_n converges almost surely and hence in probability as well as in r-mean.

We finally relate the concept of convergence in distribution to the others.

IV. $X_n \xrightarrow{p} X$ as $n \to \infty$ $\implies$ $X_n \xrightarrow{d} X$ as $n \to \infty$.

Let $\varepsilon > 0$. Then

$$
\begin{aligned}
F_{X_n}(x) &= P(X_n \leq x) \\
&= P(\{X_n \leq x\} \cap \{|X_n - X| \leq \varepsilon\}) + P(\{X_n \leq x\} \cap \{|X_n - X| > \varepsilon\}) \\
&\leq P(\{X \leq x + \varepsilon\} \cap \{|X_n - X| \leq \varepsilon\}) + P(|X_n - X| > \varepsilon) \\
&\leq P(X \leq x + \varepsilon) + P(|X_n - X| > \varepsilon),
\end{aligned}
$$

that is,

$$
F_{X_n}(x) \leq F_X(x + \varepsilon) + P(|X_n - X| > \varepsilon). \tag{3.8}
$$

By switching X_n to X, x to $x - \varepsilon$, X to X_n, and $x + \varepsilon$ to x, it follows, analogously, that

$$
F_X(x - \varepsilon) \leq F_{X_n}(x) + P(|X_n - X| > \varepsilon). \tag{3.9}
$$

Since $X_n \xrightarrow{p} X$ as $n \to \infty$, we obtain, by letting $n \to \infty$ in (3.8) and (3.9),

$$
F_X(x - \varepsilon) \leq \liminf_{n \to \infty} F_{X_n}(x) \leq \limsup_{n \to \infty} F_{X_n}(x) \leq F_X(x + \varepsilon). \tag{3.10}
$$

This relation holds for all x and for all $\varepsilon > 0$. To prove convergence in distribution, we finally suppose that $x \in C(F_X)$ and let $\varepsilon \to 0$. It follows that

$$
F_X(x) \leq \liminf_{n \to \infty} F_{X_n}(x) \leq \limsup_{n \to \infty} F_{X_n}(x) \leq F_X(x), \tag{3.11}
$$

that is,

$$
\lim_{n \to \infty} F_{X_n}(x) = F_X(x).
$$

Since $x \in C(F_X)$ was arbitrary, the conclusion follows. $\qquad \square$

Remark 3.6. We observe that if F_X has a jump at x, then we can only conclude that

$$
F_X(x-) \leq \liminf_{n \to \infty} F_{X_n}(x) \leq \limsup_{n \to \infty} F_{X_n}(x) \leq F_X(x). \tag{3.12}
$$

Here $F_X(x) - F_X(x-)$ equals the size of the jump. This explains why only continuity points are involved in the definition of distributional convergence.$\square$

Since, as was mentioned earlier, distributional convergence does not require jointly distributed random variables, whereas the other concepts do, it is clear that distributional convergence is the weakest concept. The following example shows that there exist jointly distributed random variables that converge in distribution only.

Example 3.3. Suppose that X is a random variable with a symmetric, continuous, nondegenerate distribution, and let X_1, X_2, ... be such that $X_{2n} = X$ and $X_{2n-1} = -X$, $n = 1, 2, \ldots$. Since $X_n \overset{d}{=} X$ for all n, we have, in particular, $X_n \overset{d}{\longrightarrow} X$ as $n \to \infty$. Further, since X has a nondegenerate distribution, there exists $a > 0$ such that $P(|X| > a) > 0$ (why?). It follows that for every ε, $0 < \varepsilon < 2a$,

$$P(|X_n - X| > \varepsilon) = \begin{cases} 0, & \text{for } n \text{ even,} \\ P(|X| > \varepsilon/2) > 0, & \text{for } n \text{ odd.} \end{cases}$$

This shows that X_n cannot converge in probability to X as $n \to \infty$, and thus neither almost surely nor in r-mean. $\qquad \square$

The following theorem collects our findings from this section so far:

Theorem 3.1. *Let X and X_1, X_2, ... be random variables. The following implications hold as $n \to \infty$:*

$$X_n \overset{a.s.}{\longrightarrow} X \quad \Longrightarrow \quad X_n \overset{p}{\longrightarrow} X \quad \Longrightarrow \quad X_n \overset{d}{\longrightarrow} X$$

$$X_n \overset{r}{\longrightarrow} X.$$

All implications are strict. $\qquad \square$

In addition to this general result, we have the following one, which states that convergence in probability and convergence in distribution are equivalent if the limiting random variable is degenerate.

Theorem 3.2. *Let X_1, X_2, ... be random variables and c be a constant. Then*

$$X_n \overset{d}{\longrightarrow} \delta(c) \quad as \quad n \to \infty \quad \Longleftrightarrow \quad X_n \overset{p}{\longrightarrow} c \quad as \quad n \to \infty.$$

Proof. Since the implication $\Longleftarrow$ always holds (Theorem 3.1), we only have to prove the converse.

Thus, assume that $X_n \overset{d}{\longrightarrow} \delta(c)$ as $n \to \infty$, and let $\varepsilon > 0$. Then

$$\begin{aligned} P(|X_n - c| > \varepsilon) &= 1 - P(c - \varepsilon \leq X_n \leq c + \varepsilon) \\ &= 1 - F_{X_n}(c + \varepsilon) + F_{X_n}(c - \varepsilon) - P(X_n = c - \varepsilon) \\ &\leq 1 - F_{X_n}(c + \varepsilon) + F_{X_n}(c - \varepsilon) \to 1 - 1 + 0 \\ &= 0 \quad \text{as} \quad n \to \infty, \end{aligned}$$

since $F_{X_n}(c + \varepsilon) \to F_X(c + \varepsilon) = 1$, $F_{X_n}(c - \varepsilon) \to F_X(c - \varepsilon) = 0$, and $c + \varepsilon$ and $c - \varepsilon \in C(F_X) = \{x : x \neq c\}$. $\qquad \square$

Recall, once again, that only the continuity points of the limiting distribution function were involved in Definition 1.4. The following example shows that this is necessary for the definition to make sense.

Example 3.4. Let $X_n \in \delta(1/n)$ for all n. Then, clearly, $X_n \xrightarrow{p} 0$ as $n \to \infty$. It follows from Theorem 3.1 that we also have $X_n \xrightarrow{d} \delta(0)$ as $n \to \infty$. However, $F_{X_n}(0) = 0$ for all n, whereas $F_{\delta(0)}(0) = 1$, that is, the sequence of distribution functions does not converge to the corresponding value of the distribution function of the limiting random variable at *every* point (but at all continuity points).

If, instead $X_n \in \delta(-1/n)$ for all n, then, similarly, $X_n \xrightarrow{p} 0$ and $X_n \xrightarrow{d} \delta(0)$ as $n \to \infty$. However, in this case $F_{X_n}(0) = 1$ for all n, so that the distribution functions converge properly at *every* point.

Given the similarity of the two cases it would obviously be absurd if one would have convergence in distribution in the first case but not in the second one. Luckily the requirement that convergence is only required at continuity points saves the situation. □

Remark 3.7. For a.s. convergence, convergence in probability, and convergence in r-mean, one can show that Cauchy convergence actually implies convergence. For a.s. convergence this follows from the corresponding result for real numbers, but for the other concepts this is much harder to prove.

Remark 3.8. The uniqueness theorems in Section 2 for a.s. convergence and convergence in r-mean may actually be obtained as corollaries of the uniqueness theorem for convergence in probability via Theorem 3.1. Explicitly, suppose, for example, that $X_n \xrightarrow{a.s.} X$ and that $X_n \xrightarrow{a.s.} Y$ as $n \to \infty$. According to Theorem 3.1, we then also have $X_n \xrightarrow{p} X$ and $X_n \xrightarrow{p} Y$ as $n \to \infty$ and hence, by uniqueness, that $P(X = Y) = 1$. □

Exercise 3.2. Show that $X_n \xrightarrow{r} X$ and $X_n \xrightarrow{a.s.} Y$ as $n \to \infty$ implies that $P(X = Y) = 1$.

Exercise 3.3. Toss a symmetric coin and set $X = 1$ for heads and $X = 0$ for tails. Let $X_1, X_2, \ldots$ be random variables such that $X_{2n} = X$ and $X_{2n-1} = 1 - X$, $n = 1, 2, \ldots$. Show that $X_n \xrightarrow{d} X$ as $n \to \infty$, but that $X_n \xrightarrow{p} \!\!\!\!\!/ \; X$ as $n \to \infty$. □

4 Convergence via Transforms

In Chapter 3 we found that transforms are very useful for determining the distribution of new random variables, particularly for sums of independent random variables. In this section we shall see that transforms may also be used to prove convergence in distribution; it turns out that in order to prove that $X_n \xrightarrow{d} X$ as $n \to \infty$, it suffices to assert that the transform of X_n converges to the corresponding transform of X. Theorems of this kind are called *continuity theorems*.

Two important applications will be given in the next section, where we prove two fundamental results on the convergence of normalized sums of i.i.d. random variables: the law of large numbers and the central limit theorem.

Theorem 4.1. *Let* X, X_1, X_2, ... *be nonnegative, integer-valued random variables, and suppose that*

$$g_{X_n}(t) \to g_X(t) \quad as \quad n \to \infty.$$

Then

$$X_n \xrightarrow{d} X \quad as \quad n \to \infty. \qquad \square$$

Theorem 4.2. *Let* X_1, X_2, ... *be random variables such that* $\psi_{X_n}(t)$ *exists for* $|t| < h$ *for some* $h > 0$ *and for all* n. *Suppose further that* X *is a random variable whose moment generating function* $\psi_X(t)$ *exists for* $|t| \leq h_1 < h$ *for some* $h_1 > 0$ *and that*

$$\psi_{X_n}(t) \to \psi_X(t) \quad as \quad n \to \infty, \quad for \quad |t| \leq h_1.$$

Then

$$X_n \xrightarrow{d} X \quad as \quad n \to \infty. \qquad \square$$

Theorem 4.3. *Let* X, X_1, X_2, ... *be random variables, and suppose that*

$$\varphi_{X_n}(t) \to \varphi_X(t) \quad as \quad n \to \infty, \quad for \quad -\infty < t < \infty.$$

Then

$$X_n \xrightarrow{d} X \quad as \quad n \to \infty. \qquad \square$$

Remark 4.1. Theorem 4.3 can be sharpened; we need only to assume that $\varphi_{X_n}(t) \to \varphi(t)$ as $n \to \infty$, where φ is some function that is continuous at $t = 0$. The conclusion then is that X_n converges in distribution as $n \to \infty$ to some random variable X whose characteristic function is φ. The formulation of Theorem 4.3 implicitly presupposes the knowledge that the limit is, indeed, a characteristic function and, moreover, the characteristic function of a known (to us) random variable X. In the sharper formulation we can answer the weaker question of whether or not X_n converges in distribution as $n \to \infty$; we have an *existence theorem* in this case.

Remark 4.2. The converse problem is also of interest. Namely, one can show that if X_1, X_2, ... is a sequence of random variables such that

$$X_n \xrightarrow{d} X \quad as \quad n \to \infty$$

for some random variable X, then

$$\varphi_{X_n}(t) \to \varphi_X(t) \quad as \quad n \to \infty \quad for \quad -\infty < t < \infty,$$

that is, the characteristic functions converge. $\qquad \square$

A particular case of interest is when the limiting random variable X is degenerate. We then know from Theorem 3.2 that convergence in probability and distributional convergence are equivalent. The following result is a useful consequence of this fact:

Corollary 4.3.1. *Let X_1, X_2, ... be random variables, and suppose that, for some real number c,*

$$\varphi_{X_n}(t) \to e^{itc} \quad as \quad n \to \infty, \quad for \quad -\infty < t < \infty.$$

Then

$$X_n \xrightarrow{p} c \quad as \quad n \to \infty. \qquad \square$$

Exercise 4.1. Prove Corollary 4.3.1. $\qquad \square$

Example 4.1. Show that $X_n \xrightarrow{p} 1$ as $n \to \infty$ in Example 3.1.

In order to apply the corollary, we prove that the characteristic function of X_n converges as desired as n tends to infinity:

$$\varphi_{X_n}(t) = e^{it \cdot 1}\left(1 - \frac{1}{n}\right) + e^{it \cdot n}\frac{1}{n} = e^{it} + \frac{e^{itn} - e^{it}}{n} \to e^{it} \quad as \quad n \to \infty,$$

since $|e^{itn} - e^{it}|/n \le 2/n \to 0$ as $n \to \infty$. And since e^{it} is the characteristic function of the $\delta(1)$-distribution, Corollary 4.3.1 finally implies that $X_n \xrightarrow{p} 1$ as $n \to \infty$.

Remark 4.3. The earlier, direct proof is the obvious one; the purpose here was merely to illustrate the method. Note also that Theorem 4.3, which lies behind this method, was stated without proof.

Remark 4.4. The analogous computation using moment generating functions collapses:

$$\psi_{X_n}(t) = e^{t \cdot 1}\left(1 - \frac{1}{n}\right) + e^{t \cdot n}\frac{1}{n} \to \begin{cases} e^t, & \text{for } t \le 0, \\ +\infty, & \text{for } t > 0, \end{cases} \quad as \quad n \to \infty.$$

The reason for the collapse is that, as we found in Example 3.1, the moments of order greater than or equal to 1 do not converge properly—recall that $E X_n \to 2$ and $\text{Var} X_n \to \infty$ as $n \to \infty$—and, hence, the moment generating functions cannot converge either. $\qquad \square$

Example 4.2. (Another solution of Example 1.1) Since $X_n \in \Gamma(n, 1/n)$, we have

$$\varphi_{X_n}(t) = \left(\frac{1}{1 - \frac{it}{n}}\right)^n = \frac{1}{(1 - \frac{it}{n})^n} \to \frac{1}{e^{-it}} = e^{it} = \varphi_{\delta(1)}(t) \quad as \quad n \to \infty,$$

and the desired conclusion follows from Corollary 4.3.1. $\qquad \square$

Exercise 4.2. Use transforms to solve the problem given in Exercise 1.2. □

As another, deeper example we reconsider Example 1.4 concerning the Poisson approximation of the binomial distribution.

Example 4.3. We were given $X_n \in \text{Bin}(n, \lambda/n)$ and wanted to show that $X_n \xrightarrow{d} \text{Po}(\lambda)$ as $n \to \infty$.

To verify this we compute the generating function of X_n:

$$g_{X_n}(t) = \left(1 - \frac{\lambda}{n} + \frac{\lambda}{n}t\right)^n = \left(1 + \frac{\lambda(t-1)}{n}\right)^n \to e^{\lambda(t-1)}$$

$$= g_{\text{Po}(\lambda)}(t) \quad \text{as} \quad n \to \infty,$$

and the conclusion follows from Theorem 4.1. □

In proving uniqueness for distributional convergence (Step IV in the proof of Theorem 2.1), we had some trouble with the continuity points. Here we provide a proof using characteristic functions in which we do not have to worry about such matters. (The reasons are that uniqueness theorems and continuity theorems for transforms imply distributional uniqueness and distributional convergence of random variables and also that events on sets with probability zero do not matter in theorems for transforms.)

We thus assume that $X_1, X_2, \ldots$ are random variables such that $X_n \xrightarrow{d} X$ and $X_n \xrightarrow{d} Y$ as $n \to \infty$. Then (Remark 4.2)

$$\varphi_{X_n}(t) \to \varphi_X(t) \quad \text{and} \quad \varphi_{X_n}(t) \to \varphi_Y(t) \quad \text{as} \quad n \to \infty,$$

whence

$$|\varphi_X(t) - \varphi_Y(t)| \leq |\varphi_X(t) - \varphi_{X_n}(t)| + |\varphi_{X_n}(t) - \varphi_Y(t)| \to 0 + 0 = 0$$

as $n \to \infty$. This shows that $\varphi_X(t) = \varphi_Y(t)$, which, by Theorem 3.4.2, proves that $X \stackrel{d}{=} Y$, and we are done.

5 The Law of Large Numbers and the Central Limit Theorem

The two most fundamental results in probability theory are the law of large numbers (LLN) and the central limit theorem (CLT). In a first course in probability the law of large numbers is usually proved with the aid of Chebyshev's inequality under the assumption of finite variance, and the central limit theorem is normally given without a proof. Here we shall formulate and prove both theorems under minimal conditions (in the case of i.i.d. summands).

Theorem 5.1. *(The weak law of large numbers) Let X_1, X_2, ... be i.i.d. random variables with finite expectation μ, and set $S_n = X_1 + X_2 + \cdots + X_n$, $n \geq 1$. Then*

$$\bar{X}_n = \frac{S_n}{n} \xrightarrow{p} \mu \quad as \quad n \to \infty.$$

Proof. According to Corollary 4.3.1, it suffices to show that

$$\varphi_{\bar{X}_n}(t) \to e^{it\mu} \quad as \quad n \to \infty, \quad for \quad -\infty < t < \infty.$$

By Theorem 3.4.9 and Corollary 3.4.6.1 we have

$$\varphi_{\bar{X}_n}(t) = \varphi_{S_n}\left(\frac{t}{n}\right) = \left(\varphi_{X_1}\left(\frac{t}{n}\right)\right)^n, \tag{5.1}$$

which, together with Theorem 3.4.7, yields

$$\varphi_{\bar{X}_n}(t) = \left(1 + i\frac{t}{n}\mu + o\left(\frac{t}{n}\right)\right)^n \to e^{it\mu} \quad as \quad n \to \infty$$

for all t. $\square$

Remark 5.1. With different methods one can in fact prove that $\bar{X}_n \xrightarrow{a.s.} \mu$ as $n \to \infty$ and that the assumption about finite mean is necessary. This result is called the *strong* law of large numbers in contrast to Theorem 5.1, which is called the *weak* law of large numbers. For more on this, we refer to Appendix A, where some references are given, and to the end of Subsection 7.7.3 for some remarks on complete convergence and its relation to the strong law. $\square$

Exercise 5.1. Let X_1, X_2, ... be i.i.d. random variables such that $E|X|^k < \infty$. Show that

$$\frac{X_1^k + X_2^k + \cdots + X_n^k}{n} \xrightarrow{p} E\,X^k \quad as \quad n \to \infty.$$ $\square$

Theorem 5.2. *(The central limit theorem) Let X_1, X_2, ... be i.i.d. random variables with finite expectation μ and finite variance σ^2, and set $S_n = X_1 + X_2 + \cdots + X_n$, $n \geq 1$. Then*

$$\frac{S_n - n\mu}{\sigma\sqrt{n}} \xrightarrow{d} N(0,1) \quad as \quad n \to \infty.$$

Proof. In view of the continuity theorem for characteristic functions (Theorem 4.3), it suffices to prove that

$$\varphi_{\frac{S_n - n\mu}{\sigma\sqrt{n}}}(t) \to e^{-t^2/2} \quad as \quad n \to \infty, \quad for \quad -\infty < t < \infty. \tag{5.2}$$

The relation

$$\frac{S_n - n\mu}{\sigma\sqrt{n}} = \frac{\sum_{k=1}^{n}\left(\frac{X_k - \mu}{\sigma}\right)}{\sqrt{n}}$$

shows that it is no restriction to assume, throughout the proof, that $\mu = 0$ and $\sigma = 1$. With the aid of Theorems 3.4.9, 3.4.6, and 3.4.7 (in particular, Remark 3.4.3), we then obtain

$$\varphi_{\frac{S_n - n\mu}{\sigma\sqrt{n}}}(t) = \varphi_{\frac{S_n}{\sqrt{n}}}(t) = \varphi_{S_n}\left(\tfrac{t}{\sqrt{n}}\right) = \left(\varphi_{X_1}\left(\tfrac{t}{\sqrt{n}}\right)\right)^n$$

$$= \left(1 - \frac{t^2}{2n} + o\left(\frac{t^2}{n}\right)\right)^n \to e^{-t^2/2} \quad \text{as} \quad n \to \infty. \qquad \Box$$

Remark 5.2. The centering ($\mu = 0$) in the proof has a simplifying effect. Otherwise, one would have

$$\varphi_{\frac{S_n - n\mu}{\sigma\sqrt{n}}}(t) = \exp\left\{-\frac{i\mu\sqrt{n}}{\sigma}t\right\} \cdot \varphi_{\frac{S_n}{\sigma\sqrt{n}}}(t)$$

$$= \exp\left\{-\frac{i\mu\sqrt{n}}{\sigma}t\right\} \cdot \left(\varphi_{X_1}\left(\tfrac{t}{\sigma\sqrt{n}}\right)\right)^n \qquad (5.3)$$

$$= \exp\left\{-\frac{i\mu\sqrt{n}}{\sigma}t\right\}\left(1 + i\frac{t}{\sigma\sqrt{n}}\mu - \frac{t^2}{2\sigma^2 n}E X_1^2 + o\left(\frac{t^2}{n}\right)\right)^n.$$

By exploiting the relation $x = \exp\{\log x\}$ and Taylor expansion of the function $\log(1 + z)$, which is valid for all complex z with $|z| < 1$, the last expression in (5.3) becomes

$$\exp\left\{-\frac{i\mu\sqrt{n}}{\sigma}t\right\} \cdot \exp\left\{n \cdot \log\left[1 + \frac{it\mu}{\sigma\sqrt{n}} - \frac{t^2(\sigma^2 + \mu^2)}{2\sigma^2 n} + o\left(\frac{t^2}{n}\right)\right]\right\}$$

$$= \exp\left\{-\frac{i\mu\sqrt{n}}{\sigma}t + n\left[\frac{it\mu}{\sigma\sqrt{n}} - \frac{t^2(\sigma^2 + \mu^2)}{2\sigma^2 n} + o\left(\frac{t^2}{n}\right)\right.\right.$$

$$\left.\left. - \frac{1}{2} \cdot \left(\frac{it\mu}{\sigma\sqrt{n}} - \frac{t^2(\sigma^2 + \mu^2)}{2\sigma^2 n} + o\left(\frac{t^2}{n}\right)\right)^2 + o\left(\frac{t^2}{n}\right)\right]\right\}$$

$$= \exp\left\{-\frac{i\mu\sqrt{n}}{\sigma}t + n\left[\frac{it\mu}{\sigma\sqrt{n}} - \frac{t^2(\sigma^2 + \mu^2)}{2\sigma^2 n} + \frac{1}{2}\frac{t^2\mu^2}{\sigma^2 n} + o\left(\frac{t^2}{n}\right)\right]\right\}$$

$$= e^{-t^2/2 + n \cdot o(t^2/n)} \to e^{-t^2/2} \quad \text{as} \quad n \to \infty.$$

The troublemaker is the factor $\exp\{-i\mu\sqrt{n}t/\sigma\}$, which must be annihilated by a corresponding piece in the second factor.

Remark 5.3. One may, alternatively, prove the central limit theorem with the aid of moment generating functions. However, the theorem is then only verified for random variables that actually possess a moment generating function. $\Box$

One important application of the preceding results is to the empirical distribution function.

Example 5.1. Let $X_1, X_2, \ldots, X_n$ be a sample of the random variable X. Suppose that the distribution function of X is F, and let F_n denote the *empirical distribution function* of the sample, that is,

$$F_n(x) = \frac{\#\text{ observations} \leq x}{n}.$$

Show that, for every fixed x,

(a) $F_n(x) \xrightarrow{p} F(x)$ as $n \to \infty$,

(b) $\sqrt{n}(F_n(x) - F(x)) \xrightarrow{d} N(0, \sigma^2(x))$ as $n \to \infty$, and determine $\sigma^2(x)$.

Since $\{\# \text{ observations } \leq x\} \in \text{Bin}(n, F(x))$ (recall Section 4.1), we introduce the indicators

$$I_k(x) = \begin{cases} 1, & \text{if } X_k \leq x, \\ 0, & \text{otherwise.} \end{cases}$$

The law of large numbers (i.e., Theorem 5.1) then immediately yields

$$F_n(x) = \frac{1}{n} \sum_{k=1}^{n} I_k(x) \xrightarrow{p} E\, I_1(x) = F(x) \quad \text{as} \quad n \to \infty,$$

which proves (a). To prove (b) we note that Theorem 5.2, similarly, yields

$$\sqrt{n}(F_n(x) - F(x)) \xrightarrow{d} N(0, \sigma^2(x)) \quad \text{as} \quad n \to \infty,$$

where $\sigma^2(x) = \text{Var}\, I_1(x) = F(x)(1 - F(x))$. □

Remark 5.4. Using the strong law cited in Remark 5.1, one can in fact show that $F_n(x) \xrightarrow{a.s.} F(x)$ as $n \to \infty$. A further strengthening is the Glivenko–Cantelli theorem, which states that

$$\sup_x |F_n(x) - F(x)| \xrightarrow{a.s.} 0 \quad \text{as} \quad n \to \infty.$$

Remark 5.5. The empirical distribution function is a useful tool for estimating the true (unknown) distribution function. More precisely, part (a) shows that the empirical distribution at some point x is close to the true value $F(x)$ for large samples. Part (b) gives an estimate of the deviation from the true value. Another use of the empirical distribution is to test the hypothesis that a sample or a series of observations actually has been taken from some pre-specified distribution. One such test is the Kolmogorov test, which is based on the quantity of the left-hand side in the Glivenko–Cantelli theorem cited above.

A related test quantity, which is useful for testing whether two samples of equal size have been taken from the same distribution or population is $\sup_x |F_n^{(1)}(x) - F_n^{(2)}(x)|$, where $F_n^{(1)}$ and $F_n^{(2)}$ are the empirical distribution functions of the two samples. □

The law of large numbers states that $P(|\bar{X}_n - \mu| > \varepsilon) \to 0$ as $n \to \infty$ for any $\varepsilon > 0$, which means that $\bar{X}_n - \mu$ is "small" (with high probability) when n is "large." This can be interpreted as a *qualitative* statement. A natural question now is *how* small? The central limit theorem states that $\sigma^{-1}\sqrt{n}(\bar{X}_n - \mu) \xrightarrow{d} N$ as $n \to \infty$, where $N \in N(0, 1)$, which means that "$\sigma^{-1}\sqrt{n}(\bar{X}_n - \mu) \approx$

N," or, equivalently, that "$\bar{X}_n - \mu \approx N\sigma/\sqrt{n}$" when n is "large" (provided the variance is finite). This is a *quantitative* statement. Alternatively, we may say that the central limit theorem provides information on the *rate of convergence* in the law of large numbers.

For example, if $X_1, X_2, \ldots$ is a sequence of independent, $U(0,1)$-distributed random variables, the law of large numbers only provides the information that

$$P(|\bar{X}_n - \tfrac{1}{2}| > \tfrac{1}{10}) \to 0 \quad \text{as} \quad n \to \infty,$$

whereas the central limit theorem yields the numerical result

$$P(|\bar{X}_n - \tfrac{1}{2}| > \tfrac{1}{10}) \approx 2\big(1 - \Phi(\tfrac{\sqrt{12n}}{10})\big),$$

which may be computed for any given sample size n.

Remark 5.6. The obvious next step would be to ask for rates of convergence in the central limit theorem, that is, to ask for a more detailed explanation of the statement that "$\bar{X}_n$ is approximately normally distributed when n is large," the corresponding qualitative statement of which is "$F_{(S_n-n\mu)/\sigma\sqrt{n}}(x) - \Phi(x)$ is small when n is large." The following is a quantitative result meeting this demand: Suppose, in addition, that $E|X_1|^3 < \infty$. Then

$$\sup_x |F_{\frac{S_n-n\mu}{\sigma\sqrt{n}}}(x) - \Phi(x)| \leq C \cdot \frac{E|X_1|^3}{\sigma^3\sqrt{n}}, \tag{5.4}$$

where C is a constant (0.7655 is the current best estimate).   □

We close this section with an example and an exercise. The example also provides a solution to Problem 3.8.13(a), so the reader who has not (yet) solved that problem should skip it.

Example 5.2. Let $X_1, X_2, \ldots$ be independent, $C(0,1)$-distributed random variables. Then the fact that $\varphi_X(t) = e^{-|t|}$ and formula (5.1) tell us that

$$\varphi_{\bar{X}_n}(t) = \big(\varphi_{X_1}(t/n)\big)^n = \big(e^{-|t/n|}\big)^n = e^{-|t|} = \varphi_{X_1}(t).$$

It follows from the uniqueness theorem for characteristic functions that

$$\bar{X}_n \stackrel{d}{=} X_1, \quad \text{for all} \quad n. \tag{5.5}$$

In particular, the law of large numbers does not hold. However, this is no contradiction, because the mean of the Cauchy distribution does not exist.   □

6 Convergence of Sums of Sequences of Random Variables

Let $X_1, X_2, \ldots$ and $Y_1, Y_2, \ldots$ be sequences of random variables. Suppose that $X_n \to X$ and that $Y_n \to Y$ as $n \to \infty$ in one of the four senses defined above. In this section we shall determine to what extent we may conclude that $X_n + Y_n \to X + Y$ as $n \to \infty$ (in the same sense).

Again, it is instructive to recall the corresponding proof for sequences of real numbers. Thus, assume that $a_1, a_2, \ldots$ and $b_1, b_2, \ldots$ are sequences of reals such that

$$a_n \to a \quad \text{and} \quad b_n \to b \quad \text{as} \quad n \to \infty. \tag{6.1}$$

The conclusion that $a_n + b_n \to a + b$ as $n \to \infty$ follows from the triangle inequality:

$$|a_n + b_n - (a + b)| = |(a_n - a) + (b_n - b)| \leq |a_n - a| + |b_n - b| \to 0 \tag{6.2}$$

as $n \to \infty$.

Alternatively, we could argue as follows. Given $\varepsilon > 0$, we have

$$|a_n - a| < \varepsilon \text{ for } n > n_1(\varepsilon) \quad \text{and} \quad |b_n - b| < \varepsilon \text{ for } n > n_2(\varepsilon),$$

from which it follows that

$$|a_n + b_n - (a + b)| < 2\varepsilon \quad \text{for} \quad n > \max\{n_1(\varepsilon), n_2(\varepsilon)\},$$

which yields the assertion.

Yet another proof is obtained by assuming the opposite in order to obtain a contradiction. Suppose that

$$a_n + b_n \nrightarrow a + b \quad \text{as} \quad n \to \infty. \tag{6.3}$$

We can then find infinitely many values of n such that, for some $\varepsilon > 0$,

$$|a_n + b_n - (a + b)| > \varepsilon, \tag{6.4}$$

from which we conclude that for every such n

$$|a_n - a| > \frac{\varepsilon}{2} \quad \text{or} \quad |b_n - b| > \frac{\varepsilon}{2}. \tag{6.5}$$

It follows that there must exist infinitely many n such that (at least) one of the inequalities in (6.5) holds. This shows that (at least) one of the statements $a_n \to a$ as $n \to \infty$ or $b_n \to b$ as $n \to \infty$ cannot hold, in contradiction to (6.1).

Now let us turn our attention to the corresponding problem for sums of sequences of random variables.

Theorem 6.1. *Let $X_1, X_2, \ldots$ and $Y_1, Y_2, \ldots$ be sequences of random variables such that*

$$X_n \xrightarrow{a.s.} X \quad \text{and} \quad Y_n \xrightarrow{a.s.} Y \quad \text{as} \quad n \to \infty.$$

Then

$$X_n + Y_n \xrightarrow{a.s.} X + Y \quad \text{as} \quad n \to \infty.$$

Proof. We introduce the sets N_X and N_Y from Theorem 2.1 and choose $\omega \in (N_X \cup N_Y)^c$. The conclusion now follows by modifying part (i) in the proof of Theorem 2.1 in the obvious manner (cf. (6.2)).   □

The corresponding results for convergence in probability and mean convergence follow by analogous modifications of the proof of Theorem 2.1, parts (ii) and (iii), respectively.

Theorem 6.2. *Let X_1, X_2, ... and Y_1, Y_2, ... be sequences of random variables such that*

$$X_n \xrightarrow{p} X \quad and \quad Y_n \xrightarrow{p} Y \quad as \quad n \to \infty.$$

Then

$$X_n + Y_n \xrightarrow{p} X + Y \quad as \quad n \to \infty.$$ □

Theorem 6.3. *Let X_1, X_2, ... and Y_1, Y_2, ... be sequences of random variables such that, for some $r > 0$,*

$$X_n \xrightarrow{r} X \quad and \quad Y_n \xrightarrow{r} Y \quad as \quad n \to \infty.$$

Then

$$X_n + Y_n \xrightarrow{r} X + Y \quad as \quad n \to \infty.$$ □

Exercise 6.1. Complete the proof of Theorem 6.1 and prove Theorems 6.2 and 6.3.   □

As for convergence in distribution, a little more care is needed, in that some additional assumption is required. We first prove a positive result under the additional assumption that one of the limiting random variables is degenerate, and in Theorem 6.6 we prove a result under extra independence conditions.

Theorem 6.4. *Let X_1, X_2, ... and Y_1, Y_2, ... be sequences of random variables such that*

$$X_n \xrightarrow{d} X \quad and \quad Y_n \xrightarrow{p} a \quad as \quad n \to \infty,$$

where a is a constant. Then

$$X_n + Y_n \xrightarrow{d} X + a \quad as \quad n \to \infty.$$

Proof. The proof is similar to that of Step IV in the proof of Theorem 3.1.
Let $\varepsilon > 0$ be given. Then

$$
\begin{aligned}
F_{X_n+Y_n}(x) &= P(X_n + Y_n \leq x) \\
&= P(\{X_n + Y_n \leq x\} \cap \{|Y_n - a| \leq \varepsilon\}) \\
&\quad + P(\{X_n + Y_n \leq x\} \cap \{|Y_n - a| > \varepsilon\}) \\
&\leq P(\{X_n \leq x - a + \varepsilon\} \cap \{|Y_n - a| \leq \varepsilon\}) \\
&\quad + P(|Y_n - a| > \varepsilon) \\
&\leq P(X_n \leq x - a + \varepsilon) + P(|Y_n - a| > \varepsilon) \\
&= F_{X_n}(x - a + \varepsilon) + P(|Y_n - a| > \varepsilon),
\end{aligned}
$$

from which it follows that

$$\limsup_{n\to\infty} F_{X_n+Y_n}(x) \leq F_X(x-a+\varepsilon) \tag{6.6}$$

for $x - a + \varepsilon \in C(F_X)$. A similar argument shows that

$$\liminf_{n\to\infty} F_{X_n+Y_n}(x) \geq F_X(x-a-\varepsilon) \tag{6.7}$$

for $x-a-\varepsilon \in C(F_X)$; we leave that as an exercise. Since $\varepsilon > 0$ may be arbitrarily small (and since F_X has only at most a countable number of discontinuity points), we finally conclude that

$$F_{X_n+Y_n}(x) \to F_X(x-a) = F_{X+a}(x) \quad \text{as} \quad n \to \infty$$

for $x - a \in C(F_X)$, that is, for $x \in C(F_{X+a})$. $\qquad\square$

Remark 6.1. The strength of the results so far is that no assumptions about independence have been made. $\qquad\square$

The assertions above also hold for differences, products, and ratios. We leave the formulations and proofs as an exercise, except for the result corresponding to Theorem 6.4, which is formulated next.

Theorem 6.5. *Let X_1, X_2, ... and Y_1, Y_2, ... be sequences of random variables. Suppose that*

$$X_n \xrightarrow{d} X \quad \text{and} \quad Y_n \xrightarrow{p} a \quad \text{as} \quad n \to \infty,$$

where a is a constant. Then

$$X_n + Y_n \xrightarrow{d} X + a,$$
$$X_n - Y_n \xrightarrow{d} X - a,$$
$$X_n \cdot Y_n \xrightarrow{d} X \cdot a,$$
$$\frac{X_n}{Y_n} \xrightarrow{d} \frac{X}{a}, \quad \text{for} \quad a \neq 0,$$

as $n \to \infty$. $\qquad\square$

Remark 6.2. Theorem 6.5 is frequently called Cramér's theorem or Slutsky's theorem. $\qquad\square$

Example 6.1. Let X_1, X_2, ... be independent, $U(0,1)$-distributed random variables. Show that

$$\frac{X_1 + X_2 + \cdots + X_n}{X_1^2 + X_2^2 + \cdots + X_n^2} \xrightarrow{p} \frac{3}{2} \quad \text{as} \quad n \to \infty.$$

Solution. When we multiply the numerator and denominator by $1/n$, the ratio turns into

$$\frac{(X_1 + X_2 + \cdots + X_n)/n}{(X_1^2 + X_2^2 + \cdots + X_n^2)/n}.$$

The numerator converges, according to the law of large numbers, to $E\,X_1 = 1/2$ as $n \to \infty$. Since X_1^2, X_2^2, ... are independent, equidistributed random variables with finite mean, another application of the law of large numbers shows that the denominator converges to $E\,X_1^2 = 1/3$ as $n \to \infty$. An application of Theorem 6.5 finally shows that the ratio under consideration converges to the ratio of the limits, that is, to $(1/2)/(1/3) = 3/2$ as $n \to \infty$.

Example 6.2. Let $\dot{X}_1$, X_2, ... be independent, $L(1)$-distributed random variables. Show that

$$\sqrt{n}\,\frac{X_1 + X_2 + \cdots + X_n}{X_1^2 + X_2^2 + \cdots + X_n^2} \xrightarrow{d} N(0, \sigma^2) \quad \text{as} \quad n \to \infty,$$

and determine σ^2.

Solution. By beginning as in the previous example, the left-hand side becomes

$$\frac{(X_1 + X_2 + \cdots + X_n)/\sqrt{n}}{(X_1^2 + X_2^2 + \cdots + X_n^2)/n}.$$

By the central limit theorem the numerator converges in distribution to the $N(0, 2)$-distribution as $n \to \infty$; by the law of large numbers, the denominator converges to $E\,X_1^2 = 2$ as $n \to \infty$. It follows from Cramér's theorem (Theorem 6.5) that the ratio converges in distribution to $Y \overset{d}{=} 1/2 \cdot N(0, 2) \overset{d}{=} N(0, 1/2)$ as $n \to \infty$. □

Next we present the announced result for sums of sequences of random variables under certain independence assumptions.

Theorem 6.6. *Let X_1, X_2, ... and Y_1, Y_2, ... be sequences of random variables such that*

$$X_n \xrightarrow{d} X \quad \text{and} \quad Y_n \xrightarrow{d} Y \quad \text{as} \quad n \to \infty.$$

Suppose further that X_n and Y_n are independent for all n and that X and Y are independent. Then

$$X_n + Y_n \xrightarrow{d} X + Y \quad \text{as} \quad n \to \infty.$$

Proof. The independence assumption suggests the use of transforms.

It follows from the continuity theorem for characteristic functions, Theorem 4.3, that it suffices to show that

$$\varphi_{X_n + Y_n}(t) \to \varphi_{X + Y}(t) \quad \text{as} \quad n \to \infty \quad \text{for} \quad -\infty < t < \infty. \tag{6.8}$$

In view of Theorem 3.4.6, it suffices to show that

$$\varphi_{X_n}(t)\varphi_{Y_n}(t) \to \varphi_X(t)\varphi_Y(t) \quad \text{as} \quad n \to \infty \quad \text{for} \quad -\infty < t < \infty.$$

This, however, is a simple consequence of the fact that the individual sequences of characteristic functions converge (and of Remark 4.2). $\qquad\square$

In Sections 1 and 4 we showed that a binomial distribution with large n and small $p = \lambda/n$ for some $\lambda > 0$ may be approximated with a suitable Poisson distribution. As an application of Theorem 6.6 we prove, in the following example, an addition theorem for binomial distributions with large sample sizes and small success probabilities.

Example 6.3. Let $X_n \in \text{Bin}(n_x, p_x(n))$, let $Y_n \in \text{Bin}(n_y, p_y(n))$, and suppose that X_n and Y_n are independent for all $n \geq 1$. Suppose in addition that $n_x \to \infty$ and $p_x(n) \to 0$ such that $n_x p_x(n) \to \lambda_x$ as $n_x \to \infty$, and that $n_y \to \infty$ and $p_y(n) \to 0$ such that $n_y p_y(n) \to \lambda_y$ as $n_y \to \infty$.

For the case $p_x(n) = \lambda_x/n_x$ and $p_y(n) = \lambda_y/n_y$, we know from Examples 1.4 and 4.3 that $X_n \xrightarrow{d} \text{Po}(\lambda_x)$ as $n_x \to \infty$ and that $Y_n \xrightarrow{d} \text{Po}(\lambda_y)$ as $n_y \to \infty$; for the general case see Problem 8.10(a). Furthermore, it is clear that the two limiting random variables are independent. It therefore follows from Theorem 6.6 and the addition theorem for the Poisson distribution that

$$X_n + Y_n \xrightarrow{d} \text{Po}(\lambda_x + \lambda_y) \quad \text{as} \quad n_x \text{ and } n_y \to \infty.$$

In particular, this is true if the sample sizes are equal, that is, when $n_x = n_y \to \infty$. $\qquad\square$

A common mathematics problem is whether or not one may interchange various operations, for example, taking limits and integrating. The final result of this section provides a useful answer in one simple case to the following problem.

Suppose that $X_1, X_2, \ldots$ is a sequence of random variables that converges in some sense to the random variable X and that g is a real-valued function. Is it true that the sequence $g(X_1), g(X_2), \ldots$ converges (in the same sense)? If so, does the limiting random variable equal $g(X)$?

Theorem 6.7. *Let $X_1, X_2, \ldots$ be random variables such that*

$$X_n \xrightarrow{p} a \quad as \quad n \to \infty.$$

Suppose, further, that g is a function that is continuous at a. Then

$$g(X_n) \xrightarrow{p} g(a) \quad as \quad n \to \infty.$$

Proof. The assumption is that

$$P(|X_n - a| > \delta) \to 0 \quad \text{as} \quad n \to \infty \quad \forall \delta > 0, \tag{6.9}$$

and we wish to prove that

$$P(|g(X_n) - g(a)| > \varepsilon) \to 0 \quad \text{as} \quad n \to \infty \quad \forall \varepsilon > 0. \tag{6.10}$$

The continuity of g at a implies that

$$\forall \varepsilon > 0 \; \exists \delta > 0 \quad \text{such that} \quad |x - a| < \delta \Longrightarrow |g(x) - g(a)| < \varepsilon,$$

or, equivalently, that

$$\forall \varepsilon > 0 \; \exists \delta > 0 \quad \text{such that} \quad |g(x) - g(a)| > \varepsilon \Longrightarrow |x - a| > \delta. \tag{6.11}$$

From (6.11) we conclude that

$$\{\omega : |g(X_n(\omega)) - g(a)| > \varepsilon\} \subset \{\omega : |X_n(\omega) - a| > \delta\},$$

that is, $\forall \varepsilon > 0 \; \exists \delta > 0$ such that

$$P(|g(X_n) - g(a)| > \varepsilon) \le P(|X_n - a| > \delta). \tag{6.12}$$

Since the latter probability tends to zero for *all* δ, this is, in particular, true for the very δ we chose in (6.12) as a partner of the arbitrary $\varepsilon > 0$ with which we started. $\square$

Example 6.4. Let $Y_1, Y_2, \ldots$ be random variables such that $Y_n \xrightarrow{p} 2$ as $n \to \infty$. Then $Y_n^2 \xrightarrow{p} 4$ as $n \to \infty$, since the function $g(x) = x^2$ is continuous at $x = 2$.

Example 6.5. Let $X_1, X_2, \ldots$ be i.i.d. random variables with finite mean $\mu \ge 0$. Show that $\sqrt{\bar{X}_n} \xrightarrow{p} \sqrt{\mu}$ as $n \to \infty$.

To see, this we first note that by the law of large numbers we have $\bar{X}_n \xrightarrow{p} \mu$ as $n \to \infty$, and since the function $g(x) = \sqrt{x}$ is continuous, in particular at $x = \mu$, the conclusion follows. $\square$

Exercise 6.2. It is a little harder to show that if, instead, we assume that $X_n \xrightarrow{p} X$ as $n \to \infty$ and that g is continuous, then $g(X_n) \xrightarrow{p} g(X)$ as $n \to \infty$. Generalize the proof of Theorem 6.7, and try to find out why this is harder. $\square$

We conclude this section with some further examples, which aim to combine Theorems 6.5 and 6.7.

Example 6.6. Let $\{X_n, n \ge 1\}$ be independent, $N(0,1)$-distributed random variables. Show that

$$\frac{X_1}{\sqrt{\frac{1}{n} \sum_{k=1}^{n} X_k^2}} \xrightarrow{d} N(0,1) \quad \text{as} \quad n \to \infty.$$

Since the X_1 in the numerator is standard normal, it follows in particular that $X_1 \xrightarrow{d} N(0,1)$ as $n \to \infty$. As for the denominator, $\frac{1}{n}\sum_{k=1}^{n} X_k^2 \xrightarrow{p} E X^2 = 1$ as $n \to \infty$ in view of the law of large numbers (recall Exercise 5.1). It follows from Theorem 6.7 (cf. also Example 6.5) that

$$\sqrt{\frac{1}{n}\sum_{k=1}^{n} X_k^2} \xrightarrow{p} 1 \quad \text{as} \quad n \to \infty.$$

An application of Cramér's theorem finally proves the conclusion.

Example 6.7. Let $Z_n \in N(0,1)$ and $V_n \in \chi^2(n)$ be independent random variables, and set

$$T_n = \frac{Z_n}{\sqrt{\frac{V_n}{n}}}, \quad n = 1, 2, \ldots.$$

Show that

$$T_n \xrightarrow{d} N(0,1) \quad \text{as} \quad n \to \infty.$$

Solution. Since $E V_n = n$ and $\operatorname{Var} V_n = 2n$, it follows from Chebyshev's inequality that

$$P(|\frac{V_n}{n} - 1| > \varepsilon) \le \frac{\operatorname{Var}(\frac{V_n}{n})}{\varepsilon^2} = \frac{2n}{n^2 \cdot \varepsilon^2} = \frac{2}{n\varepsilon^2} \to 0 \quad \text{as} \quad n \to \infty$$

and hence that $V_n/n \xrightarrow{p} 1$ as $n \to \infty$. Since $g(x) = \sqrt{x}$ is continuous at $x = 1$, it further follows, from Theorem 6.7, that $\sqrt{V_n/n} \xrightarrow{p} 1$ as $n \to \infty$. An application of Cramér's theorem finishes the proof.   □

Remark 6.3. From statistics we know that $T_n \in t(n)$ and that the t-distribution, for example, is used in order to obtain confidence intervals for μ when σ is unknown in the normal distribution. When n, the number of degrees of freedom, is large, one approximates the t-percentile with the corresponding percentile of the standard normal distribution. Example 6.7 shows that this is reasonable in the sense that $t(n) \xrightarrow{d} N(0,1)$ as $n \to \infty$. In this case the percentiles converge too, namely, $t_\alpha(n) \to \lambda_\alpha$ as $n \to \infty$ (since the normal distribution function is strictly increasing).

Remark 6.4. It is not necessary that V_n and Z_n are independent for the conclusion to hold. It is, however, necessary in order for T_n to be t-distributed, which is of statistical importance; cf. Remark 6.3.   □

The following exercise deals with the analogous problem of the success probability in Bernoulli trials or coin tossing experiments being unknown:

Exercise 6.3. Let $X_1, X_2, \ldots, X_n$ be independent, Be(p)-distributed random variables, $0 < p < 1$, and set $\bar{X}_n = (1/n)\sum_{k=1}^{n} X_k$. The interval

$\bar{X}_n \pm \lambda_{\alpha/2}\sqrt{\bar{X}_n(1 - \bar{X}_n)/n}$ is commonly used as an approximate confidence interval for p on the confidence level $1 - \alpha$. Show that this is acceptable in the sense that

$$\frac{\bar{X}_n - p}{\sqrt{\bar{X}_n(1 - \bar{X}_n)/n}} \xrightarrow{d} N(0, 1) \quad \text{as} \quad n \to \infty. \qquad \square$$

Finally, in this section we provide an example where a sequence of random variables converges in distribution to a standard normal distribution, but the variance tends to infinity.

Example 6.8. Let $X_2, X_3, \ldots$ be as described in Example 3.1 and, furthermore, independent of $X \in N(0, 1)$. Set

$$Y_n = X \cdot X_n, \quad n \geq 2.$$

Show that

$$Y_n \xrightarrow{d} N(0, 1) \quad \text{as} \quad n \to \infty, \tag{6.13}$$

that

$$E Y_n = 0, \tag{6.14}$$

and (but) that

$$\text{Var}\, Y_n \to +\infty \quad \text{as} \quad n \to \infty. \tag{6.15}$$

Solution. Since $X_n \xrightarrow{p} 1$ as $n \to \infty$ and $X \in N(0, 1)$, an application of Cramér's theorem proves (6.13).

Furthermore, by independence,

$$E Y_n = E(X \cdot X_n) = E X \cdot E X_n = 0 \cdot E X_n = 0,$$

and

$$\text{Var}\, Y_n = E Y_n^2 = E(X \cdot X_n)^2 = E X^2 \cdot E X_n^2$$
$$= 1 \cdot \left(1^2\left(1 - \frac{1}{n}\right) + n^2 \cdot \frac{1}{n}\right) = 1 - \frac{1}{n} + n \to +\infty \quad \text{as} \quad n \to \infty. \qquad \square$$

7 The Galton–Watson Process Revisited

In Section 3.7 we encountered branching processes, more precisely, *Galton–Watson* processes defined as follows:

At time $t = 0$ there exists an initial population, which we suppose consists of one individual: $X(0) = 1$. In the following, every individual gives birth to a random number of children, who during their lifespans give birth to a random number of children, and so on. The reproduction rules for the Galton–Watson process are that all individuals give birth according to the same probability law, independently of each other, and that the number of children produced by an individual is independent of the number of individuals in their generation.

We further introduced the random variables

$$X(n) = \# \text{ individuals in generation } n, \quad n \geq 1,$$

and used Y and $\{Y_k, k \geq 1\}$ as generic random variables to denote the number of children obtained by individuals. We also excluded the case $P(Y = 1) = 1$, and for asymptotics the case $P(Y = 0) = 0$ (since otherwise the population can never die out).

In Problem 3.8.46 we considered the total population "so far", that is, we let T_n, $n \geq 1$, denote the total progeny up to and including the nth generation, viz.,

$$T_n = 1 + X(1) + \cdots + X(n), \quad n \geq 1.$$

With $g(t)$ and $G_n(t)$ being the generating functions of Y and T_n, respectively, the task was to establish the relation

$$G_n(t) = t \cdot g\big(G_{n-1}(t)\big). \tag{7.1}$$

The trick to see that this is true is to rewrite T_n as

$$T_n = 1 + \sum_{k=0}^{X(1)} T_{n-1}(k),$$

where $T_{n-1}(k)$ are i.i.d. random variables corresponding to the total progeny up to and including generation $n - 1$ of the children in the first generation.

Now, suppose that $m = EY \leq 1$. We then know from Theorem 3.7.3 that the probability of extinction is equal to 1 $(\eta = 1)$. This means that there will be a random variable T that describes the total population, where

$$T = \lim_{n \to \infty} T_n.$$

More precisely the family of random variable $T_n \nearrow T$ as $n \to \infty$, which, in particular, implies that the generating functions converge. Letting $n \to \infty$ in equation (7.1) we obtain

$$G(t) = tg\big(G(t)\big), \tag{7.2}$$

where thus $G(t) = g_T(t)$.

Letting $t \nearrow 1$ we find that $G(1) = g(G(1))$, that is, $G(1)$ is a root of the equation $t = g(t)$. But since there is no root to this equation in $[0, 1)$ we conclude that $G(1) = 1$, which shows that T is a bona fide random variable; $P(T < \infty) = 1$.

By differentiating and recalling our formulas in Section 3.2 that relate derivatives and moments (provided they exist) we now may derive expressions for the mean and the variance of T, the total progeny of the process.

If $m = 1$ we have $EX(n) = 1$ for all $n \geq 1$, so that $ET = +\infty$. We therefore suppose that $m < 1$.

Differentiating (7.2) twice we obtain

$$G'(t) = g(G(t)) + tg'(G(t))G'(t),$$
$$G''(t) = 2g'(G(t))G'(t) + tg''(G(t)) \cdot (G'(t))^2 + tg'(G(t))G''(t).$$

Letting $t \nearrow 1$ in the first derivative now yields

$$ET = G'(1) = g(G(1)) + g'(G(1))G'(1) = g(1) + g'(1)G'(1) = 1 + m \cdot ET,$$

so that

$$ET = \frac{1}{1-m}, \tag{7.3}$$

in agreement with Problem 3.8.44b.

If, in addition $\operatorname{Var} Y = \sigma^2 < \infty$ we may let $t \nearrow 1$ in the expression of second derivative, which yields

$$\begin{aligned} ET(T-1) = G''(1) &= 2g'(G(1))G'(1) + g''(G(1)) \cdot (G'(1))^2 \\ &\quad + g'(G(1))G''(1) \\ &= 2mET + EY(Y-1) \cdot (ET)^2 + m \cdot ET(T-1), \end{aligned}$$

which, after rearranging and joining with (7.3), tells us that

$$\operatorname{Var} T = \frac{\sigma^2}{(1-m)^3}. \tag{7.4}$$

We conclude by mentioning the particular case when $Y \in \operatorname{Po}(m)$, that is, when every individual produces a Poisson distributed number of children. Given our results above it is clear that we must have $m \leq 1$.

In this case G is implicitly given via the relation

$$G(t) = t \cdot e^{\lambda(G(t)-1)}, \tag{7.5}$$

from which one can show that the corresponding distribution is given by

$$P(T = k) = \frac{1}{k!}(\lambda k)^{k-1} e^{-\lambda k}, \quad k = 1, 2, \ldots.$$

This particular distribution has a name; it is called the *Borel distribution*.

Since $EY = \operatorname{Var} Y = m$ when $Y \in \operatorname{Po}(m)$, we have

$$ET = \frac{1}{1-m} \quad \text{and} \quad \operatorname{Var} T = \frac{m}{(1-m)^3}. \tag{7.6}$$

in this case (provided, of course, that $m < 1$).

Exercise 7.1. Check (7.6) by differentiating (7.2).

Another special case is when $Y \in Ge(p)$, where $p \geq 1/2$ in order for $m = p/q \leq 1$. Then

$$G(t) = t \cdot \frac{p}{1 - qG(t)}. \tag{7.7}$$

By solving equation (7.7), which is a second degree equation in $G(t)$, one finds that

$$G(t) = \frac{1 - \sqrt{1 - 4pqt}}{2q}. \tag{7.8}$$

8 Problems

1. Let $X_1, X_2, \ldots$ be $U(0, 1)$-distributed random variables. Show that
   (a) $\max_{1 \leq k \leq n} X_k \xrightarrow{p} 1$ as $n \to \infty$,
   (b) $\min_{1 \leq k \leq n} X_k \xrightarrow{p} 0$ as $n \to \infty$.
2. Let $\{X_n, n \geq 1\}$ be a sequence of i.i.d. random variables with density

$$f(x) = \begin{cases} e^{-(x-a)}, & \text{for } x \geq a, \\ 0, & \text{for } x < a. \end{cases}$$

Set $Y_n = \min\{X_1, X_2, \ldots, X_n\}$. Show that

$$Y_n \xrightarrow{p} a \quad \text{as} \quad n \to \infty.$$

Remark. This is a translated exponential distribution; technically, if X is distributed as above, then $X - a \in Exp(1)$. We may interpret this as X having age a and a remaining lifetime $X - a$, which is standard exponential.

3. Let X_k, $k \geq 1$, be i.i.d. random variables with finite variance σ^2, and let, for $n \geq 1$,

$$\bar{X}_n = \frac{1}{n} \sum_{k=1}^{n} X_k \quad \text{and} \quad s_n^2 = \frac{1}{n-1} \sum_{k=1}^{n} (X_k - \bar{X}_n)^2$$

denote the arithmetic mean and sample variance, respectively. It is well known(?) that s_n^2 is an unbiased estimator of σ^2, that is, that $E\,s_n^2 = \sigma^2$.
   (a) Prove this well known fact.
   (b) Prove that, moreover, $s_n^2 \xrightarrow{p} \sigma^2$ as $n \to \infty$.
4. Let (X_k, Y_k), $1 \leq k \leq n$, be a sample from a two-dimensional distribution with mean vector and covariance matrix

$$\mu = \begin{pmatrix} \mu_x \\ \mu_y \end{pmatrix}, \qquad \Lambda = \begin{pmatrix} \sigma_x^2 & \rho \\ \rho & \sigma_y^2 \end{pmatrix},$$

respectively, and let

$$\bar{X}_n = \frac{1}{n}\sum_{k=1}^{n}X_k, \qquad s_{n,x}^2 = \frac{1}{n-1}\sum_{k=1}^{n}(X_k - \bar{X}_n)^2,$$

$$\bar{Y}_n = \frac{1}{n}\sum_{k=1}^{n}Y_k, \qquad s_{n,y}^2 = \frac{1}{n-1}\sum_{k=1}^{n}(Y_k - \bar{Y}_n)^2,$$

denote the arithmetic means and sample variances of the respective components. The *empirical correlation coefficient* is defined as

$$r_n = \frac{\sum_{k=1}^{n}(X_k - \bar{X}_n)(Y_k - \bar{Y}_n)}{\sqrt{\sum_{k=1}^{n}(X_k - \bar{X}_n)^2 \sum_{k=1}^{n}(Y_k - \bar{Y}_n)^2}}.$$

Prove that

$$r_n \xrightarrow{p} \rho \quad \text{as} \quad n \to \infty.$$

Hint. $s_{n,x}^2 \xrightarrow{p} ??$, $\frac{1}{n}\sum_{k=1}^{n}X_kY_k \xrightarrow{p} ??$.

5. Let $X_1, X_2, \ldots$ be independent, $C(0,1)$-distributed random variables. Determine the limit distribution of

$$Y_n = \frac{1}{n}\cdot\max\{X_1, X_2, \ldots, X_n\}$$

as $n \to \infty$.

Remark. It may be helpful to know that $\arctan x + \arctan 1/x = \pi/2$ and that $\arctan y = y - y^3/3 + y^5/5 - y^7/7 + \cdots$.

6. Suppose that $X_1, X_2, \ldots$ are independent, $\mathrm{Pa}(1,2)$-distributed random variables, and set $Y_n = \max\{X_1, X_2, \ldots, X_n\}$.

   (a) Show that $Y_n \xrightarrow{p} 1$ as $n \to \infty$.

   It thus follows that $Y_n \approx 1$ with a probability close to 1 when n is large. One might therefore suspect that there exists a limit theorem to the effect that $Y_n - 1$, suitably rescaled, converges in distribution as $n \to \infty$ (note that $Y_n > 1$ always).

   (b) Show that $n(Y_n - 1)$ converges in distribution as $n \to \infty$, and determine the limit distribution.

7. Let $X_1, X_2, \ldots$ be i.i.d. random variables, and let

$$\tau(t) = \min\{n : X_n > t\}, \quad t \geq 0.$$

   (a) Determine the distribution of $\tau(t)$.

   (b) Show that, if $p_t = P(X_1 > t) \to 0$ as $t \to \infty$, then

$$p_t\tau(t) \xrightarrow{d} \mathrm{Exp}(1) \quad \text{as} \quad t \to \infty.$$

8. Suppose that $X_n \in \mathrm{Ge}(\lambda/(n+\lambda))$, $n = 1, 2, \ldots$, where λ is a positive constant. Show that X_n/n converges in distribution to an exponential distribution as $n \to \infty$, and determine the parameter of the limit distribution.

9. Let X_1, X_2, ... be a sequence of random variables such that

$$P\left(X_n = \frac{k}{n}\right) = \frac{1}{n}, \quad \text{for} \quad k = 1, 2, \ldots, n.$$

Determine the limit distribution of X_n as $n \to \infty$.

10. Let $X_n \in \text{Bin}(n, p_n)$.

(a) Suppose that $n \cdot p_n \to m$ as $n \to \infty$. Show that

$$X_n \xrightarrow{d} \text{Po}(m) \quad \text{as} \quad n \to \infty.$$

(b) Suppose that $p_n \to 0$ and that $np_n \to \infty$ as $n \to \infty$. Show that

$$\frac{X_n - np_n}{\sqrt{np_n}} \xrightarrow{d} N(0, 1) \quad \text{as} \quad n \to \infty.$$

(c) Suppose that $np_n(1 - p_n) \to \infty$ as $n \to \infty$. Show that

$$\frac{X_n - np_n}{\sqrt{np_n(1 - p_n)}} \xrightarrow{d} N(0, 1) \quad \text{as} \quad n \to \infty.$$

Remark. These results, which usually are presented without proofs in a first probability course, verify the common approximations of the binomial distribution with the Poisson and normal distributions.

11. Let $X_n \in \text{Bin}(n^2, m/n)$, $m > 0$. Show that

$$\frac{X_n - n \cdot m}{\sqrt{nm}} \xrightarrow{d} N(0, 1) \quad \text{as} \quad n \to \infty.$$

12. Let X_{n1}, X_{n2}, ..., X_{nn} be independent random variables with a common distribution given as follows:

$$P(X_{nk} = 0) = 1 - \frac{1}{n} - \frac{1}{n^2}, \quad P(X_{nk} = 1) = \frac{1}{n}, \quad P(X_{nk} = 2) = \frac{1}{n^2},$$

where $k = 1, 2, \ldots, n$ and $n = 2, 3, \ldots$. Set $S_n = X_{n1} + X_{n2} + \cdots + X_{nn}$, $n \geq 2$. Show that

$$S_n \xrightarrow{d} \text{Po}(1) \quad \text{as} \quad n \to \infty.$$

13. Let X_1, X_2, ... be independent, equidistributed random variables with characteristic function

$$\varphi(t) = \begin{cases} 1 - \sqrt{|t|(2 - |t|)}, & \text{for} \quad |t| \leq 1, \\ 0, & \text{for} \quad |t| \geq 1. \end{cases}$$

Set $S_n = \sum_{k=1}^n X_k$, $n \geq 1$. Show that S_n/n^2 converges in distribution as $n \to \infty$, and determine the limit distribution.

14. Let $X_{n1}, X_{n2}, \ldots, X_{nn}$ be independent random variables, with a common distribution given, and set

$$P(X_{nk} = 0) = 1 - \frac{2}{n} - \frac{4}{n^3}, \quad P(X_{nk} = 1) = \frac{2}{n}, \quad P(X_{nk} = 2) = \frac{4}{n^3},$$

and $S_n = X_{n1} + X_{n2} + \cdots + X_{nn}$ for $k = 1, 2, \ldots, n$ and $n \geq 2$. Show that

$$S_n \xrightarrow{d} \text{Po}(\lambda) \quad \text{as} \quad n \to \infty,$$

and determine λ.

15. Let X and Y be random variables such that

$$Y \mid X = x \in N(0, x) \quad \text{with} \quad X \in \text{Po}(\lambda).$$

(a) Find the characteristic function of Y.
(b) Show that

$$\frac{Y}{\sqrt{\lambda}} \xrightarrow{d} N(0, 1) \quad \text{as} \quad \lambda \to \infty.$$

16. Let $X_1, X_2, \ldots$ be independent, $L(a)$-distributed random variables, and let $N \in \text{Po}(m)$ be independent of $X_1, X_2, \ldots$. Determine the limit distribution of $S_N = X_1 + X_2 + \cdots + X_N$ (where $S_0 = 0$) as $m \to \infty$ and $a \to 0$ in such a way that $m \cdot a^2 \to 1$.

17. Let $N, X_1, X_2, \ldots$ be independent random variables such that $N \in \text{Po}(\lambda)$ and $X_k \in \text{Po}(\mu)$, $k = 1, 2, \ldots$. Determine the limit distribution of $Y = X_1 + X_2 + \cdots + X_N$ as $\lambda \to \infty$ and $\mu \to 0$ such that $\lambda \cdot \mu \to \gamma > 0$. (The sum is zero for $N = 0$.)

18. Let $X_1, X_2, \ldots$ be independent $\text{Po}(m)$-distributed random variables, suppose that $N \in \text{Ge}(p)$ is independent of $X_1, X_2, \ldots$, and set $S_N = X_1 + X_2 + \cdots + X_N$ (and $S_0 = 0$ for $N = 0$). Let $m \to 0$ and $p \to 0$ in such a way that $p/m \to \alpha$, where α is a given positive number. Show that S_N converges in distribution, and determine the limit distribution.

19. Suppose that the random variables $N_n, X_1, X_2, \ldots$ are independent, that $N_n \in \text{Ge}(p_n)$, $0 < p_n < 1$, and that $X_1, X_2, \ldots$ are equidistributed with finite mean μ. Show that if $p_n \to 0$ as $n \to \infty$ then $p_n(X_1 + X_2 + \cdots + X_{N_n})$ converges in distribution as $n \to \infty$, and determine the limit distribution.

20. Suppose that $X_1, X_2, \ldots$ are i.i.d. symmetric random variables with finite variance σ^2, let $N_p \in \text{Fs}(p)$ be independent of $X_1, X_2, \ldots$, and set $Y_p = \sum_{k=1}^{N_p} X_k$. Show that

$$\sqrt{p} Y_p \xrightarrow{d} L(a) \quad \text{as} \quad p \to 0,$$

and determine a.

21. Let X_1, X_2, ... be independent, $U(0,1)$-distributed random variables, and let $N_m \in \text{Po}(m)$ be independent of X_1, X_2, ... Set

$$V_m = \max\{X_1, \ldots, X_{N_m}\}$$

($V_m = 0$ when $N_m = 0$). Determine
   (a) the distribution function of V_m,
   (b) the moment generating function of V_m.
       It is reasonable to believe that V_m is "close" to 1 when m is "large" (cf. Problem 8.1). The purpose of parts (c) and (d) is to show how this can be made more precise.
   (c) Show that $E V_m \to 1$ as $m \to \infty$.
   (d) Show that $m(1 - V_m)$ converges in distribution as $m \to \infty$, and determine the limit distribution.

22. Let X_{1n}, X_{2n}, ..., X_{nn} be independent random variables such that $X_{kn} \in \text{Be}(p_{k,n})$, $k = 1, 2, \ldots, n$, $n \geq 1$. Suppose, further, that $\sum_{k=1}^{n} p_{k,n} \to \lambda < \infty$ and that $\max_{1 \leq k \leq n} p_{k,n} \to 0$ as $n \to \infty$. Show that $\sum_{k=1}^{n} X_{kn}$ converges in distribution as $n \to \infty$, and determine the limit distribution.

23. Show that

$$\lim_{n \to \infty} e^{-n} \sum_{k=0}^{n} \frac{n^k}{k!} = \frac{1}{2}$$

by applying the central limit theorem to suitably chosen, independent, Poisson-distributed random variables.

24. Let X_1, X_2, ... be independent, $U(-1,1)$-distributed random variables.
   (a) Show that

$$Y_n = \frac{\sum_{k=1}^{n} X_k}{\sum_{k=1}^{n} X_k^2 + \sum_{k=1}^{n} X_k^3}$$

converges in probability as $n \to \infty$, and determine the limit.
   (b) Show that Y_n, suitably normalized, converges in distribution as $n \to \infty$, and determine the limit distribution.

25. Let $X_n \in \Gamma(n, 1)$, and set

$$Y_n = \frac{X_n - n}{\sqrt{X_n}}.$$

Show that $Y_n \xrightarrow{d} N(0, 1)$ as $n \to \infty$.

26. Let X_1, X_2, ... be positive, i.i.d. random variables with mean μ and finite variance σ^2, and set $S_n = \sum_{k=1}^{n} X_k$, $n \geq 1$. Show that

$$\frac{S_n - n\mu}{\sqrt{S_n}} \xrightarrow{d} N(0, b^2) \quad \text{as} \quad n \to \infty,$$

and determine b^2.

27. Let X_1, X_2, ... be i.i.d. random variables with finite mean $\mu \neq 0$, and set $S_n = \sum_{k=1}^{n} X_k$, $n \geq 1$.

(a) Show that

$$\frac{S_n - n\mu}{S_n + n\mu} \quad \text{converges in probability as} \quad n \to \infty,$$

and determine the limit.

(b) Suppose in addition that $0 < \operatorname{Var} X = \sigma^2 < \infty$. Show that

$$\sqrt{n} \frac{S_n - n\mu}{S_n + n\mu} \xrightarrow{d} N(0, a^2) \quad \text{as} \quad n \to \infty,$$

and determine a^2.

28. Suppose that $X_1, X_2, \ldots$ are i.i.d. random variables with mean 0 and variance 1. Show that

$$\frac{\sum_{k=1}^n X_k}{\sqrt{\sum_{k=1}^n X_k^2}} \xrightarrow{d} N(0, 1) \quad \text{as} \quad n \to \infty.$$

29. Let $X_1, X_2, \ldots$ and $Y_1, Y_2, \ldots$ be two (not necessarily independent) sequences of random variables and suppose that $X_n \xrightarrow{d} X$ and that $P(X_n \neq Y_n) \to 0$ as $n \to \infty$. Prove that (also)

$$Y_n \xrightarrow{d} X \quad \text{as} \quad n \to \infty.$$

30. Let $\{Y_k, k \geq 1\}$ be independent, $U(-1, 1)$-distributed random variables, and set

$$X_n = \frac{\sum_{k=1}^n Y_k}{\sqrt{n} \cdot \max_{1 \leq k \leq n} Y_k}.$$

Show that $X_n \xrightarrow{d} N(0, 1/3)$ as $n \to \infty$.

31. Let $X_1, X_2, \ldots$ be independent, $U(-a, a)$-distributed random variables $(a > 0)$. Set

$$S_n = \sum_{k=1}^n X_k, \quad Z_n = \max_{1 \leq k \leq n} X_k, \quad \text{and} \quad V_n = \min_{1 \leq k \leq n} X_k.$$

Show that $S_n Z_n / V_n$, suitably normalized, converges in distribution as $n \to \infty$, and determine the limit distribution.

32. Let $X_1, X_2, \ldots$ be independent, $U(0, 1)$-distributed random variables, and set

$$Z_n = \max_{1 \leq k \leq n} X_k \quad \text{and} \quad V_n = \min_{1 \leq k \leq n} X_k.$$

Determine the limit distribution of nV_n/Z_n as $n \to \infty$.

33. Let $X_1, X_2, \ldots$ be independent random variables such that $X_k \in \operatorname{Exp}(k!)$, $k = 1, 2 \ldots$, and set $S_n = \sum_{k=1}^n X_k$, $n \geq 1$. Show that

$$\frac{S_n}{n!} \xrightarrow{d} \operatorname{Exp}(1) \quad \text{as} \quad n \to \infty.$$

Hint. What is the distribution of $X_n/n!$?

34. Let X_1, X_2, ... be i.i.d. random variables with expectation 1 and finite variance σ^2, and set $S_n = X_1 + X_2 + \cdots + X_n$, for $n \geq 1$. Show that

$$\sqrt{S_n} - \sqrt{n} \xrightarrow{d} N(0, b^2) \quad \text{as} \quad n \to \infty,$$

and determine the constant b^2.

35. Let X_1, X_2, ... be i.i.d. random variables with mean μ and positive, finite variance σ^2, and set $S_n = \sum_{k=1}^{n} X_k$, $n \geq 1$. Finally, suppose that g is twice continuously differentiable, and that $g'(\mu) \neq 0$. Show that

$$\sqrt{n}\left(g\left(\frac{S_n}{n}\right) - g(\mu)\right) \xrightarrow{d} N(0, b^2) \quad \text{as} \quad n \to \infty,$$

and determine b^2.

Hint. Try Taylor expansion.

36. Let X_1, X_2, ... and Y_1, Y_2, ... be independent sequences of independent random variables. Suppose that there exist sequences $\{a_n, n \geq 1\}$ of real numbers and $\{b_n, n \geq 1\}$ of positive real numbers tending to infinity such that

$$\frac{X_n - a_n}{b_n} \xrightarrow{d} Z_1 \quad \text{and} \quad \frac{Y_n - a_n}{b_n} \xrightarrow{d} Z_2 \quad \text{as} \quad n \to \infty,$$

where Z_1 and Z_2 are independent random variables. Show that

$$\frac{\max\{X_n, Y_n\} - a_n}{b_n} \xrightarrow{d} \max\{Z_1, Z_2\} \quad \text{as} \quad n \to \infty,$$

$$\frac{\min\{X_n, Y_n\} - a_n}{b_n} \xrightarrow{d} \min\{Z_1, Z_2\} \quad \text{as} \quad n \to \infty.$$

37. Suppose that $\{U_t, t \geq 0\}$ and $\{V_t, t \geq 0\}$ are families of random variables, such that

$$U_t \xrightarrow{p} a \quad \text{and} \quad V_t \xrightarrow{d} V \quad \text{as} \quad t \to \infty,$$

for some finite constant a, and random variable V. Prove that

$$P(\max\{U_t, V_t\} \leq y) \to \begin{cases} 0, & \text{for } y < a, \\ P(V \leq y), & \text{for } y > a, \end{cases}$$

as $t \to \infty$.

Remark. This is a kind of Cramér theorem for the maximum.

38. Let X_1, X_2, ... be independent random variables such that, for some fixed positive integer m, X_1, ..., X_m are equidistributed with mean μ_1 and variance σ_1^2, and X_{m+1}, X_{m+2}, ... are equidistributed with mean μ_2 and variance σ_2^2. Set $S_n = \sum_{k=1}^{n} X_k$, $n \geq 1$. Show that the central limit theorem (still) holds.

Remark. Begin with the case $m = 1$.

39. Let X_1, X_2, ... be $U(-1,1)$-distributed random variables, and set

$$Y_n = \begin{cases} X_n, & \text{for } |X_n| \leq 1 - \dfrac{1}{n}, \\ n, & \text{otherwise.} \end{cases}$$

   (a) Show that Y_n converges in distribution as $n \to \infty$, and determine the limit distribution.
   (b) Let Y denote the limiting random variable. Consider the statements $E Y_n \to E Y$ and $\operatorname{Var} Y_n \to \operatorname{Var} Y$ as $n \to \infty$. Are they true or false?

40. Let $Z \in U(0,1)$ be independent of Y_1, Y_2, ..., where

$$P(Y_n = 1) = 1 - \frac{1}{n^\alpha} \quad \text{and} \quad P(Y_n = n) = \frac{1}{n^\alpha}, \quad n \geq 2, \quad (\alpha > 0),$$

   and set

$$X_n = Z \cdot Y_n, \quad n \geq 2.$$

   (a) Show that X_n converges in distribution as $n \to \infty$ and determine the limit distribution.
   (b) What about $E X_n$ and $\operatorname{Var} X_n$ as $n \to \infty$?

41. Let X_1, X_2, ... be identically distributed random variables converging in distribution to the random variable X, let $\{a_n, n \geq 1\}$ and $\{b_n, n \geq 1\}$ be sequences of positive reals $\nearrow +\infty$, and set

$$Y_n = \begin{cases} X_n, & \text{when } X_n \leq a_n, \\ b_n, & \text{when } X_n > a_n. \end{cases}$$

   (a) Show that $Y_n \xrightarrow{d} X$ as $n \to \infty$.
   (b) Suppose, in addition, that $E|X| < \infty$. Provide some sufficient condition to ensure that $E Y_n \to E X$ as $n \to \infty$.

42. The following example shows that a sequence of continuous random variables may converge in distribution without the sequence of densities being convergent. Namely, let X_n have a distribution function given by

$$F_n(x) = \begin{cases} x - \dfrac{\sin(2n\pi x)}{2n\pi}, & \text{for } 0 < x < 1, \\ 0, & \text{otherwise.} \end{cases}$$

   Show that $X_n \xrightarrow{d} U(0,1)$ as $n \to \infty$, but that $f_n(x)$ does not converge to the density of the $U(0,1)$-distribution.

43. Let Y_1, Y_2, ... be a sequence of random variables such that

$$Y_n \xrightarrow{d} Y \quad \text{as } n \to \infty,$$

   and let $\{N_n, n \geq 1\}$ be a sequence of nonnegative, integer-valued random variables such that

$$N_n \xrightarrow{p} +\infty \quad \text{as} \quad n \to \infty.$$

Finally, suppose that the sequences $\{N_n, n \geq 1\}$ and $Y_1, Y_2, \ldots$ are independent of each other. Prove (for example, with the aid of characteristic functions) that

$$Y_{N_n} \xrightarrow{d} Y \quad \text{as} \quad n \to \infty.$$

Hint. Consider the cases $\{N_n \leq M\}$ and $\{N_n > M\}$ where M is some suitably chosen "large" integer.

44. Let $X_1, X_2, \ldots$ and X be normal random variables. Show that

$$X_n \xrightarrow{d} X \quad \text{as} \quad n \to \infty \iff$$
$$E\,X_n \to E\,X \quad \text{and} \quad \text{Var}\,X_n \to \text{Var}\,X \quad \text{as} \quad n \to \infty.$$

45. Suppose that $X_n \in \text{Exp}(a_n)$, $n \geq 1$. Show that

$$X_n \xrightarrow{d} X \in \text{Exp}(a) \quad \text{as} \quad n \to \infty \iff a_n \to a \quad \text{as} \quad n \to \infty.$$

46. Prove that if $\{X_n, n \geq 1\}$ are Poissonian random variables such that X_n converges to X in square mean, then X must be Poissonian too.

47. Suppose that $\{Z_k, k \geq 1\}$ is a sequence of branching processes—all of them starting with one single individual at time 0. Suppose, furthermore, that $Z_k \xrightarrow{d} Z$ as $k \to \infty$, where Z is some nonnegative integer-valued random variable. Show that the corresponding sequence $\{\eta_k, k \geq 1\}$ of extinction probabilities converges as $k \to \infty$.

48. Let $X_1, X_2, \ldots$ be independent random variables such that $X_k \in \text{Po}(k)$, $k = 1, 2, \ldots$, and set $Z_n = \frac{1}{n}\{\sum_{k=1}^{n} X_k - n^2/2\}$. Show that Z_n converges in distribution as $n \to \infty$, and determine the limit distribution.

    Hint. Note that $X_k \overset{d}{=} \sum_{j=1}^{k} Y_{j,k}$ for every $k \geq 1$, where $\{Y_{j,k}, 1 \leq j \leq k, k \geq 1\}$ are independent $\text{Po}(1)$-distributed random variables.

49. Suppose that $N \in \text{Po}(\lambda)$ independent observations of a random variable, X, with mean 0 and variance 1, are performed. Moreover, assume that N is independent of $X_1, X_2, \ldots$. Show that

$$\frac{X_1 + X_2 + \cdots + X_N}{\sqrt{N}} \xrightarrow{d} N(0,1) \quad \text{as} \quad \lambda \to \infty.$$

50. The purpose of this problem is to show that one can obtain a (kind of) central limit theorem even if the summands have infinite variance (if the variance does not exist). A short introduction to the general topic of possible limit theorems for normalized sums without finite variance is given in Section 7.3.

    Let $X_1, X_2, \ldots$ be independent random variables with the following symmetric Pareto distribution:

$$f_X(x) = \begin{cases} \dfrac{1}{|x|^3}, & \text{for } |x| > 1, \\ 0, & \text{otherwise.} \end{cases}$$

Set $S_n = \sum_{k=1}^n X_k$, $n \geq 1$. Show via the following steps that

$$\frac{S_n}{\sqrt{n \log n}} \xrightarrow{d} N(0,1) \quad \text{as} \quad n \to \infty.$$

(Note that we do not normalize by $\sqrt{n}$ as in the standard case.)
Fix n and consider, for $k = 1, 2, \ldots, n$, the truncated random variables

$$Y_{nk} = \begin{cases} X_k, & \text{when } |X_k| \leq \sqrt{n}, \\ 0, & \text{otherwise,} \end{cases}$$

and

$$Z_{nk} = \begin{cases} X_k, & \text{when } |X_k| > \sqrt{n}, \\ 0, & \text{otherwise,} \end{cases}$$

and note that $Y_{nk} + Z_{nk} = X_k$. Further, set $S'_n = \sum_{k=1}^n Y_{nk}$ and $S''_n = \sum_{k=1}^n Z_{nk}$.

(a) Show that

$$E\left| \frac{S''_n}{\sqrt{n \log n}} \right| \to 0 \quad \text{as} \quad n \to \infty,$$

and conclude that

$$\frac{S''_n}{\sqrt{n \log n}} \to 0 \quad \text{in 1-mean} \quad \text{as} \quad n \to \infty,$$

and hence in probability (why?).

(b) Show that it remains to prove that

$$\frac{S'_n}{\sqrt{n \log n}} \xrightarrow{d} N(0,1) \quad \text{as} \quad n \to \infty.$$

(c) Let φ denote a characteristic function. Show that

$$\varphi_{Y_{nk}}(t) = 1 - 2 \int_1^{\sqrt{n}} \frac{1 - \cos tx}{x^3} \, dx,$$

and hence that

$$\varphi_{\frac{S'_n}{\sqrt{n \log n}}}(t) = \left(1 - 2 \int_1^{\sqrt{n}} \frac{1 - \cos \frac{tx}{\sqrt{n \log n}}}{x^3} \, dx \right)^n.$$

(d) Show that it remains to prove

$$2 \int_1^{\sqrt{n}} \frac{1 - \cos \frac{tx}{\sqrt{n \log n}}}{x^3} \, dx = \frac{t^2}{2n} + o\left(\frac{1}{n} \right) \quad \text{as} \quad n \to \infty.$$

(e) Prove this relation.

Remark. Note that (a) and (b) together show that S_n and S'_n have the same asymptotic distributional behavior, that $\operatorname{Var} Y_{nk} = \log n$ for $1 \leq k \leq n$ and $n \geq 1$, and hence that we have used the "natural" normalization $\sqrt{\operatorname{Var} S'_n}$ for S'_n.

7

An Outlook on Further Topics

Probability theory is, of course, much more than what one will find in this book (so far). In this chapter we provide an outlook on some extensions and further areas and concepts in probability theory. For more we refer to the more advanced literature cited in Appendix A.

We begin, in the first section, by presenting some extensions of the classical limit theorems, that is, the law of large numbers and the central limit theorem, to cases where one relaxes the assumptions of independence and equidistribution.

Another question in this context is whether there exist (other) limit distributions if the variance of the summands does not exist (is infinite). This leads, in the case of i.i.d. summands, to the class of stable distributions and their, what is called, *domains of attraction*. Sections 2 and 3 are devoted to this problem.

In connection with the convergence concepts in Section 6.3, it was mentioned that convergence in r-mean was, in general, not implied by the other convergence concepts. In Section 4 we define *uniform integrability*, which is the precise condition one needs in order to assure that moments converge whenever convergence almost surely, in probability, or in distribution holds. As a pleasant illustration we prove Stirling's formula with the aid of the exponential distribution.

There exists an abundance of situations where *extremes* rather than sums are relevant; earthquakes, floods, storms, and many others. Analogous to "limit theory for sums" there exists a "limit theory for extremes," that is for $Y_n = \max\{X_1, X_2, \ldots, X_n\}$, $n \geq 1$, where (in our case) $X_1, X_2, \ldots, X_n$ are i.i.d. random variables. Section 5 provides an introduction to the what is called *extreme value theory*. We also mention the closely related *records*, which are extremes at first appearance.

Section 7 introduces the Borel–Cantelli lemmas, which are a useful tool for studying the limit superior and limit inferior of sequences of events, and, as an extension, in order to decide whether some special event will occur infinitely many times or not. As a toy example we prove the intuitively obvious fact that

A. Gut, *An Intermediate course in Probabilty*, Springer Texts in Statistics,
DOI: 10.1007/978-1-4419-0162-0_7,
© Springer Science + Business Media, LLC 2009

if one tosses a coin an infinite number of times there will appear infinitely many heads and infinitely many tails. For a fair coin this is trivial due to symmetry, but what about an unfair coin? We also revisit Examples 6.3.1 and 6.3.2, and introduce the concept of *complete convergence*.

The final section, preceding some problems for solution, serves as an introduction to one of the most central tools in probability theory and the theory of stochastic processes, namely the theory of *martingales*, which, as a very rough definition, may be thought of as an extension of the theory of sums of independent random variables with mean zero and of fair games. In order to fully appreciate the theory one needs to base it on measure theory. Nevertheless, the basic flavor of the topic can be understood with our more elementary approach.

1 Extensions of the Main Limit Theorems

Several generalizations of the central limit theorem seem natural, such as:

1. the summands have (somewhat) different distributions;
2. the summands are *not independent*;
3. the variance does not exist.

In the first two subsections we provide some hints on the law of large numbers and the central limit theorem for the case of independent but not identically distributed summands. In the third subsection a few comments are given in the case of dependent summands. Possible (other) limit theorems when the variance is infinite (does not exist) is a separate issue, to which we return in Sections 2 and 3 for a short introduction.

1.1 The Law of Large Numbers: The Non-i.i.d. Case

It is intuitively reasonable to expect that the law of large numbers remains valid if the summands have different distributions—within limits.

We begin by presenting two extensions of this result.

Theorem 1.1. *Let X_1, X_2, ... be independent random variables with $E\,X_k = \mu_k$ and $\mathrm{Var}\,X_k = \sigma_k^2$, and suppose that*

$$\frac{1}{n}\sum_{k=1}^{n}\mu_k \to \mu \quad \text{and that} \quad \frac{1}{n}\sum_{k=1}^{n}\sigma_k^2 \to \sigma^2 \quad \text{as} \quad n \to \infty,$$

(where $|\mu| < \infty$ and $\sigma^2 < \infty$). Then

$$\frac{1}{n}\sum_{k=1}^{n}X_k \xrightarrow{p} \mu \quad \text{as} \quad n \to \infty.$$

Proof. Set $S_n = \sum_{k=1}^n X_k$, $m_n = \sum_{k=1}^n \mu_k$, and $s_n^2 = \sum_{k=1}^n \sigma_k^2$, and let $\varepsilon > 0$. By Chebyshev's inequality we then have

$$P\left(\left|\frac{S_n - m_n}{n}\right| > \varepsilon\right) \leq \frac{s_n^2}{n^2 \varepsilon^2} \to 0 \quad \text{as} \quad n \to \infty,$$

which tells us that

$$\frac{S_n - m_n}{n} \xrightarrow{p} 0 \quad \text{as} \quad n \to \infty,$$

which implies that

$$\frac{S_n}{n} = \frac{S_n - m_n}{n} + \frac{m_n}{n} \xrightarrow{p} 0 + \mu = \mu \quad \text{as} \quad n \to \infty$$

via Theorem 6.6.2. $\qquad\square$

The next result is an example of *the law of large numbers for weighted sums.*

Theorem 1.2. *Let X_1, X_2, $\ldots$ be i.i.d. random variables with finite mean μ, and let $\{(a_{nk}, 1 \leq k \leq n), n \geq 1\}$ be "weights," that is, suppose that $a_{nk} \geq 0$ and $\sum_{k=1}^n a_{nk} = 1$ for $n = 1, 2, \ldots$. Suppose, in addition, that*

$$\max_{1 \leq k \leq n} a_{nk} \leq \frac{C}{n} \quad \text{for all} \quad n,$$

for some positive constant C, and set

$$S_n = \sum_{k=1}^n a_{nk} X_k, \quad n = 1, 2, \ldots.$$

Then

$$S_n \xrightarrow{p} \mu \quad \text{as} \quad n \to \infty.$$

Proof. The proof follows very much the lines of the previous one. We first note that

$$E\,S_n = \mu \sum_{k=1}^n a_{nk} = \mu \quad \text{and that} \quad \text{Var}\,S_n = \sigma^2 \sum_{k=1}^n a_{nk}^2 = \sigma^2 A_n,$$

where thus

$$A_n = \sum_{k=1}^n a_{nk}^2 \leq \max_{1 \leq k \leq n} a_{nk} \sum_{k=1}^n a_{nk} \leq \frac{C}{n} \cdot 1 = \frac{C}{n}.$$

By Chebyshev's inequality we now obtain

$$P\left(|S_n - \mu| > \varepsilon\right) \leq \frac{\text{Var}\,S_n}{\varepsilon^2} = \frac{\sigma^2 A_n}{\varepsilon^2} \leq \sigma^2 \frac{C}{n} \to 0 \quad \text{as} \quad n \to \infty,$$

and the conclusion follows. $\qquad\square$

1.2 The Central Limit Theorem: The Non-i.i.d. Case

An important criterion pertaining to the central limit theorem is the *Lyapounov condition*. It should be said, however, that more than finite variance is necessary in order for the condition to apply. This is the price one pays for relaxing the assumption of equidistribution. For the proof we refer to the literature cited in Appendix A.

Theorem 1.3. *Suppose that $X_1, X_2, \ldots$ are independent random variables, set, for $k \geq 1$, $\mu_k = E X_k$ and $\sigma_k^2 = \operatorname{Var} X_k$, and suppose that $E|X_k|^r < \infty$ for all k and some $r > 2$. If*

$$\beta(n,r) = \frac{\sum_{k=1}^n E|X_k - \mu_k|^r}{\left(\sum_{k=1}^n \sigma_k^2\right)^{r/2}} \to 0 \quad as \quad n \to \infty, \tag{1.1}$$

then

$$\frac{\sum_{k=1}^n (X_k - \mu_k)}{\sqrt{\sum_{k=1}^n \sigma_k^2}} \xrightarrow{d} N(0,1) \quad as \quad n \to \infty. \qquad \square$$

If, in particular, $X_1, X_2, \ldots$ are identically distributed and, for simplicity, with mean zero, then Lyapounov's condition turns into

$$\beta(n,r) = \frac{n E|X_1|^r}{(n\sigma^2)^{r/2}} = \frac{E|X_1|^r}{\sigma^r} \cdot \frac{1}{n^{1-r/2}} \to 0 \quad \text{as} \quad n \to \infty, \tag{1.2}$$

which proves the central limit theorem under this slightly stronger assumption.

1.3 Sums of Dependent Random Variables

There exist many notions of dependence. One of the first things one learns in probability theory is that the outcomes of repeated drawings of balls *with replacement* from an urn of balls with different colors are independent, whereas the drawings *without replacement* are not. *Markov dependence* means, vaguely speaking, that the future of a process depends on the past only through the present. Another important dependence concept is *martingale dependence*, which is the topic of Section 8. Generally speaking, a typical dependence concept is defined via some kind of decay, in the sense that the further two elements are apart in time or index, the weaker is the dependence.

A simple such concept is m-dependence.

Definition 1.1. *The sequence $X_1, X_2, \ldots$ is m-dependent if X_i and X_j are independent whenever $|i - j| > m$.* $\qquad \square$

Remark 1.1. Independence is the same as 0-dependence.[1]

---

[1] In Swedish this looks fancier: "Oberoende" is the same as "0-beroende."

Example 1.1. $Y_1, Y_2, \ldots$ be i.i.d. random variables, and set

$$X_1 = Y_1 \cdot Y_2, \quad X_2 = Y_2 \cdot Y_3, \quad \ldots, \quad X_k = Y_k \cdot Y_{k+1}, \quad \ldots.$$

The sequence $X_1, X_2, \ldots$ clearly is a 1-dependent sequence; neighboring X variables are dependent, but X_i and X_j with $|i - j| > 1$ are independent.   □

A common example of m-dependent sequences are the so-called $(m + 1)$-block factors defined by

$$X_n = g(Y_n, Y_{n+1}, \ldots, Y_{n+m-1}, Y_{n+m}), \quad n \geq 1,$$

where $Y_1, Y_2, \ldots$ are independent random variables, and $g : \mathbb{R}^{m+1} \to \mathbb{R}$. Note that our example is a 2-block factor with $g(y_1, y_2) = y_1 \cdot y_2$.

The law of large numbers and the central limit theorem are both valid in this setting. Following is the law of large numbers.

Theorem 1.4. *Suppose that $X_1, X_2, \ldots$ is a sequence of m-dependent random variables with finite mean μ and set $S_n = \sum_{k=1}^{n} X_k$, $n \geq 1$. Then*

$$\frac{S_n}{n} \xrightarrow{p} \mu \quad as \quad n \to \infty.$$

Proof. For simplicity we confine ourselves to proving the theorem for $m = 1$. We then separate S_n into the sums over the odd and even summands, respectively.

Since the even as well as the odd summands are independent, the law of large numbers for independent summands, Theorem 6.5.1 tells us that

$$\frac{\sum_{k=1}^{m} X_{2k}}{m} \xrightarrow{p} \mu \quad and \quad \frac{\sum_{k=1}^{m} X_{2k-1}}{m} \xrightarrow{p} \mu \quad as \quad m \to \infty,$$

so that an application of Theorem 6.6.2 yields

$$\frac{S_{2m}}{2m} = \frac{1}{2}\frac{\sum_{k=1}^{m} X_{2k-1}}{m} + \frac{1}{2}\frac{\sum_{k=1}^{m} X_{2k}}{m} \xrightarrow{p} \frac{1}{2}\mu + \frac{1}{2}\mu = \mu \quad as \quad m \to \infty,$$

when $n = 2m$ is even. For $n = 2m + 1$ odd we similarly obtain

$$\frac{S_{2m+1}}{2m+1} = \frac{m+1}{2m+1} \cdot \frac{\sum_{k=1}^{m+1} X_{2k-1}}{m+1} + \frac{m}{2m+1} \cdot \frac{\sum_{k=1}^{m} X_{2k}}{m}$$

$$\xrightarrow{p} \frac{1}{2}\mu + \frac{1}{2}\mu = \mu \quad as \quad m \to \infty,$$

which finishes the proof.   □

Exercise 1.1. Complete the proof of the theorem for general m.   □

In the m-dependent case the dependence stops abruptly. A natural generalization would be to allow the dependence to drop gradually. This introduces the concept of *mixing*. There are variations with different names. We refer to the more advanced literature for details.

2 Stable Distributions

Let $X, X_1, X_2, \ldots$ be i.i.d. random variables with partial sums S_n, $n \geq 1$. The law of large numbers states that $S_n/n \xrightarrow{p} \mu$ as $n \to \infty$ if the mean μ is finite. The central limit theorem states that $(S_n - n\mu)/(\sigma\sqrt{n}) \xrightarrow{d} N(0,1)$ as $n \to \infty$, provided the mean μ and the variance σ^2 exist. A natural question is whether there exists something "in between," that is, can we obtain some (other) limit by normalizing with n to some other power than 1 or $1/2$? In this section and the next one we provide a glimpse into more general limit theorems for sums of i.i.d. random variables.

Before addressing the question just raised, here is another observation. If, in particular, we assume that the random variables are $C(0,1)$-distributed, then we recall from Remark 6.5.2 that, for any $n \geq 1$,

$$\varphi_{\frac{S_n}{n}}(t) = \left(\varphi_X\left(\frac{t}{n}\right)\right)^n = \left(e^{-|t/n|}\right)^n = e^{-|t|} = \varphi_X(t),$$

that

$$\frac{S_n}{n} \overset{d}{=} X \quad \text{for all} \quad n,$$

and, hence, that law of large numbers does not hold, which was no contradiction, because the mean does not exist.

Now, if, instead the random variables are $N(0, \sigma^2)$-distributed, then the analogous computation shows that

$$\varphi_{\frac{S_n}{\sqrt{n}}}(t) = \left(\varphi_X\left(\frac{t}{\sqrt{n}}\right)\right)^n = \left(\exp\left\{-\frac{1}{2}\left(\frac{t}{\sqrt{n}}\right)^2\right\}\right)^n = e^{-t^2/2} = \varphi_X(t),$$

that is,

$$\frac{S_n}{\sqrt{n}} \overset{d}{=} X \quad \text{for all} \quad n,$$

in view of the uniqueness theorem for characteristic functions.

Returning to our question above it seems, with this in mind, reasonable to try a distribution whose characteristic function equals $\exp\{-|t|^\alpha\}$ for $\alpha > 0$ (provided this is really a characteristic function also when $\alpha \neq 1$ and $\neq 2$). By modifying the computations above we similarly find that

$$\frac{S_n}{n^{1/\alpha}} \overset{d}{=} X \quad \text{for all} \quad n, \tag{2.1}$$

where, thus, $\alpha = 1$ corresponds to the Cauchy distribution and $\alpha = 2$ to the normal distribution.

Distributions with a characteristic function of the form

$$\varphi(t) = e^{-c|t|^\alpha}, \quad \text{where } 0 < \alpha \leq 2 \text{ and } c > 0, \tag{2.2}$$

are called *symmetric stable*. However, φ as defined in (2.2) is *not* a characteristic function for any $\alpha > 2$.

The general definition of stable distributions, stated in terms of random variables is as follows.

Definition 2.1. *Let X_1, X_2, ... be i.i.d. random variables, and set $S_n = X_1 + X_2 + \cdots + X_n$. The distribution of the random variables is* stable (in the broad sense) *if there exist sequences $a_n > 0$ and b_n such that*

$$S_n \overset{d}{=} a_n X + b_n.$$

The distribution is strictly stable *if $b_n = 0$ for all n.* □

Remark 2.1. The stability pertains to the fact that the sum of any number of random variables has the same distribution as the individual summands themselves (after scaling and translation).

Remark 2.2. One can show that if X has a stable distribution, then, necessarily, $a_n = n^{1/\alpha}$ for some $\alpha > 0$, which means that our first attempt to investigate possible characteristic functions was exhaustive (except for symmetry) and that, once again, only the case $0 < \alpha \leq 2$ is possible. Moreover, α is called the *index*.

Exercise 2.1. Another fact is that if X has a stable distribution with index α, $0 < \alpha < 2$, then

$$E|X|^r \begin{cases} < \infty, & \text{for } 0 < r < \alpha, \\ = \infty, & \text{for } r \geq \alpha. \end{cases}$$

This implies, in particular, that the law of large numbers must hold for stable distributions with $\alpha > 1$. Prove directly via characteristic functions that this is the case. Recall also, from above, that the case $\alpha = 1$ corresponds to the Cauchy distribution for which the law of large numbers does not hold.

We close this section by mentioning that there exist characterizations in terms of characteristic functions for the general class of stable distributions (not just the symmetric ones), but that is beyond the present outlook.

3 Domains of Attraction

We now return to the question posed in the introduction of Section 2, namely whether there exist limit theorems "in between" the law of large numbers and the central limit theorem. With the previous section in mind it is natural to guess that the result is positive, that such results would be connected with the stable distributions, and that the variance is not necessarily assumed to exist.

In order to discuss this problem we introduce the notion of *domains of attraction*.

Definition 3.1. *Let $X, X_1, X_2, \ldots$ be i.i.d. random variables with partial sums S_n, $n \geq 1$. We say that X, or, equivalently, the distribution F_X, belongs to the domain of attraction of the (non-degenerate) distribution G if there exist normalizing sequences $\{a_n > 0, n \geq 1\}$ and $\{b_n, n \geq 1\}$ such that*

$$\frac{S_n - b_n}{a_n} \xrightarrow{d} G \quad as \quad n \to \infty.$$

The notation is $F_X \in \mathcal{D}(G)$; alternatively, $X \in \mathcal{D}(Z)$ if $Z \in G$.   □

If $\operatorname{Var} X < \infty$, the central limit theorem tells us that X belongs to the domain of attraction of the normal distribution; choose $b_n = nE\,X$, and $a_n = \sqrt{n\operatorname{Var} X}$. In particular, the normal distribution belongs to its own domain of attraction. Recalling Section 2 we also note that the stable distributions belong to their own domain of attraction.

In fact, the stable distributions are the only possible limit distributions.

Theorem 3.1. *Only the stable distributions or random variables possess a domain of attraction.*

With this information the next problem of interest would be to exhibit criteria for a distribution to belong to the domain of attraction of some given (stable) distribution. In order to state such results we need some facts about what is called *regular* and *slow variation*.

Definition 3.2. *Let $a > 0$. A positive measurable function u on $[a, \infty)$ varies regularly at infinity with exponent ρ, $-\infty < \rho < \infty$, denoted $u \in \mathcal{RV}(\rho)$, iff*

$$\frac{u(tx)}{u(t)} \to x^\rho \quad as \quad t \to \infty \quad for\ all \quad x > 0.$$

If $\rho = 0$ the function is slowly varying *at infinity; $u \in \mathcal{SV}$.*   □

Typical examples of regularly varying functions are

$$x^\rho, \quad x^\rho \log^+ x, \quad x^\rho \log^+ \log^+ x, \quad x^\rho \frac{\log^+ x}{\log^+ \log^+ x}, \quad \text{and so on.}$$

Typical slowly varying functions are the above when $\rho = 0$. Every positive function with a positive finite limit as $x \to \infty$ is slowly varying.

Exercise 3.1. Check that the typical functions behave as claimed.

Here is now the main theorem.

Theorem 3.2. *A random variable X with distribution function F belongs to the domain of attraction of a stable distribution iff there exists $L \in \mathcal{SV}$ such that*

$$U(x) = E\,X^2 I\{|X| \leq x\} \sim x^{2-\alpha}L(x) \quad as \quad x \to \infty, \tag{3.1}$$

and, moreover, for $\alpha \in (0,2)$, *that*

$$\frac{P(X > x)}{P(|X| > x)} \to p \quad and \quad \frac{P(X < -x)}{P(|X| > x)} \to 1 - p \quad as \quad x \to \infty. \quad (3.2)$$

By partial integration and properties of regularly varying functions one can show that (3.1) is equivalent to

$$\frac{x^2 P(|X| > x)}{U(x)} \to \frac{2 - \alpha}{\alpha} \quad \text{as} \quad x \to \infty, \quad \text{for} \quad 0 < \alpha \le 2, \quad (3.3)$$

$$P(|X| > x) \sim \frac{2 - \alpha}{\alpha} \cdot \frac{L(x)}{x^\alpha} \quad \text{as} \quad x \to \infty, \quad \text{for} \quad 0 < \alpha < 2, \quad (3.4)$$

which, in view of Theorem 3.1 yields the following alternative formulation of Theorem 3.2.

Theorem 3.3. *A random variable X with distribution function F belongs to the domain of attraction of*
(a) *the normal distribution iff $U \in \mathcal{SV}$;*
(b) *a stable distribution with index $\alpha \in (0,2)$ iff (3.4) and (3.2) hold.*

Let us, as a first illustration, look at the simplest example.

Example 3.1. Let $X, X_1, X_2, \ldots$ be independent random variables with common density

$$f(x) = \begin{cases} \dfrac{1}{2x^2}, & \text{for} \quad |x| > 1, \\ 0, & \text{otherwise.} \end{cases}$$

Note that the distribution is symmetric and that the mean is infinite.
Now, for $x > 1$,

$$P(X > x) = \frac{1}{2x}, \quad P(X < -x) = \frac{1}{2|x|}, \quad P(|X| > x) = \frac{1}{x}, \quad U(x) = x - 1,$$

so that (3.1)–(3.4) are satisfied ($p = 1/2$ and $L(x) = 1$).

Our second example is a boundary case in that the variance does not exist, but the asymptotic distribution is still the normal one.

Example 3.2. Suppose that $X, X_1, X_2, \ldots$ are independent random variables with common density

$$f(x) = \begin{cases} \dfrac{1}{|x|^3}, & \text{for} \quad |x| > 1, \\ 0, & \text{otherwise.} \end{cases}$$

The distribution is symmetric again, the mean is finite and the variance is infinite $-\int_1^\infty (x^2/x^3)\, dx = +\infty$. As for (3.1) we find that

$$U(x) = \int_{|y|\leq x} y^2 f(y)\, dy = 2\int_1^x \frac{dy}{y} = 2\log x,$$

so that $U \in \mathcal{SV}$ as $x \to \infty$, that is, X belongs to the domain of attraction of the normal distribution.

This means that, for a suitable choice of normalizing constants $\{a_n,\ n \geq 1\}$ (no centering because of symmetry), we have

$$\frac{S_n}{a_n} \xrightarrow{d} N(0,1) \quad \text{as} \quad n \to \infty.$$

More precisely, omitting all details, we just mention that one can show that, in fact,

$$\frac{S_n}{\sqrt{n\log n}} \xrightarrow{d} N(0,1) \quad \text{as} \quad n \to \infty.$$

Remark 3.1. The object of Problem 6.8.50 was to prove this result with the aid of characteristic functions, that is, directly, without using the theory of domains of attraction.

4 Uniform Integrability

We found in Section 6.3 that convergence in probability does not necessarily imply convergence of moments. A natural question is whether there exists some condition that guarantees that a sequence that converges in probability (or almost surely or in distribution) also converges in r-mean. It turns out that *uniform integrability* is the adequate concept for this problem.

Definition 4.1. *A sequence* $X_1, X_2, \ldots$ *is called* uniformly integrable *if*

$$E|X_n|I\{|X_n| > a\} \to 0 \qquad \text{as} \quad a \to \infty \quad \text{uniformly in} \quad n. \qquad \square$$

Remark 4.1. If, for example, all distributions involved are continuous, this is the same as

$$\int_{|x|>a} |x| f_{X_n}(x)\, dx \to 0 \quad \text{as} \quad a \to \infty \quad \text{uniformly in} \quad n. \qquad \square$$

The following result shows why uniform integrability is the correct concept. For a proof and much more on uniform integrability, we refer to the literature cited in Appendix A.

Theorem 4.1. *Let* $X, X_1, X_2, \ldots$ *be random variables such that* $X_n \xrightarrow{p} X$ *as* $n \to \infty$. *Let* $r > 0$, *and suppose that* $E|X_n|^r < \infty$ *for all* n. *The following are equivalent:*

(a) $\{|X_n|^r,\ n \geq 1\}$ *is uniformly integrable;*
(b) $X_n \xrightarrow{r} X$ *as* $n \to \infty$;
(c) $E|X_n|^r \to E|X|^r$ *as* $n \to \infty$. $\qquad \square$

The immediate application of the theorem is manifested in the following exercise.

Exercise 4.1. Show that if $X_n \xrightarrow{p} X$ as $n \to \infty$ and $X_1, X_2, \ldots$ is uniformly integrable, then $E X_n \to E X$ as $n \to \infty$.   □

Example 4.1. A uniformly bounded sequence of random variables is uniformly integrable. Technically, if the random variables $X_1, X_2, \ldots$ are uniformly bounded, there exists some constant $A > 0$ such that $P(|X_n| \le A) = 1$ for all n. This implies that the expectation in the definition, in fact, *equals* zero as soon as $a > A$.

Example 4.2. In Example 6.3.1 we found that X_n converges in probability as $n \to \infty$ and that X_n converges in r-mean as $n \to \infty$ when $r < 1$ but not when $r \ge 1$. In view of Theorem 4.1 it must follow that $\{|X_n|^r, n \ge 1\}$ is uniformly integrable when $r < 1$ but not when $r \ge 1$.

Indeed, it follows from the definition that (for $a > 1$)

$$E|X_n|^r I\{|X_n| > a\} = n^r \cdot \frac{1}{n} \cdot I\{a < n\} \to 0 \quad \text{as} \quad a \to \infty$$

uniformly in n iff $r < 1$, which verifies the desired conclusion.   □

Exercise 4.2. State and prove an analogous statement for Example 6.3.2.

Exercise 4.3. Consider the following modification of Example 6.3.1. Let $X_1, X_2, \ldots$ be random variables such that

$$P(X_n = 1) = 1 - \frac{1}{n} \quad \text{and} \quad P(X_n = 1000) = \frac{1}{n}, \quad n \ge 2.$$

Show that $X_n \xrightarrow{p} 1$ as $n \to \infty$, that $\{|X_n|^r, n \ge 1\}$ is uniformly integrable for all $r > 0$, and hence that $X_n \xrightarrow{r} 1$ as $n \to \infty$ for all $r > 0$.   □

Remark 4.2. Since $X_1, X_2, \ldots$ are uniformly bounded, the latter part follows immediately from Example 4.1, but it is instructive to verify the conclusion directly via the definition.

Note also that the difference between Exercise 4.3 and Example 6.3.1 is that there the "rare" value n drifts off to infinity, whereas here it is a fixed constant (1000).   □

It is frequently difficult to verify uniform integrability of a sequence directly. The following result provides a convenient sufficient criterion.

Theorem 4.2. *Let $X_1, X_2, \ldots$ be random variables, and suppose that*

$$\sup_n E|X_n|^r < \infty \quad \text{for some} \quad r > 1.$$

Then $\{X_n, n \ge 1\}$ is uniformly integrable. In particular, this is the case if $\{|X_n|^r, n \ge 1\}$ is uniformly integrable for some $r > 1$.

Proof. We have

$$E|X_n|I\{|X_n| > a\} \le a^{1-r}E|X_n|^r I\{|X_n| > a\} \le a^{1-r}E|X_n|^r$$
$$\le a^{1-r}\sup_n E|X_n|^r \to 0 \quad \text{as} \quad a \to \infty,$$

independently, hence uniformly, in n.

The particular case is immediate since more is assumed. □

Remark 4.3. The typical case is when one wishes to prove convergence of the sequence of expected values and knows that the sequence of variances is uniformly bounded. □

We close this section with an illustration of how one can prove Stirling's formula via the central limit theorem with the aid of the exponential distribution and Theorems 4.1 and 4.2.

Example 4.3. Let $X_1, X_2, \ldots$ be independent Exp(1)-distributed random variables, and set $S_n = \sum_{k=1}^n X_k$, $n \ge 1$. From the central limit theorem we know that

$$\frac{S_n - n}{\sqrt{n}} \xrightarrow{d} N(0,1) \quad \text{as} \quad n \to \infty,$$

and, since, for example, the variances of the normalized partial sums are equal to 1 for all n (so that the second moments are uniformly bounded), it follows from Theorems 4.2 and 4.1 that

$$\lim_{n\to\infty} E\left|\frac{S_n - n}{\sqrt{n}}\right| = E|N(0,1)| = \sqrt{\frac{2}{\pi}}. \tag{4.1}$$

Since we know that $S_n \in \Gamma(n,1)$ the expectation can be spelled out exactly and we can rewrite (4.1) as

$$\int_0^\infty \left|\frac{x-n}{\sqrt{n}}\right| \frac{1}{\Gamma(n)} x^{n-1} e^{-x}\, dx \to \sqrt{\frac{2}{\pi}} \quad \text{as} \quad n \to \infty. \tag{4.2}$$

By splitting the integral at $x = n$, and making the change of variable $u = x/n$ one arrives after some additional computations at the relation

$$\lim_{n\to\infty} \frac{\left(\frac{n}{e}\right)^n \sqrt{2n\pi}}{n!} = 1,$$

which is—Stirling's formula. □

Exercise 4.4. Carry out the details of the program. □

5 An Introduction to Extreme Value Theory

Suppose that X_1, X_2, ... is a sequence of i.i.d. distributed random variables. What are the possible limit distributions of the normalized partial sums? If the variance is finite the answer is the normal distribution in view of the central limit theorem. In the general case, we found in Section 3 that the possible limit distributions are the stable distributions.

This section is devoted to the analogous problem for extremes. Thus, let, for $n \geq 1$,

$$Y_n = \max\{X_1, X_2, \ldots, X_n\}.$$

What are the possible limit distributions of Y_n, after suitable normalization, as $n \to \infty$?

The following definition is the analog of Definition 3.1 (which concerned sums) for extremes.

Definition 5.1. *Let X, X_1, X_2, ... be i.i.d. random variables, and set $Y_n = \max_{1 \leq k \leq n} X_k$, $n \geq 1$. We say that X, or, equivalently, the distribution function F_X, belongs to the domain of attraction of the extremal distribution G if there exist normalizing sequences $\{a_n > 0, n \geq 1\}$ and $\{b_n, n \geq 1\}$, such that*

$$\frac{Y_n - b_n}{a_n} \xrightarrow{d} G \quad as \quad n \to \infty. \qquad \square$$

Example 5.1. Let X_1, X_2, ... be independent $\mathrm{Exp}(1)$-distributed random variables, and set $Y_n = \max\{X_1, X_2, \ldots, X_n\}$, $n \geq 1$. Then,

$$F(x) = 1 - e^{-x} \quad for \quad x > 0,$$

(and 0 otherwise), so that

$$P(Y_n \leq x) = \left(1 - e^{-x}\right)^n.$$

Aiming at something like $(1 - u/n)^n \to e^u$ as $n \to \infty$ suggests that we try $a_n = 1$ and $b_n = \log n$ to obtain

$$F_{Y_n - \log n}(x) = P(Y_n \leq x + \log n) = \left(1 - e^{-x - \log n}\right)^n$$

$$= \left(1 - \frac{e^{-x}}{n}\right)^n \to e^{-e^{-x}} \quad as \quad n \to \infty,$$

for all $x \in \mathbb{R}$.

Example 5.2. Let X_1, X_2, ... be independent $\mathrm{Pa}(\beta, \alpha)$-distributed random variables, and set $Y_n = \max\{X_1, X_2, \ldots, X_n\}$, $n \geq 1$. Then,

$$F(x) = \int_\beta^x \frac{\alpha \beta^\alpha}{y^{\alpha+1}} \, dy = 1 - \left(\frac{\beta}{x}\right)^\alpha \quad for \quad x > \beta,$$

(and 0 otherwise), so that

$$P(Y_n \leq x) = \left(1 - \left(\frac{\beta}{x}\right)^\alpha\right)^n.$$

An inspection of this relation suggests the normalization $a_n = n^{1/\alpha}$ and $b_n = 0$, which, for $x > 0$ and n large, yields

$$F_{n^{-1/\alpha}Y_n}(x) = P(Y_n \leq xn^{1/\alpha}) = \left(1 - \left(\frac{\beta}{xn^{1/\alpha}}\right)^\alpha\right)^n$$

$$= \left(1 - \frac{(\beta/x)^\alpha}{n}\right)^n \to e^{-(\beta/x)^\alpha} \quad \text{as} \quad n \to \infty.$$

Remark 5.1. For $\beta = 1$ the example reduces to Example 6.1.2.

Example 5.3. Let X_1, X_2, ... be independent $U(0,\theta)$-distributed random variables ($\theta > 0$), and set $Y_n = \max\{X_1, X_2, \ldots, X_n\}$, $n \geq 1$. Thus, $F(x) = x/\theta$ for $x \in (0,\theta)$ and 0 otherwise, so that,

$$P(Y_n \leq x) = \left(\frac{x}{\theta}\right)^n.$$

Now, since $Y_n \overset{p}{\to} \theta$ as $n \to \infty$ (this is intuitively "obvious," but check Problem 6.8.1), it is more convenient to study $\theta - Y_n$, viz.,

$$P(\theta - Y_n \leq x) = P(Y_n \geq \theta - x) = 1 - \left(1 - \frac{x}{\theta}\right)^n.$$

The usual approach now suggests $a_n = 1/n$ and $b_n = \theta$. Using this we obtain, for any $x < 0$,

$$P(n(Y_n - \theta) \leq x) = P\left(\theta - Y_n \geq \frac{(-x)}{n}\right) = \left(1 - \frac{(-x)}{\theta n}\right)^n$$

$$\to e^{-(-x)/\theta} \quad \text{as} \quad n \to \infty. \qquad \square$$

• Looking back at the examples we note that the limit distributions have different expressions and that their domains vary; they are $x > 0$, $x \in \mathbb{R}$, and $x < 0$, respectively. It seems that the possible limits may be of at least three kinds. The following result tells us that this is indeed the case. More precisely, there are exactly three so-called *types*, meaning those mentioned in the theorem below, together with linear transformations of them.

Theorem 5.1. *There exist three types of extremal distributions:*

Fréchet: $\quad \Phi_\alpha(x) = \begin{cases} 0, & \text{for } x < 0, \\ \exp\{-x^{-\alpha}\}, & \text{for } x \geq 0, \end{cases} \quad \alpha > 0;$

Weibull: $\quad \Psi_\alpha(x) = \begin{cases} \exp\{-(-x)^\alpha\}, & \text{for } x < 0, \\ 1, & \text{for } x \geq 0, \end{cases} \quad \alpha > 0;$

Gumbel: $\quad \Lambda(x) = \exp\{-e^{-x}\}, \quad \text{for } x \in \mathbb{R}.$

The proof is beyond the scope of this book, let us just mention that the so-called convergence of types theorem is a crucial ingredient.

Remark 5.2. Just as the normal and stable distributions belong to their own domain of attraction (recall relation (2.1) above), it is natural to expect that the three extreme value distributions of the theorem belong to their domain of attraction. This is more formally spelled out in Problem 9.10 below.

6 Records

Let $X, X_1, X_2, \ldots$ be i.i.d. continuous random variables. The *record times* are $L(1) = 1$ and, recursively,

$$L(n) = \min\{k : X_k > X_{L(n-1)}\}, \quad n \geq 2,$$

and the *record values* are

$$X_{L(n)}, \quad n \geq 1.$$

The associated *counting process* $\{\mu(n), n \geq 1\}$ is defined by

$$\mu(n) = \#\text{records among } X_1, X_2, \ldots, X_n = \max\{k : L(k) \leq n\}.$$

The reason for assuming continuity is that we wish to avoid ties.

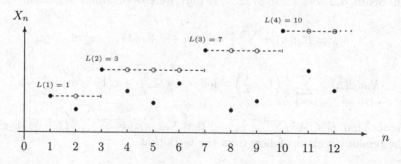

Fig. 7.1. Partial maxima ∘

Whereas the sequence of partial maxima, Y_n, $n \geq 1$, describe "the largest value so far," the record values pick these values the first time they appear. The sequence of record values thus constitutes a subsequence of the partial maxima. Otherwise put, the sequence of record values behaves like a compressed sequence of partial maxima, as is depicted in the above figure.

We begin by noticing that the record times and the number of records are distribution independent (under our continuity assumption). This is due to the fact that for a given random variable X with distribution function F, it follows that $F(X) \in U(0,1)$. This implies that there is a 1-to-1 map from every random variable to every other one, which preserves the record *times*, and therefore also the number of records—but not the record *values*.

Next, set

$$I_k = \begin{cases} 1, & \text{if } X_k \text{ is a record,} \\ 0, & \text{otherwise,} \end{cases}$$

so that $\mu(n) = \sum_{k=1}^{n} I_k$, $n \geq 1$.

By symmetry, all permutations between $X_1, X_2, \ldots, X_n$ are equally likely, from which we conclude that

$$P(I_k = 1) = 1 - P(I_k = 0) = \frac{1}{k}, \quad k = 1, 2, \ldots, n.$$

In addition one can show that the random variables $\{I_k, k \geq 1\}$ are independent. We collect these facts in the following result.

Theorem 6.1. *Let $X_1, X_2, \ldots, X_n$, $n \geq 1$, be i.i.d. continuous random variables. Then*

(a) *the indicators $I_1, I_2, \ldots, I_n$ are independent;*

(b) $P(I_k = 1) = 1/k$ *for $k = 1, 2, \ldots, n$.*

As a corollary it is now a simple task to compute the mean and the variance of $\mu(n)$ and their asymptotics.

Theorem 6.2. *Let $\gamma = 0.5772\ldots$ denote Euler's constant. We have*

$$m_n = E\,\mu(n) = \sum_{k=1}^{n} \frac{1}{k} = \log n + \gamma + o(1) \quad \text{as} \quad n \to \infty;$$

$$\operatorname{Var}\mu(n) = \sum_{k=1}^{n} \frac{1}{k}\left(1 - \frac{1}{k}\right) = \log n + \gamma - \frac{\pi^2}{6} + o(1) \quad \text{as} \quad n \to \infty.$$

Proof. That $E\,\mu(n) = \sum_{k=1}^{n} \frac{1}{k}$, and that $\operatorname{Var}\mu(n) = \sum_{k=1}^{n} \frac{1}{k}(1 - \frac{1}{k})$, is clear. The remaining claims follow from the facts that

$$\sum_{k=1}^{n} \frac{1}{k} = \log n + \gamma + o(1) \text{ as } n \to \infty \quad \text{and} \quad \sum_{n=1}^{\infty} \frac{1}{n^2} = \frac{\pi^2}{6}. \qquad \square$$

Next we present the weak laws of large numbers for the counting process.

Theorem 6.3. *We have*

$$\frac{\mu(n)}{\log n} \xrightarrow{p} 1 \quad \text{as} \quad n \to \infty.$$

Proof. Chebyshev's inequality together with Theorem 6.2 yields

$$P\left(\frac{\mu(n) - E\,\mu(n)}{\operatorname{Var}(\mu(n))} > \varepsilon\right) \leq \frac{1}{\varepsilon^2 \operatorname{Var}(\mu(n))} \to 0 \quad \text{as} \quad n \to \infty,$$

which tells us that

$$\frac{\mu(n) - E\,\mu(n)}{\text{Var}\,(\mu(n))} \xrightarrow{p} 0 \quad \text{as} \quad n \to \infty.$$

Finally,

$$\frac{\mu(n)}{\log n} = \frac{\mu(n) - E\,\mu(n)}{\text{Var}\,(\mu(n))} \cdot \frac{\text{Var}\,(\mu(n))}{\log n} + \frac{E\,\mu(n)}{\log n} \xrightarrow{p} 0 \cdot 1 + 1 = 1 \quad \text{as} \quad n \to \infty,$$

in view of Theorem 6.2 (and Theorem 6.6.2). □

The central limit theorem for the counting process runs as follows.

Theorem 6.4. *We have*

$$\frac{\mu(n) - \log n}{\sqrt{\log n}} \xrightarrow{d} N(0,1) \quad \text{as} \quad n \to \infty.$$

Proof. We check the Lyapounov condition (1.1) with $r = 3$:

$$E|I_k - E\,I_k|^3 = \left|0 - \frac{1}{k}\right|^3 \cdot \left(1 - \frac{1}{k}\right) + \left|1 - \frac{1}{k}\right|^3 \frac{1}{k}$$

$$= \left(1 - \frac{1}{k}\right)\frac{1}{k} \cdot \left(\frac{1}{k^2} + \left(1 - \frac{1}{k}\right)^2\right) \leq 2\left(1 - \frac{1}{k}\right)\frac{1}{k},$$

so that

$$\beta(n,3) = \frac{\sum_{k=1}^{n} E|X_k - \mu_k|^3}{\left(\sum_{k=1}^{n} \sigma_k^2\right)^{3/2}} \leq 2\frac{\sum_{k=1}^{n} \left(1 - \frac{1}{k}\right)\frac{1}{k}}{\left(\sum_{k=1}^{n} \left(1 - \frac{1}{k}\right)\frac{1}{k}\right)^{3/2}}$$

$$= 2\left(\sum_{k=1}^{n} \left(1 - \frac{1}{k}\right)\frac{1}{k}\right)^{-1/2} \to 0 \quad \text{as} \quad n \to \infty,$$

since

$$\sum_{k=1}^{n} \left(1 - \frac{1}{k}\right)\frac{1}{k} \geq \frac{1}{2}\sum_{k=2}^{n}\frac{1}{k} \to \infty \quad \text{as} \quad n \to \infty. \qquad \square$$

Exercise 6.1. Another way to prove this is via characteristic functions or moment generating functions; note, in particular, that $|I_k - \frac{1}{k}| \leq 1$ for all k.□

The analogous results for record times state that

$$\frac{\log L(n)}{n} \xrightarrow{p} 1 \quad \text{as} \quad n \to \infty,$$

$$\frac{\log L(n) - n}{\sqrt{n}} \xrightarrow{d} N(0,1) \quad \text{as} \quad n \to \infty.$$

In the opening of this section we found that the *record values*, $\{X_{L(n)}, n \geq 1\}$, seemed to behave like a compressed sequence of partial maxima, which makes it reasonable to believe that there exist three possible limit distributions for $X_{L(n)}$ as $n \to \infty$, which are somehow connected with the the three limit theorems for extremes. The following theorem shows that this is, indeed, the case.

Theorem 6.5. *Suppose that F is absolutely continuous. The possible types of limit distributions for record values are*

$$\Phi(-\log(-\log G(x))),$$

where G is an extremal distribution and Φ the distribution function of the standard normal distribution. More precisely, the three classes or types of limit distributions are

$$\Phi_\alpha^{(R)}(x) = \begin{cases} 0, & \text{for } x < 0, \\ \Phi(\alpha \log x), & \text{for } x \geq 0, \end{cases} \quad \alpha > 0;$$

$$\Psi_\alpha^{(R)}(x) = \begin{cases} \Phi(-\alpha \log(-x)), & \text{for } x < 0, \\ 1, & \text{for } x \geq 0, \end{cases} \quad \alpha > 0;$$

$$\Lambda^{(R)}(x) = \Phi(x), \qquad\qquad \text{for } x \in \mathbb{R}.$$

7 The Borel–Cantelli Lemmas

The aim of this section is to provide some additional material on a.s. convergence. Although the reader cannot be expected to appreciate the concept fully at this level, we add here some additional facts and properties to shed somewhat light on it. The main results or tools are the Borel–Cantelli lemmas. We begin, however, with the following definition:

Definition 7.1. *Let $\{A_n, n \geq 1\}$ be a sequence of events (subsets of Ω). We define*

$$A_* = \liminf_{n \to \infty} A_n = \bigcup_{n=1}^{\infty} \bigcap_{m=n}^{\infty} A_m,$$

$$A^* = \limsup_{n \to \infty} A_n = \bigcap_{n=1}^{\infty} \bigcup_{m=n}^{\infty} A_m. \qquad\qquad \square$$

Thus, if $\omega \in \Omega$ belongs to the set $\liminf_{n \to \infty} A_n$, then ω belongs to $\bigcap_{m=n}^{\infty} A_m$ for some n, that is, there exists an n such that $\omega \in A_m$ for *all* $m \geq n$. In particular, if A_n is the event that something special occurs at "time" n, then $\liminf_{n \to \infty} A_n^c$ means that from some n onward this property never occurs.

Similarly, if $\omega \in \Omega$ belongs to the set $\limsup_{n \to \infty} A_n$, then ω belongs to $\bigcup_{m=n}^{\infty} A_m$ for every n, that is, no matter how large we choose m there is always some $n \geq m$ such that $\omega \in A_n$, or, equivalently, $\omega \in A_n$ for infinitely many values of n or, equivalently, for arbitrarily large values of n. A convenient way to express this is

$$\omega \in \{A_n \text{ infinitely often (i.o.)}\} \quad \Longleftrightarrow \quad \omega \in A^*. \tag{7.1}$$

Example 7.1. Let $X_1, X_2, \ldots$ be a sequence of random variables and let $A_n = \{|X_n| > \varepsilon\}$, $n \geq 1$, $\varepsilon > 0$. Then $\omega \in \liminf_{n\to\infty} A_n^c$ means that ω is such that $|X_n(\omega)| \leq \varepsilon$ for all sufficiently large n, and $\omega \in \limsup_{n\to\infty} A_n$ means that ω is such that there exist arbitrarily large values of n such that $|X_n(\omega)| > \varepsilon$. In particular, every ω for which $X_n(\omega) \to 0$ as $n \to \infty$ must be such that, for every $\varepsilon > 0$, only finitely many of the real numbers $X_n(\omega)$ exceed ε in absolute value. Hence,

$$X_n \overset{a.s.}{\to} 0 \text{ as } n \to \infty \quad \Longleftrightarrow \quad P(|X_n| > \varepsilon \text{ i.o.}) = 0 \quad \text{for all} \quad \varepsilon > 0. \quad (7.2)$$

We shall return to this example later. □

Here is the first Borel–Cantelli lemma.

Theorem 7.1. *Let $\{A_n, n \geq 1\}$ be arbitrary events. Then*

$$\sum_{n=1}^{\infty} P(A_n) < \infty \quad \Longrightarrow \quad P(A_n \text{ i.o.}) = 0.$$

Proof. We have

$$P(A_n \text{ i.o.}) = P(\limsup_{n\to\infty} A_n) = P(\bigcap_{n=1}^{\infty} \bigcup_{m=n}^{\infty} A_m)$$

$$\leq P(\bigcup_{m=n}^{\infty} A_m) \leq \sum_{m=n}^{\infty} P(A_m) \to 0 \quad \text{as} \quad n \to \infty. \quad \square$$

The converse does not hold in general—one example is given at the very end of this section. However, with an additional assumption of independence, the following, second Borel–Cantelli lemma, holds true.

Theorem 7.2. *Let $\{A_n, n \geq 1\}$ be independent events. Then*

$$\sum_{n=1}^{\infty} P(A_n) = \infty \quad \Longrightarrow \quad P(A_n \text{ i.o.}) = 1.$$

Proof. By the De Morgan formula and independence we obtain

$$P(A_n \text{ i.o.}) = P\left(\bigcap_{n=1}^{\infty} \bigcup_{m=n}^{\infty} A_m\right) = 1 - P\left(\bigcup_{n=1}^{\infty} \bigcap_{m=n}^{\infty} A_m^c\right)$$

$$= 1 - \lim_{n\to\infty} P\left(\bigcap_{m=n}^{\infty} A_m^c\right) = 1 - \lim_{n\to\infty} \lim_{N\to\infty} P\left(\bigcap_{m=n}^{N} A_m^c\right)$$

$$= 1 - \lim_{n\to\infty} \lim_{N\to\infty} \prod_{m=n}^{N} \left(1 - P(A_m)\right).$$

Now, since for $0 < x < 1$ we have $e^{-x} \geq 1 - x$, it follows that

$$\prod_{m=n}^{N} (1 - P(A_m)) \leq \exp\left\{ - \sum_{m=n}^{N} P(A_m) \right\} \to 0 \quad \text{as} \quad N \to \infty \,.$$

for every n, since, by assumption, $\sum_{m=1}^{\infty} P(A_m) = \infty$.   □

Remark 7.1. There exist more general versions of this result that allow for some dependence between the events (i.e., independence is not necessary for the converse to hold).   □

As a first application, let us reconsider Examples 6.3.1 and 6.3.2.

Example 7.2. Thus, X_2, X_3, ... is a sequence of random variables such that

$$P(X_n = 1) = 1 - \frac{1}{n^{\alpha}} \quad \text{and} \quad P(X_n = n) = \frac{1}{n^{\alpha}}, \quad n \geq 2,$$

where α is some positive number. Under the additional assumption that the random variables are independent, it was claimed in Remark 6.3.5 that $X_n \overset{a.s.}{\to} 1$ as $n \to \infty$ when $\alpha = 2$ and proved in Example 6.3.1 that this is not the case when $\alpha = 1$.

Now, in view of the first Borel–Cantelli lemma, it follows immediately that $X_n \overset{a.s.}{\to} 1$ as $n \to \infty$ for *all* $\alpha > 1$, even without any assumption about independence! To see this we first recall Example 7.1, according to which

$$X_n \overset{a.s.}{\to} 1 \quad \text{as} \quad n \to \infty \quad \Longleftrightarrow \quad P(|X_n - 1| > \varepsilon \text{ i.o.}) = 0 \quad \text{for all} \quad \varepsilon > 0.$$

The desired conclusion now follows from Theorem 7.1 since, for $\alpha > 1$,

$$\sum_{n=1}^{\infty} P(|X_n - 1| > \varepsilon) = \sum_{n=1}^{\infty} \frac{1}{n^{\alpha}} < \infty \quad \text{for all} \quad \varepsilon > 0.$$

It follows, moreover, from the second Borel–Cantelli lemma that if, in addition, we assume that X_1, X_2, ... are independent, then we do not have almost-sure convergence for any $\alpha \leq 1$. In particular, almost-sure convergence holds if and only if $\alpha > 1$ in that case.   □

A second look at the arguments above shows (please check!) that, in fact, the following, more general result holds true.

Theorem 7.3. *Let X_1, X_2, ... be a sequence of independent random variables. Then*

$$X_n \overset{a.s.}{\to} 0 \text{ as } \tilde{n} \to \infty \quad \Longleftrightarrow \quad \sum_{n=1}^{\infty} P(|X_n| > \varepsilon) < \infty \quad \text{for all} \quad \varepsilon > 0. \quad □$$

Let us now comment on formula(s) (6.3.1) (and (6.3.2)), which were presented before without proof, and show, at least, that almost-sure convergence implies their validity. Toward this end, let $X_1, X_2, \ldots$ be a sequence of random variables and $A = \{\omega : X_n(\omega) \to X(\omega) \text{ as } n \to \infty\}$ for some random variable X. Then (why?)

$$A = \bigcap_{n=1}^{\infty} \bigcup_{m=1}^{\infty} \bigcap_{i=m}^{\infty} \left\{ |X_i - X| \leq \frac{1}{n} \right\}. \tag{7.3}$$

Thus, assuming that almost-sure convergence holds, we have $P(A) = 1$, from which it follows that

$$P\left(\bigcup_{m=1}^{\infty} \bigcap_{i=m}^{\infty} \left\{ |X_i - X| \leq \frac{1}{n} \right\} \right) = 1$$

for all n. Furthermore, the sets $\{ \bigcap_{i=m}^{\infty} \{ |X_i - X| \leq 1/n \}, m \geq 1 \}$ are monotone increasing as $m \to \infty$, which, in view of Lemma 6.3.1, implies that, for all n,

$$\lim_{m \to \infty} P\left(\bigcap_{i=m}^{\infty} \left\{ |X_i - X| \leq \frac{1}{n} \right\} \right) = P\left(\bigcup_{m=1}^{\infty} \bigcap_{i=m}^{\infty} \left\{ |X_i - X| \leq \frac{1}{n} \right\} \right).$$

However, the latter probability was just seen to equal 1, from which it follows that $P(\bigcap_{i=m}^{\infty} \{ |X_i - X| \leq 1/n \})$ can be made arbitrary close to 1 by choosing m large enough. Therefore, since n was arbitrary we have shown (why?) that if $X_n \overset{a.s.}{\to} X$ as $n \to \infty$ then, for every $\varepsilon > 0$ and δ, $0 < \delta < 1$, there exists m_0 such that for all $m > m_0$ we have

$$P\left(\bigcap_{i=m}^{\infty} \{ |X_i - X| < \varepsilon \} \right) > 1 - \delta,$$

which is exactly (6.3.1) (which was equivalent to (6.3.2)).

7.1 Patterns

We begin with an example of a different and simpler nature.

Example 7.3. Toss a regular coin repeatedly (independent tosses) and let $A_n = \{$the nth toss yields a head$\}$ for $n \geq 1$. Then

$$P(A_n \text{ i.o.}) = 1.$$

To see this we note that $\sum_{n=1}^{\infty} P(A_n) = \sum_{n=1}^{\infty} 1/2 = \infty$, and the conclusion follows from Theorem 7.2.

In words, if we toss a regular coin repeatedly, we obtain only finitely many heads with probability zero. Intuitively, this is obvious since, by symmetry, if this were not true, the same would not be true for tails either, which is impossible, since at least one of them must appear infinitely often.

However, for a biased coin, one could imagine that if the probability of obtaining heads is "very small," then it might happen that, with some "very small" probability, only finitely many heads appear. To treat that case, suppose that $P(\text{heads}) = p$, where $0 < p < 1$. Then $\sum_{n=1}^{\infty} P(A_n) = \sum_{n=1}^{\infty} p = \infty$. We thus conclude, from the second Borel–Cantelli lemma, that $P(A_n \text{ i.o.}) = 1$ for *any* coin (unless it has two heads and no tails, or vice versa). $\square$

The following exercise can be solved similarly, but a little more care is required, since the corresponding events are no longer independent; recalling Subsection 1.3 we find that the events form a 1-dependent sequence.

Exercise 7.1. Toss a coin repeatedly as before and let $A_n = \{\text{the } (n-1)\text{th}$ and the nth toss both yield a head$\}$ for $n \geq 2$. Then

$$P(A_n \text{ i.o.}) = 1.$$

In other words, the event "two heads in a row" will occur infinitely often with probability 1.

Exercise 7.2. Toss another coin as above. Show that any finite pattern occurs infinitely often with probability 1. $\square$

Remark 7.2. There exists a theorem, called Kolmogorov's 0-1 law, according to which, for independent events $\{A_n, n \geq 1\}$, the probability $P(A_n \text{ i.o.})$ can only assume the values 0 or 1. Example 7.3 above is of this kind, and, by exploiting the fact that the events $\{A_{2n}, n \geq 1\}$ are independent, one can show that the law also applies to Exercise 7.1. The problem is, of course, to decide which of the values is the true one for the problem at hand. $\square$

The previous problem may serve as an introduction to *patterns*. In some vague sense we may formulate this by stating that given a finite alphabet, any finite sequence of letters, such that the letters are selected uniformly at random, will appear infinitely often with probability 1. A natural question is to ask how long one has to wait for the appearance of a given sequence. That this problem is more sophisticated than one might think at first glance is illustrated by the following example.

Example 7.4. Let $X, X_1, X_2, \ldots$ be i.i.d. random variables, such that $P(X = 0) = P(X = 1) = 1/2$.

(a) Let N_1 be the number of 0's and 1's until the first appearance of the pattern 10. Find $E\, N_1$.
(b) Let N_2 be the number of 0's and 1's until the first appearance of the pattern 11. Find $E\, N_2$.

Before we try to solve this problem it seems pretty obvious that the answers are the same for (a) and (b). However, this is not true!

(a) Let N_1 be the required number. A realization of the game would run as follows: We start off with a random number of 0's (possibly none) which at some point are followed by a 1, after which we are done as soon as a 0 appears. Technically, the pattern 10 appears after the following sequence

$$\underbrace{000\ldots0001}_{M_1}\underbrace{111\ldots1110}_{M_2},$$

where thus M_1 and M_2 are independent Fs(1/2)-distributed random variables, which implies that

$$E\,N_1 = E(M_1 + M_2) = E\,M_1 + E\,M_2 = 2 + 2 = 4.$$

(b) Let N_2 be the required number. This case is different, because when the first 1 has appeared we are done *only if the next digit equals* 1. If this is not the case we start over again. This means that there will be a geometric number of M_1 blocks followed by 0, after which the sequence is finished off with another M_1 block followed by 1:

$$\underbrace{000\ldots0001}_{M_1(1)}0\underbrace{000\ldots0001}_{M_1(2)}0\ldots\underbrace{000\ldots0001}_{M_1(Y)}0\underbrace{000\ldots0001}_{M_1^*}1\,,$$

that is,

$$N_2 = \sum_{k=1}^{Y}(M_1(k) + 1) + (M_1^* + 1),$$

where, thus $Y \in \mathrm{Ge}(1/2)$, $M_1(k)$, and M_1^* all are distributed as M_1 and all random variables are independent. Thus,

$$E\,N_2 = E\,(Y + 1) \cdot E(M_1 + 1) = (1 + 1) \cdot (2 + 1) = 6.$$

Alternatively, and as the mathematics reveals, we may consider the experiment as consisting of $Z\,(= Y + 1)$ blocks of size $M_1 + 1$, where the last block is a success and the previous ones are failures. With this viewpoint we obtain

$$N_2 = \sum_{k=1}^{Z}(M_1(k) + 1),$$

and the expected value turns out the same as before, since $Z \in \mathrm{Fs}(1/2)$.

Another solution that we include because of its beauty is to condition on the outcome of the first digit(s) and see how the process evolves after that using the law of total probability. A similar kind of argument was used in the early part of the proof of Theorem 3.7.3 concerning the probability of extinction in a branching process.

There are three ways to start off:

1. the first digit is a 0, after which we start from scratch;
2. the first two digits are 10, after which we start from scratch;
3. the first two digits are 11, after which we are done.

It follows that

$$N_2 = \frac{1}{2}(1 + N_2') + \frac{1}{4}(2 + N_2'') + \frac{1}{4} \cdot 2,$$

where N_2' and N_2'' are distributed as N_2. Taking expectation yields

$$EN_2 = \frac{1}{2} \cdot (1 + E\,N_2) + \frac{1}{4} \cdot (2 + E\,N_2) + \frac{1}{4} \cdot 2 = \frac{3}{2} + \frac{3}{4}E\,N_2,$$

from which we conclude that $E\,N_2 = 6$.

To summarize, for the sequence "10" the expected number was 4 and for the sequence "11" it was 6. By symmetry it follows that for "01" and "00" the answers must also be 4 and 6, respectively.

The reason for the different answers is that beginning and end are overlapping in 11 and 00, but not in 10 and 01. The overlapping makes it harder to obtain the desired sequence. This may also be observed in the different solutions. Whereas in (a) once the first 1 has appeared we simply have to wait for a 0, in (b) the 0 *must appear immediately* after the 1, otherwise we start from scratch again. Note how this is reflected in the last solution of (b).

7.2 Records Revisited

For another application of the Borel–Cantelli lemmas we recall the records from Section 6. For a sequence $X_1, X_2, \ldots$ of i.i.d. continuous random variables the record times were $L(1) = 1$ and $L(n) = \min\{k : X_k > X_{L(n-1)}\}$ for $n \geq 2$. We also introduced the indicator variables $\{I_k, k \geq 1\}$, which equal 1 if a record is observed and 0 otherwise, and the counting process $\{\mu(n), n \geq 1\}$ is defined by

$$\mu(n) = \sum_{k=1}^{n} I_k = \#\,\text{records among } X_1, X_2, \ldots, X_n = \max\{k : L(k) \leq n\}.$$

Since $P(I_k = 1) = 1/k$ for all k we conclude that

$$\sum_{n=1}^{\infty} P(I_k = 1) = \infty,$$

so that, because of the independence of the indicators, the second Borel–Cantelli lemma tells us that there will be infinitely many records with probability 1. This is not surprising, since, intuitively, there is always room for a new observation that is bigger than all others so far.

After this it is tempting to introduce *double records*, which appear whenever there are two records immediately following each other. Intuition this time might suggest once more that there is always room for two records in a row. So, let us check this.

Let $D_n = 1$ if X_n produces a double record, that is, if X_{n-1} and X_n both are records, and let $D_n = 0$ otherwise. Then, for $n \geq 2$,

$$P(D_n = 1) = P(I_n = 1, I_{n-1} = 1) = P(I_n = 1) \cdot P(I_{n-1} = 1) = \frac{1}{n} \cdot \frac{1}{n-1}.$$

We also note that the random variables $\{D_n, n \geq 2\}$ are *not* independent (more precisely, they are 1-dependent), which causes no problem. Namely,

$$\sum_{n=2}^{\infty} P(D_n = 1) = \sum_{n=2}^{\infty} \frac{1}{n(n-1)} = \lim_{m \to \infty} \sum_{n=2}^{m} \left(\frac{1}{n-1} - \frac{1}{n} \right) = \lim_{m \to \infty} \left(1 - \frac{1}{m} \right) = 1,$$

so that by the first Borel–Cantelli lemma—which does not require independence—we conclude that

$$P(D_n = 1 \text{ i.o.}) = 0,$$

that is, the probability of infinitely many double records is equal to zero.

Moreover, the expected number of double records is

$$E \sum_{n=2}^{\infty} D_n = \sum_{n=2}^{\infty} E D_n = \sum_{n=2}^{\infty} P(D_n = 1) = 1;$$

in other words, we can expect *one* double record. A detailed analysis shows that, in fact, the total number of double records is

$$\sum_{n=2}^{\infty} D_n \in \text{Po}(1).$$

7.3 Complete Convergence

We close this section by introducing another convergence concept, which, as will be seen, is closely related to the Borel–Cantelli lemmas.

Definition 7.2. *A sequence* $\{X_n, n \geq 1\}$ *of random variables* converges completely *to the constant* θ *if*

$$\sum_{n=1}^{\infty} P(|X_n - \theta| > \varepsilon) < \infty \quad \text{for all} \quad \varepsilon > 0.$$ □

Two immediate observations are that complete convergence always implies a.s. convergence in view of the first Borel–Cantelli lemma and that complete convergence and almost-sure convergence are equivalent for sequences of independent random variables.

Theorem 7.4. *Let X_1, X_2, ... be random variables and θ be some constant. The following implications hold as $n \to \infty$:*

$$X_n \to \theta \text{ completely} \implies X_n \overset{a.s.}{\to} \theta.$$

If, in addition, X_1, X_2, ... are independent, then

$$X_n \to \theta \text{ completely} \iff X_n \overset{a.s.}{\to} \theta. \qquad \square$$

Example 7.5. Another inspection of Example 6.3.1 tells us that it follows immediately from the definition of complete convergence that $X_n \to 1$ completely as $n \to \infty$ when $\alpha > 1$ and that complete convergence does not hold if X_1, X_2, ... are independent and $\alpha \le 1$. $\qquad \square$

The concept was introduced in the late 1940s in connection with the following result:

Theorem 7.5. *Let X_1, X_2, ... be a sequence of i.i.d. random variables, and set $S_n = \sum_{k=1}^{n} X_k$, $n \ge 1$. Then*

$$\frac{S_n}{n} \to 0 \text{ completely as } n \to \infty \iff EX = 0 \text{ and } EX^2 < \infty,$$

or, equivalently,

$$\sum_{n=1}^{\infty} P(|S_n| > n\varepsilon) < \infty \text{ for all } \varepsilon > 0 \iff EX = 0 \text{ and } EX^2 < \infty. \quad \square$$

Remark 7.3. A first naive attempt to prove the sufficiency would be to use Chebyshev's inequality. The attack fails, however, since the harmonic series diverges; more sophisticated tools are required. $\qquad \square$

We mentioned in Remark 6.5.1 that the so-called strong law of large numbers, which states that S_n/n converges almost surely as $n \to \infty$, is equivalent to the existence of the mean, EX. Consequently, if the mean exists and/but the variance (or any moment of higher order than the first one) does not exist, then almost-sure convergence holds. In particular, if the mean equals 0, then

$$P(|S_n| > n\varepsilon \text{ i.o.}) = 0 \quad \text{for all} \quad \varepsilon > 0,$$

whereas Theorem 7.5 tells us that the corresponding Borel–Cantelli sum *diverges* in this case. This is the example we promised just before stating Theorem 7.2. Note also that the events $\{|S_n| > n\varepsilon, n \ge 1\}$ are definitely not independent.

8 Martingales

One of the most important modern concepts in probability is the concept of martingales. A rigorous treatment is beyond the scope of this book. The purpose of this section is to give the reader a flavor of martingale theory in a slightly simplified way.

Definition 8.1. Let X_1, X_2, ... be a sequence of random variables with finite expectations. We call X_1, X_2, ... a martingale if

$$E(X_{n+1} \mid X_1, X_2, \ldots, X_n) = X_n \quad \text{for all} \quad n \geq 1. \qquad \square$$

The term *martingale* originates in gambling theory. The famous game *double or nothing*, in which the gambling strategy is to double one's stake as long as one loses and leave as soon as one wins, is called a "martingale." That it is, indeed, a martingale in the sense of our definition will be seen below.

Exercise 8.1. Use Theorem 2.2.1 to show that X_1, X_2, ... is a martingale if and only if

$$E(X_n \mid X_1, X_2, \ldots, X_m) = X_m \quad \text{for all} \quad n \geq m \geq 1. \qquad \square$$

In general, consider a game such that X_n is the gambler's fortune after n plays, $n \geq 1$. If the game satisfies the martingale property, it means that the expected fortune of the player, given the history of the game, equals the current fortune. Such games may be considered to be fair, since on average neither the player nor the bank loses any money.

Example 8.1. The canonical example of a martingale is a sequence of partial sums of independent random variables with mean zero. Namely, let Y_1, Y_2, ... be independent random variables with mean zero, and set

$$X_n = Y_1 + Y_2 + \cdots + Y_n, \quad n \geq 1.$$

Then

$$\begin{aligned}
E(X_{n+1} \mid X_1, X_2, \ldots, X_n) &= E(X_n + Y_{n+1} \mid X_1, X_2, \ldots, X_n) \\
&= X_n + E(Y_{n+1} \mid X_1, X_2, \ldots, X_n) \\
&= X_n + E(Y_{n+1} \mid Y_1, Y_2, \ldots, Y_n) \\
&= X_n + 0 = X_n,
\end{aligned}$$

as claimed. For the second equality we used Theorem 2.2.2(a), and for the third one we used the fact that knowledge of $X_1, X_2, \ldots, X_n$ is equivalent to knowledge of $Y_1, Y_2, \ldots, Y_n$. The last equality follows from the independence of the summands; recall Theorem 2.2.2(b). $\qquad \square$

Another example is a sequence of products of independent random variables with mean 1.

Example 8.2. Suppose that $Y_1, Y_2, \ldots$ are independent random variables with mean 1, and set $X_n = \prod_{k=1}^{n} Y_k$, $n \geq 1$ (with $Y_0 = X_0 = 1$). Then

$$
\begin{aligned}
E(X_{n+1} \mid X_1, X_2, \ldots, X_n) &= E(X_n \cdot Y_{n+1} \mid X_1, X_2, \ldots, X_n) \\
&= X_n \cdot E(Y_{n+1} \mid X_1, X_2, \ldots, X_n) \\
&= X_n \cdot 1 = X_n,
\end{aligned}
$$

which verifies the martingale property of $\{X_n, n \geq 1\}$.

One application of this example is the game "double or nothing" mentioned above. To see this, set $X_0 = 1$ and, recursively,

$$
X_{n+1} = \begin{cases} 2X_n, & \text{with probability } \frac{1}{2}, \\ 0, & \text{with probability } \frac{1}{2}, \end{cases}
$$

or, equivalently,

$$
P(X_n = 2^n) = \frac{1}{2^n}, \qquad P(X_n = 0) = 1 - \frac{1}{2^n} \quad \text{for} \quad n \geq 1.
$$

Since

$$
X_n = \prod_{k=1}^{n} Y_k,
$$

where $Y_1, Y_2, \ldots$ are i.i.d. random variables such that $P(Y_k = 0) = P(Y_k = 2) = 1/2$ for all $k \geq 1$, it follows that X_n equals a product of i.i.d. random variables with mean 1, so that $\{X_n, n \geq 1\}$ is a martingale.

A problem with this game is that the expected money spent when the game is over is infinite. Namely, suppose that the initial stake is 1 euro. If the gambler wins at the nth game, she or he has spent $1+2+4+\cdots+2^{n-1} = 2^n - 1$ euros and won 2^n euros, for a total net of 1 euro. The total number of games is Fs(1/2)-distributed. This implies on the one hand that, on average, a success or win occurs after two games, and on the other hand that, on average, the gambler will have spent an amount of

$$
\sum_{n=1}^{\infty} \frac{1}{2^n} \cdot (2^n - 1) = \infty
$$

euros in order to achieve this. In practice this is therefore an impossible game. A truncated version would be to use the same strategy but to leave the game no matter what happens after (at most) a fixed number of games (to be decided before the game starts).

Another example is related to the likelihood ratio test. Let $Y_1, Y_2, \ldots, Y_n$ be independent random variables with common density f and some characterizing parameter θ of interest. In order to test the null hypothesis $H_0 : \theta = \theta_0$

against the alternative $H_1 : \theta = \theta_1$, the *Neyman–Pearson lemma* in statistics tells us that such a test should be based on the likelihood ratio statistic

$$L_n = \prod_{k=1}^{n} \frac{f(X_k; \theta_1)}{f(X_k; \theta_0)},$$

where f_{θ_0} and f_{θ_1} are the densities under the null and alternative hypotheses, respectively.

Now, the factors $f(X_k; \theta_1)/f(X_k; \theta_0)$ are i.i.d. random variables, and, under the null hypothesis, the mean equals

$$E_0\left(\frac{f(X_k; \theta_1)}{f(X_k; \theta_0)}\right) = \int_{-\infty}^{\infty} \frac{f(x; \theta_1)}{f(x; \theta_0)} f(x; \theta_0)\, dx = \int_{-\infty}^{\infty} f(x; \theta_1)\, dx = 1,$$

that is, L_n is made up as a product of i.i.d. random variables with mean 1, from which we immediately conclude that $\{L_n, n \geq 1\}$ is a martingale.

We also remark that if $=$ in the definition is replaced by $\geq$ then $X_1, X_2, \ldots$ is called a *submartingale*, and if it is replaced by $\leq$ it is called a *supermartingale*. As a typical example one can show that if $\{X_n, n \geq 1\}$ is a martingale and $E|X_n|^r < \infty$ for all $n \geq 1$ and some $r \geq 1$, then $\{|X_n|^r, n \geq 1\}$ is a submartingale.

Applying this to the martingale in Example 8.1 tells us that whereas the sums $\{X_n, n \geq 1\}$ of independent random variables with mean zero constitute a martingale, such is not the case with the sequence of sums of squares $\{X_n^2, n \geq 1\}$ (provided the variances are finite); that sequence is a submartingale. However by centering the sequence one obtains a martingale. This is the topic of Problems 9.11 and 9.12.

There also exist so-called *reversed martingales*. If we interpret n as time, then "reversing" means reversing time. Traditionally one defines reversed martingales via the relation

$$X_n = E(X_m \mid X_{n+1}, X_{n+2}, X_{n+3}, \ldots) \quad \text{for all} \quad m < n,$$

which means that one conditions on "the future." The more modern way is to let the index set be the negative integers as follows.

Definition 8.2. *Let* $\ldots, X_{-3}, X_{-2}, X_{-1}$ *be a sequence of random variables with finite expectations. We call* $\ldots, X_{-3}, X_{-2}, X_{-1}$ *a reversed martingale if*

$$E(X_{n+1} \mid \ldots, X_{n-3}, X_{n-2}, X_{n-1}, X_n) = X_n \quad \text{for all} \quad n \leq -1. \qquad \square$$

The obvious parallel to Exercise 8.1 is next.

Exercise 8.2. Use Theorem 2.2.1 to show that $\ldots, X_{-3}, X_{-2}, X_{-1}$ is a reversed martingale if and only if

$$E(X_n \mid \ldots, X_{m-3}, X_{m-2}, X_{m-1}, X_m) = X_m \quad \text{for all} \quad m \leq n \leq 0.$$

In particular, $\ldots, X_{-3}, X_{-2}, X_{-1}$ is a reversed martingale if and only if,

$$E(X_{-1} \mid \ldots, X_{m-3}, X_{m-2}, X_{m-1}, X_m) = X_m \quad \text{for all} \quad m \leq -1. \qquad \square$$

Just as the sequence of sums of independent random variables with mean zero constitutes the generic example of a martingale it turns out that the sequence of arithmetic means of i.i.d. random variables with finite mean (not necessarily equal to zero) constitutes the generic example of a reversed martingale.

To see this, suppose that $Y_1, Y_2, \ldots$ are i.i.d. random variables with finite mean μ, set $S_n = \sum_{k=1}^n Y_k$, $n \geq 1$, and

$$X_{-n} = \frac{S_n}{n} \quad \text{for} \quad n \geq 1.$$

We wish to show that

$$\{X_n, n \leq -1\} \quad \text{is a martingale.} \tag{8.1}$$

Now, knowing the arithmetic means when $k \geq n$ is the same as knowing S_n and Y_k, $k > n$, so that, due to independence,

$$E(X_{-n} \mid X_k, k \leq n-1) = E\left(\frac{S_n}{n} \mid S_{n+1}, Y_{n+2}, Y_{n+3}, \ldots\right)$$

$$= E\left(\frac{S_n}{n} \mid S_{n+1}\right) = \frac{1}{n} \sum_{k=1}^n E(Y_k \mid S_{n+1})$$

$$= \frac{1}{n} \sum_{k=1}^n \frac{S_{n+1}}{n+1} = \frac{S_{n+1}}{n+1} = X_{-n-1},$$

where, in the third to last equality we exploited the symmetry, which in turn, is due to the equidistribution.

We have thus established relation (8.1) as desired.

Remark 8.1. Reversed submartingales and *reversed supermartingales* may be defined "the obvious way."   □

Exercise 8.3. Define them!   □

We close this introduction to the theory of martingales by stating (without proof) the main convergence results. Analogous, although slightly different, results also hold for submartingales and supermartingales.

Theorem 8.1. *Suppose that $\{X_n, n \geq 1\}$ is a martingale. If*

$$\sup_n E \max\{X_n, 0\} < \infty,$$

then X_n converges almost surely as $n \to \infty$. Moreover, the following are equivalent:

(a) *$\{X_n, n \geq 1\}$ is uniformly integrable;*
(b) *X_n converges in 1-mean;*

(c) $X_n \overset{a.s.}{\to} X_\infty$ as $n \to \infty$, where $E|X_\infty| < \infty$, and X_∞ closes the sequence, that is, $\{X_n, n = 1, 2, \ldots, \infty\}$ is a martingale;

(d) there exists a random variable Y with finite mean such that

$$X_n = E(Y \mid X_1, X_2, \ldots, X_n) \quad \text{for all} \quad n \geq 1. \qquad \square$$

The analog for reversed martingales runs as follows.

Theorem 8.2. *Suppose that* $\{X_n, n \leq -1\}$ *is a reversed martingale. Then*

(a) $\{X_n, n \leq -1\}$ *is uniformly integrable;*

(b) $X_n \to X_{-\infty}$ *a.s. and in 1-mean as* $n \to -\infty$;

(c) $\{X_n, -\infty \leq n \leq -1\}$ *is a martingale.* $\qquad \square$

Note that the results differ somewhat. This is due to the fact that whereas ordinary, forward martingales always have a first element, but not necessarily a last element (which would correspond to X_∞), reversed martingales always have a last element, namely X_{-1}, but not necessarily a first element (which would correspond to $X_{-\infty}$). This, in turn, has the effect that reversed martingales "automatically" are uniformly integrable, as a consequence of which conclusions (a)–(c) are "automatic" for reversed martingales, but only hold under somewhat stronger assumptions for (forward) martingales.

Note also that the generic martingale, the sum of independent random variables with mean zero, need not be convergent at all. This is, in particular, the case if the summands are equidistributed with finite variance σ^2, in which case the sum S_n behaves, asymptotically, like $\sigma\sqrt{n} \cdot N(0,1)$, where $N(0,1)$ is a standard normal random variable.

9 Problems

1. Let $X_1, X_2, \ldots$ be independent, equidistributed random variables, and set $S_n = X_1 + \cdots + X_n$, $n \geq 1$. The sequence $\{S_n, n \geq 0\}$ (where $S_0 = 0$) is called a *random walk*. Consider the following "perturbed" random walk. Let $\{\varepsilon_n, n \geq 1\}$ be a sequence of random variables such that, for some fixed $A > 0$, we have $P(|\varepsilon_n| \leq A) = 1$ for all n, and set

$$T_n = S_n + \varepsilon_n, \quad n = 1, 2, \ldots.$$

Suppose that $E X_1 = \mu$ exists. Show that the law of large numbers holds for the perturbed random walk $\{T_n, n \geq 1\}$.

2. In a game of dice one wishes to use one of two dice A and B. A has two white and four red faces and B has two red and four white faces. A coin is tossed in order to decide which die is to be used and that die is then used throughout. Let $\{X_k, k \geq 1\}$ be a sequence of random variables defined as follows:

$$X_k = \begin{cases} 1, & \text{if red is obtained,} \\ 0, & \text{if white is obtained} \end{cases}$$

at the kth roll of the die. Show that the law of large numbers does not hold for the sequence $\{X_k,\ k \geq 1\}$. Why is this the case?

3. Suppose that $X_1, X_2, \ldots$ are independent random variables such that $X_k \in Be(p_k)$, $k \geq 1$, and set $S_n = \sum_{k=1}^{n} X_k$, $m_n = \sum_{k=1}^{n} p_k$, and $s_n^2 = \sum_{k=1}^{n} p_k(1 - p_k)$, $n \geq 1$. Show that if

$$\sum_{k=1}^{\infty} p_k(1 - p_k) = +\infty, \tag{9.1}$$

then

$$\frac{S_n - m_n}{s_n} \xrightarrow{d} N(0,1) \quad \text{as} \quad n \to \infty.$$

Remark 1. The case $p_k = 1/k$, $k \geq 1$, corresponds to the record times, and we rediscover Theorem 6.4.

Remark 2. One can show that the assumption (9.1) is necessary for the conclusion to hold.

4. Prove the following central limit theorem for a sum of independent (not identically distributed) random variables: Suppose that $X_1, X_2, \ldots$ are independent random variables such that $X_k \in U(-k, k)$, and set $S_n = \sum_{k=1}^{n} X_k$, $n \geq 1$. Show that

$$\frac{S_n}{n^{3/2}} \xrightarrow{d} N(\mu, \sigma^2) \quad \text{as} \quad n \to \infty,$$

and determine μ and σ^2.

Remark. Note that the normalization is not proportional to $\sqrt{n}$; rather, it is asymptotically proportional to $\sqrt{\text{Var}\, S_n}$.

5. Let $X_1, X_2, \ldots$ be independent, $U(0, 1)$-distributed random variables. We say that there is a *peak* at X_k if X_{k-1} and X_{k+1} are both smaller than X_k, $k \geq 2$. What is the probability of a peak at
   (a) X_2?
   (b) X_3?
   (c) X_2 and X_3?
   (d) X_2 and X_4?
   (e) X_2 and X_5?
   (f) X_i and X_j, $i, j \geq 2$?

   Remark. Letting $I_k = 1$ if there is a peak at X_k and 0 otherwise, the sequence $\{I_k,\ k \geq 1\}$ forms a 2-dependent sequence of random variables.

6. Verify formula (2.1), i.e., that if $X, X_1, X_2, \ldots$ are i.i.d. symmetric stable random variables, then

$$\frac{S_n}{n^{1/\alpha}} \overset{d}{=} X \quad \text{for all} \quad n.$$

7. Prove that the law of large numbers holds for symmetric, stable distributions with index α, $1 < \alpha \leq 2$.

8. Let $0 < \alpha < 2$ and suppose that $X, X_1, X_2, \ldots$ are independent random variables with common (two-sided Pareto) density

$$f(x) = \begin{cases} \dfrac{\alpha}{2|x|^{\alpha+1}}, & \text{for} \quad |x| > 1, \\ 0, & \text{otherwise.} \end{cases}$$

Show that the distribution belongs to the domain of attraction of a symmetric stable distribution with index α; in other words, that the sums $S_n = \sum_{k=1}^{n} X_k$, suitably normalized, converge in distribution to a symmetric stable distribution with index α.

Remark 1. More precisely, one can show that $S_n/n^{1/\alpha}$ converges in distribution to a symmetric stable law with index α.

Remark 2. This problem generalizes Examples 3.1 and 3.2.

9. The same problem as the previous one, but for the density

$$f(x) = \begin{cases} \dfrac{c \log |x|}{|x|^{\alpha+1}}, & \text{for} \quad |x| > 1, \\ 0, & \text{otherwise,} \end{cases}$$

where c is an appropriate normalizing constant.

Remark. In this case one can show that $S_n/(n \log n)^{1/\alpha}$ converges in distribution to a symmetric stable law with index α.

10. Show that the extremal distributions belong to their own domain of attraction. More precisely, let $X, X_1, X_2, \ldots$ be i.i.d. random variables, and set

$$Y_n = \max\{X_1, X_2, \ldots, X_n\}, \quad n \geq 1.$$

Show that,

(a) if X has a Fréchet distribution, then

$$\frac{Y_n}{n^{1/\alpha}} \overset{d}{=} X;$$

(b) if X has a Weibull distribution, then

$$n^{1/\alpha} Y_n \overset{d}{=} X;$$

(c) if X has a Gumbel distribution, then

$$Y_n - \log n \overset{d}{=} X.$$

11. Let $Y_1, Y_2, \ldots$ be independent random variables with mean zero and finite variances $\text{Var}\, Y_k = \sigma_k^2$. Set

$$X_n = \Big(\sum_{k=1}^{n} Y_k \Big)^2 - \sum_{k=1}^{n} \sigma_k^2, \quad n \geq 1.$$

Show that $X_1, X_2, \ldots$ is a martingale.

12. Let $Y_1, Y_2, \ldots$ be i.i.d. random variables with finite mean μ, and finite variance σ^2, and let S_n, $n \geq 1$, denote their partial sums. Set

$$X_n = (S_n - n\mu)^2 - n\sigma^2, \quad n \geq 1.$$

Show that $X_1, X_2, \ldots$ is a martingale.

13. Let $X(n)$ be the number of individuals in the nth generation of a branching process $(X(0) = 1)$ with reproduction mean m $(= E\,X(1))$. Set

$$U_n = \frac{X(n)}{m^n}, \quad n \geq 1.$$

Show that $U_1, U_2, \ldots$ is a martingale.

14. Let $Y_1, Y_2, \ldots$ are i.i.d. random variables with a finite moment generating function ψ, set $S_n = \sum_{k=1}^n Y_k$, $n \geq 1$, with $S_0 = 0$, and

$$X_n = \frac{e^{tS_n}}{(\psi(t))^n}, \quad n \geq 1.$$

(a) Show that $\{X_n, n \geq 1\}$ is a martingale (which is frequently called *the exponential martingale*).

(b) Find the relevant martingale if the common distribution is the standard normal one.

8

The Poisson Process

1 Introduction and Definitions

Suppose that an event E may occur at any point in time and that the number of occurrences of E during disjoint time intervals are independent. As examples we might think of the arrivals of customers to a store (where E means that a customer arrives), calls to a telephone switchboard, the emission of particles from a radioactive source, and accidents at a street crossing. The common feature in all these examples, although somewhat vaguely expressed, is that very many repetitions of independent Bernoulli trials are performed and that the success probability of each such trial is very small. A little less vaguely, let us imagine the time interval $(0, t]$ split into the n parts $(0, t/n]$, $(t/n, 2t/n]$, ..., $((n-1)t/n, t]$, where n is "very large." The probability of an arrival of a customer, the emission of a particle, and so forth, then is very small in every small time interval, events in disjoint time intervals are independent, and the number of time intervals is large. The Poisson approximation of the binomial distribution then tells us that the total number of occurrences in $(0, t]$ is approximately Poisson-distributed. (Observe that we have discarded the possibility of more than one occurrence in a small time interval—only one customer at a time can get through the door!)

1.1 First Definition of a Poisson Process

The discrete stochastic process in continuous time, which is commonly used to describe phenomena of the above kind, is called the *Poisson process*. We shall denote it by $\{X(t), t \geq 0\}$, where

$$X(t) = \ \# \text{ occurrences in } (0, t].$$

Definition I. *A Poisson process is a stochastic process $\{X(t), t \geq 0\}$ with independent, stationary, Poisson-distributed increments. Also, $X(0) = 0$. In other words,*

A. Gut, *An Intermediate course in Probabilty,* Springer Texts in Statistics, DOI: 10.1007/978-1-4419-0162-0_8, © Springer Science + Business Media, LLC 2009

(a) *the increments $\{X(t_k) - X(t_{k-1}), 1 \le k \le n\}$ are independent random variables for all $0 \le t_0 \le t_1 \le t_2 \le \cdots \le t_{n-1} \le t_n$ and all n;*
(b) *$X(0) = 0$ and there exists $\lambda > 0$ such that*

$$X(t) - X(s) \in \text{Po}(\lambda(t - s)), \quad \text{for} \quad 0 \le s < t.$$

The constant λ is called the intensity *of the process.*   □

By the law of large numbers (essentially) or by Chebyshev's inequality, it follows easily from the definition that $X(t)/t \xrightarrow{p} \lambda$ as $t \to \infty$ (in fact, almost-sure convergence holds). This shows that the intensity measures the average frequency or density of occurrences. A further interpretation can be made via Definition II ahead.

1.2 Second Definition of a Poisson Process

In addition to the independence between disjoint time intervals, we remarked that it is "almost impossible" that there are two or more occurrences in a small time interval. For an arbitrary time interval $((i-1)t/n, it/n]$, $i = 1, 2, \ldots, n$, it is thus reasonably probable that E occurs once and essentially impossible that E occurs more than once. We shall begin by showing that these facts hold true in a mathematical sense for the Poisson process as defined above and then see that, in fact, these properties (together with the independence between disjoint time intervals) characterize the Poisson process.

We first observe that

$$0 < 1 - e^{-x} < x, \quad \text{for} \quad x > 0,$$

from which it follows that

$$\begin{aligned}
P(E \text{ occurs once during } (t, t+h]) &= e^{-\lambda h} \lambda h \\
&= \lambda h - \lambda h(1 - e^{-\lambda h}) \\
&= \lambda h + o(h) \quad \text{as} \quad h \to 0. \quad (1.1)
\end{aligned}$$

Furthermore,

$$\sum_{k=2}^{\infty} \frac{x^k}{k!} \le \frac{1}{2} \sum_{k=2}^{\infty} x^k \le \frac{1}{2} \frac{x^2}{1 - x} \le x^2, \quad \text{for} \quad 0 < x < \frac{1}{2},$$

which implies that

$$\begin{aligned}
P(\text{at least 2 occurrences of } E &\text{ during } (t, t+h]) \\
&= \sum_{k=2}^{\infty} e^{-\lambda h} \frac{(\lambda h)^k}{k!} \le (\lambda h)^2 = o(h) \quad \text{as} \quad h \to 0. \quad (1.2)
\end{aligned}$$

Definition II. *A Poisson process* $\{X(t), t \geq 0\}$ *is a nonnegative, integer-valued stochastic process such that* $X(0) = 0$ *and*

(a) *the process has independent increments;*
(b) P*(exactly one occurrence during* $(t, t+h]) = \lambda h + o(h)$ *as* $h \to 0$ *for some* $\lambda > 0$;
(c) P*(at least two occurrences during* $(t, t+h]) = o(h)$ *as* $h \to 0$. $\qquad\square$

Remark 1.1. It follows from this definition that $\{X(t), t \geq 0\}$ is nondecreasing. This is also clear from the fact that the process counts the number of occurrences (of some event). Sometimes, however, the process is defined in terms of jumps instead of occurrences; then the assumption that the process is nondecreasing has to be incorporated in the definition (see also Problem 9.33). $\qquad\square$

Theorem 1.1. *Definitions I and II are equivalent.*

Proof. The implication Definition I $\Rightarrow$ Definition II has already been demonstrated. In order to prove the converse, we wish to show that the increments follow a Poisson distribution, that is, that

$$X(t) - X(s) \in \text{Po}(\lambda(t - s)) \quad \text{for} \quad 0 \leq s < t. \tag{1.3}$$

First let $s = 0$. Our aim is thus to show that

$$X(t) \in \text{Po}(\lambda t), \quad t > 0. \tag{1.4}$$

For $n = 0, 1, 2, \ldots$, let

$$E_n = \{\text{exactly } n \text{ occurrences during } (t, t+h]\},$$

and set

$$P_n(t) = P(X(t) = n).$$

For $n = 0$ we have

$$P_0(t + h) = P_0(t) \cdot P(X(t + h) = 0 \mid X(t) = 0) = P_0(t) \cdot P(E_0)$$
$$= P_0(t)(1 - \lambda h + o(h)) \quad \text{as} \quad h \to 0,$$

and hence

$$P_0(t + h) - P_0(t) = -\lambda h P_0(t) + o(h) \quad \text{as} \quad h \to 0.$$

Division by h and letting $h \to 0$ leads to the differential equation

$$P_0'(t) = -\lambda P_0(t). \tag{1.5a}$$

An analogous argument for $n \geq 1$ yields

$$P_n(t+h) = P\left(\bigcup_{k=0}^{n}\{X(t) = k, X(t+h) = n\}\right)$$

$$= P_n(t) \cdot P(X(t+h) = n \mid X(t) = n)$$
$$+ P_{n-1}(t) \cdot P(X(t+h) = n \mid X(t) = n-1)$$
$$+ P\left(\bigcup_{k=0}^{n-2}\{X(t) = k, X(t+h) = n\}\right)$$

$$= P_n(t)P(E_0) + P_{n-1}(t)P(E_1) + P\left(\bigcup_{k=0}^{n-2}\{X(t) = k, E_{n-k}\}\right)$$

$$= P_n(t) \cdot (1 - \lambda h + o(h)) + P_{n-1}(t) \cdot (\lambda h + o(h)) + o(h)$$

as $h \to 0$, since, by part (c) of Definition II,

$$P\left(\bigcup_{k=0}^{n-2}\{X(t) = k, E_{n-k}\}\right)$$
$$\leq P(\text{at least two occurrences during } (t, t+h]) = o(h)$$

as $h \to 0$. By moving $P_n(t)$ to the left-hand side above, dividing by h, and letting $h \to 0$, we obtain

$$P_n'(t) = -\lambda P_n(t) + \lambda P_{n-1}(t), \quad n \geq 1. \tag{1.5b}$$

Formally, we have only proved that the right derivatives exist and satisfy the differential equations in (1.5). A completely analogous argument for the interval $(t-h, t]$ shows, however, that the left derivatives exist and satisfy the same system of differential equations.

Since equation (1.5.b) contains P_n', P_n, as well as P_{n-1}, we can (only) express P_n as a function of P_{n-1}. However, (1.5.a) contains only P_0' and P_0 and is easy to solve. Once this is done, we let $n = 1$ in (1.5.b), insert our solution P_0 into (1.5.b), solve for P_1, let $n = 2$, and so forth.

To solve (1.5), we use the method of integrating factors. The condition $X(0) = 0$ amounts to the initial condition

$$P_0(0) = 1. \tag{1.5c}$$

Starting with (1.5.a), the computations run as follows:

$$P_0'(t) + \lambda P_0(t) = 0,$$
$$\frac{d}{dt}\left(e^{\lambda t} P_0(t)\right) = e^{\lambda t} P_0'(t) + \lambda e^{\lambda t} P_0(t) = 0,$$
$$e^{\lambda t} P_0(t) = c_0 = \text{constant},$$
$$P_0(t) = c_0 e^{-\lambda t},$$

which, together with (1.5.c), yields $c_0 = 1$ and hence

$$P_0(t) = e^{-\lambda t}, \tag{1.6a}$$

as desired.

Inserting (1.6.a) into (1.5.b) with $n = 1$ and arguing similarly yield

$$P_1'(t) + \lambda P_1(t) = \lambda e^{-\lambda t},$$

$$\frac{d}{dt}\left(e^{\lambda t} P_1(t)\right) = e^{\lambda t} P_1'(t) + \lambda e^{\lambda t} P_1(t) = \lambda,$$

$$e^{\lambda t} P_1(t) = \lambda t + c_1,$$

$$P_1(t) = (\lambda t + c_1)e^{-\lambda t}.$$

By (1.5.c) we must have $P_n(0) = 0$, $n \geq 1$, which leads to the solution

$$P_1(t) = \lambda t e^{-\lambda t}. \tag{1.6b}$$

For the general case we use induction. Thus, suppose that

$$P_k(t) = e^{-\lambda t}\frac{(\lambda t)^k}{k!}, \quad k = 0, 1, 2, \ldots, n-1.$$

We claim that

$$P_n(t) = e^{-\lambda t}\frac{(\lambda t)^n}{n!}. \tag{1.6c}$$

By (1.5.b) and the induction hypothesis it follows that

$$P_n'(t) + \lambda P_n(t) = \lambda P_{n-1}(t) = \lambda e^{-\lambda t}\frac{(\lambda t)^{n-1}}{(n-1)!},$$

$$\frac{d}{dt}\left(e^{\lambda t} P_n(t)\right) = \frac{\lambda^n t^{n-1}}{(n-1)!},$$

$$e^{\lambda t} P_n(t) = \frac{(\lambda t)^n}{n!} + c_n,$$

which (since $P_n(0) = 0$ yields $c_n = 0$) proves (1.6.c). This finishes the proof of (1.4) when $s = 0$. For $s > 0$, we set

$$Y(t) = X(t+s) - X(s), \quad t \geq 0, \tag{1.7}$$

and note that $Y(0) = 0$ and that the Y-process has independent increments since the X-process does. Furthermore, Y-occurrences during $(t, t+h]$ correspond to X-occurrences during $(t+s, t+s+h]$. The Y-process thus satisfies the conditions in Definition II, which, according to what has already been shown, proves that $Y(t) \in \mathrm{Po}(\lambda t)$, $t > 0$, that is, that $X(t+s) - X(s) \in \mathrm{Po}(\lambda t)$.

The proof of the theorem thus is complete. $\square$

One step in proving the equivalence of the two definitions of a Poisson process was to start with Definition II. This led to the system of differential equations (1.5). The solution above was obtained by iteration and induction. The following exercise provides another way to solve the equations.

Exercise 1.1. Let $g(t, s) = g_{X(t)}(s)$ be the generating function of $X(t)$. Multiply the equation $P'_n(t) = \cdots$ by s^n for all $n = 0, 1, 2, \ldots$ and add all equations. Show that this, together with the initial condition (1.5.c), yields

(a) $\dfrac{\partial g(t, s)}{\partial t} = \lambda(s - 1)g(t, s)$,

(b) $g(t, s) = e^{\lambda t(s-1)}$,

(c) $X(t) \in \text{Po}(\lambda t)$.   □

1.3 The Lack of Memory Property

A typical realization of a Poisson process is thus a step function that begins at 0, where it stays for a random time period, after which it jumps to 1, where it stays for a random time period, and so on. The step function thus is such that the steps have height 1 and random lengths. Moreover, the step function is right continuous.

Now let $T_1, T_2, \ldots$ be the successive time points of the occurrences of an event E. Set $\tau_1 = T_1$ and $\tau_k = T_k - T_{k-1}$ for $k \geq 2$. Figure 1.1 depicts a typical realization.

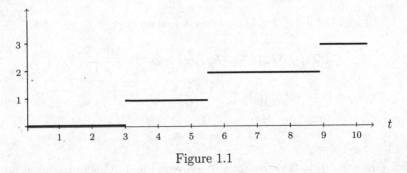

Figure 1.1

In this example, $T_1 = 3$, $T_2 = 5.5$, and $T_3 = 8.9$. Furthermore, $\tau_1 = 3$, $\tau_2 = 2.5$, and $\tau_3 = 3.4$.

Our next task is to investigate the occurrence times $\{T_n, n \geq 1\}$ and the durations $\{\tau_k, k \geq 1\}$. We first consider $T_1 \, (= \tau_1)$.

Let $t > 0$. Since

$$\{T_1 > t\} = \{X(t) = 0\}, \tag{1.8}$$

we obtain

$$1 - F_{\tau_1}(t) = 1 - F_{T_1}(t) = P(T_1 > t) = P(X(t) = 0) = e^{-\lambda t}, \tag{1.9}$$

that is, T_1 and τ_1 are $\text{Exp}(1/\lambda)$-distributed.

The exponential distribution is famous for the *lack of memory property*.

Theorem 1.2. $P(T_1 > t + s \mid T_1 > s) = e^{-\lambda t} = P(T_1 > t)$.

Proof. $P(T_1 > t + s \mid T_1 > s) = \dfrac{P(T_1 > t + s)}{P(T_1 > s)} = \dfrac{e^{-\lambda(t+s)}}{e^{-\lambda s}} = e^{-\lambda t}.$ $\square$

The significance of this result is that if, at time s, we know that there has been no occurrence, then the residual waiting time until an occurrence is, again, $\mathrm{Exp}(1/\lambda)$-distributed. In other words, the residual waiting time has the same distribution as the initial waiting time. Another way to express this fact is that an object whose lifetime has an exponential distribution does not age; once we know that it has reached a given (fixed) age, its residual lifetime has the same distribution as the original lifetime. This is the celebrated lack of memory property.

Example 1.1. Customers arrive at a store according to a Poisson process with an intensity of two customers every minute. Suddenly the cashier realizes that he has to go to the bathroom. He believes that one minute is required for this.

(a) He decides to rush away as soon as he is free in order to be back before the next customer arrives. What is the probability that he will succeed?
(b) As soon as he is free he first wonders whether or not he *dares* to leave. After 30 seconds he decides to do so. What is the probability that he will be back before the next customer arrives?

We first observe that because of the lack of memory property the answers in (a) and (b) are the same. As for part (a), the cashier succeeds if the waiting time $T \in \mathrm{Exp}(1/2)$ until the arrival of the next customer exceeds 1:

$$P(T > 1) = e^{-2 \cdot 1} = e^{-2}.$$ $\square$

The following exercise contains a slight modification of Problem 2.6.4:

Exercise 1.2. The task in that problem was to find the probability that the lifetime of a *new* lightbulb in an overhead projector was long enough for the projector to function throughout a week. What is the probability if the lightbulb is not necessarily new? For example, we know that everything was all right last week, and we ask for the probability that the lightbulb will last long enough for the projector to function this week, too. $\square$

Since the exponential distribution, and hence the Poisson process, has no memory, that is, "begins from scratch," at any given, *fixed*, observed timepoint one might be tempted to guess that the Poisson process also begins from scratch at (certain?) *random* times, for example, at T_1. If this were true, then the time until the first occurrence after time T_1 should also be $\mathrm{Exp}(1/\lambda)$-distributed, that is, we should have $\tau_2 \in \mathrm{Exp}(1/\lambda)$. Moreover, τ_1 and τ_2 should be independent, and hence $T_2 \in \Gamma(2, 1/\lambda)$. By repeating the arguments, we have made the following result plausible:

Theorem 1.3. *For $k \geq 1$, let T_k denote the time of the kth occurrence in a Poisson process, and set $\tau_1 = T_1$ and $\tau_k = T_k - T_{k-1}$, $k \geq 2$. Then*

(a) τ_k, $k \geq 1$, *are independent,* $\mathrm{Exp}(1/\lambda)$-*distributed random variables;*
(b) $T_k \in \Gamma(k, 1/\lambda)$.

Proof. For $k = 1$, we have already shown that T_1 and τ_1 are distributed as claimed. A fundamental relation in the following is

$$\{T_k \leq t\} = \{X(t) \geq k\} \tag{1.10}$$

(for $k = 1$, recall (1.8)).

Now, let $k = 2$ and $0 \leq s \leq t$. Then

$$
\begin{aligned}
P(T_1 \leq s, T_2 > t) &= P(X(s) \geq 1, X(t) < 2) \\
&= P(X(s) = 1, X(t) = 1) \\
&= P(X(s) = 1, X(t) - X(s) = 0) \\
&= P(X(s) = 1) \cdot P(X(t) - X(s) = 0) \\
&= \lambda s e^{-\lambda s} \cdot e^{-\lambda(t-s)} = \lambda s e^{-\lambda t}.
\end{aligned}
$$

Since $P(T_1 \leq s, T_2 > t) + P(T_1 \leq s, T_2 \leq t) = P(T_1 \leq s)$, it follows that

$$
\begin{aligned}
F_{T_1, T_2}(s, t) &= P(T_1 \leq s, T_2 \leq t) \\
&= 1 - e^{-\lambda s} - \lambda s e^{-\lambda t}, \quad \text{for} \quad 0 \leq s \leq t. \tag{1.11}
\end{aligned}
$$

Differentiation yields the joint density of T_1 and T_2:

$$f_{T_1, T_2}(s, t) = \lambda^2 e^{-\lambda t}, \quad \text{for} \quad 0 \leq s \leq t. \tag{1.12}$$

By the change of variable $\tau_1 = T_1$, $\tau_2 = T_2 - T_1$ (i.e., $T_1 = \tau_1$, $T_2 = \tau_1 + \tau_2$) and Theorem 1.2.1, we obtain the joint density of τ_1 and τ_2:

$$f_{\tau_1, \tau_2}(u_1, u_2) = \lambda^2 e^{-\lambda(u_1+u_2)} = \lambda e^{-\lambda u_1} \cdot \lambda e^{-\lambda u_2}, \quad u_1, u_2 > 0. \tag{1.13}$$

This proves (a) for the case $k = 2$. In the general case, (a) follows similarly, but the computations become more (and more) involved. We carry out the details for $k = 3$ below, and indicate the proof for the general case. Once (a) has been established (b) is immediate.

Thus, let $k = 3$ and $0 \leq s \leq t \leq u$. By arguing as above, we have

$$
\begin{aligned}
P(T_1 &\leq s \leq T_2 < t, T_3 > u) \\
&= P(X(s) = 1, X(t) = 2, X(u) < 3) \\
&= P(X(s) = 1, X(t) - X(s) = 1, X(u) - X(t) = 0) \\
&= P(X(s) = 1) \cdot P(X(t) - X(s) = 1) \cdot P(X(u) - X(t) = 0) \\
&= \lambda s e^{-\lambda s} \cdot \lambda(t-s) e^{-\lambda(t-s)} \cdot e^{-\lambda(u-t)} = \lambda^2 s(t-s) e^{-\lambda u},
\end{aligned}
$$

and

$$P(T_1 \le s \le T_2 \le t, T_3 \le u) + P(T_1 \le s \le T_2 \le t, T_3 > u)$$
$$= P(T_1 \le s < T_2 \le t) = P(X(s) = 1, X(t) \ge 2)$$
$$= P(X(s) = 1, X(t) - X(s) \ge 1)$$
$$= P(X(s) = 1) \cdot (1 - P(X(t) - X(s) = 0))$$
$$= \lambda s e^{-\lambda s} \cdot (1 - e^{-\lambda(t-s)}) = \lambda s(e^{-\lambda s} - e^{-\lambda t}).$$

Next we note that

$$F_{T_1,T_2,T_3}(s,t,u) = P(T_1 \le s, T_2 \le t, T_3 \le u)$$
$$= P(T_2 \le s, T_3 \le u) + P(T_1 \le s < T_2 \le t, T_3 \le u),$$

that

$$P(T_2 \le s, T_3 \le u) + P(T_2 \le s, T_3 > u)$$
$$= P(T_2 \le s) = P(X(s) \ge 2) = 1 - P(X(s) \le 1)$$
$$= 1 - e^{-\lambda s} - \lambda s e^{-\lambda s},$$

and that

$$P(T_2 \le s, T_3 > u) = P(X(s) \ge 2, X(u) < 3)$$
$$= P(X(s) = 2, X(u) - X(s) = 0)$$
$$= P(X(s) = 2) \cdot P(X(u) - X(s) = 0)$$
$$= \frac{(\lambda s)^2}{2} e^{-\lambda s} \cdot e^{-\lambda(u-s)} = \frac{(\lambda s)^2}{2} e^{-\lambda u}.$$

We finally combine the above to obtain

$$F_{T_1,T_2,T_3}(s,t,u) = P(T_2 \le s) - P(T_2 \le s, T_3 > u)$$
$$+ P(T_1 \le s < T_2 \le t) - P(T_1 \le s < T_2 \le t, T_3 > u)$$
$$= 1 - e^{-\lambda s} - \lambda s e^{-\lambda s} - \frac{(\lambda s)^2}{2} e^{-\lambda u}$$
$$+ \lambda s(e^{-\lambda s} - e^{-\lambda t}) - \lambda^2 s(t - s)e^{-\lambda u}$$
$$= 1 - e^{-\lambda s} - \lambda s e^{-\lambda t} - \lambda^2 \left(st - \frac{s^2}{2} \right) e^{-\lambda u}, \tag{1.14}$$

and, after differentiation,

$$f_{T_1,T_2,T_3}(s,t,u) = \lambda^3 e^{-\lambda u}, \quad \text{for} \quad 0 < s < t < u. \tag{1.15}$$

The change of variables $\tau_1 = T_1$, $\tau_1 + \tau_2 = T_2$, and $\tau_1 + \tau_2 + \tau_3 = T_3$ concludes the derivation, yielding

$$f_{\tau_1,\tau_2,\tau_3}(v_1, v_2, v_3) = \lambda e^{-\lambda v_1} \cdot \lambda e^{-\lambda v_2} \cdot \lambda e^{-\lambda v_3}, \tag{1.16}$$

for $v_1, v_2, v_3 > 0$, which is the desired conclusion.

Before we proceed to the general case we make the crucial observation that the probability $P(T_1 \leq s < T_2 \leq t, T_3 > u)$ was the only quantity containing *all* of s, t, and u and, hence, since differentiation is with respect to *all* variables, this probability was the only one that contributed to the density. This carries over to the general case, that is, it suffices to actually compute only the probability containing all variables.

Thus, let $k \geq 3$ and let $0 \leq t_1 \leq t_2 \leq \cdots \leq t_k$. In analogy with the above we find that the crucial probability is precisely the one in which the T_i are separated by the t_i. It follows that

$$F_{T_1, T_2, \ldots, T_k}(t_1, t_2, \ldots, t_k)$$
$$= -P(T_1 \leq t_1 < T_2 \leq t_2 < \cdots < T_{k-1} \leq t_{k-1}, T_k > t_k)$$
$$+ R(t_1, t_2, \ldots, t_k)$$
$$= -\lambda^{k-1} t_1 (t_2 - t_1)(t_3 - t_2) \cdots (t_{k-1} - t_{k-2}) e^{-\lambda t_k},$$
$$+ R(t_1, t_2, \ldots, t_k), \tag{1.17}$$

where $R(t_1, t_2, \ldots, t_k)$ is a reminder containing the probabilities of lower order, that is, those for which at least one t_i is missing.

Differentiation now yields

$$f_{T_1, T_2, \ldots, T_k}(t_1, t_2, \ldots, t_k) = \lambda^k e^{-\lambda t_k}, \tag{1.18}$$

which, after the transformation $\tau_1 = T_1$, $\tau_2 = T_2 - T_1$, $\tau_3 = T_3 - T_2, \ldots,$ $\tau_k = T_k - T_{k-1}$, shows that

$$f_{\tau_1, \tau_2, \ldots, \tau_k}(u_1, u_2, \ldots, u_k) = \prod_{i=1}^k \lambda e^{-\lambda u_i} \quad \text{for} \quad u_1, u_2, \ldots, u_k > 0, \tag{1.19}$$

and we are done. $\qquad \qquad \Box$

Remark 1.2. A simple proof of (b) can be obtained from (1.10):

$$1 - F_{T_k}(t) = P(T_k > t) = P(X(t) < k) = \sum_{j=0}^{k-1} e^{-\lambda t} \frac{(\lambda t)^j}{j!}. \tag{1.20}$$

Differentiation yields

$$-f_{T_k}(t) = \sum_{j=1}^{k-1} e^{-\lambda t} \frac{\lambda^j t^{j-1}}{(j-1)!} - \sum_{j=0}^{k-1} \lambda e^{-\lambda t} \frac{(\lambda t)^j}{j!}$$
$$= \lambda e^{-\lambda t} \left(\sum_{j=0}^{k-2} \frac{(\lambda t)^j}{j!} - \sum_{j=0}^{k-1} \frac{(\lambda t)^j}{j!} \right) = -\lambda e^{-\lambda t} \frac{(\lambda t)^{k-1}}{(k-1)!},$$

that is,

$$f_{T_k}(t) = \frac{1}{\Gamma(k)} \lambda^k t^{k-1} e^{-\lambda t}, \quad \text{for} \quad t > 0.$$

Remark 1.3. Note that we cannot deduce (a) from (b), since $T_k \in \Gamma(k, 1/\lambda)$ and $T_k = \tau_1 + \tau_2 + \cdots + \tau_k$ does not imply (a). An investigation involving joint distributions is required.

Remark 1.4. Theorem 1.3 shows that the Poisson process starts from scratch not only at fixed time points, but also at the occurrence times $\{T_k, \, k \geq 1\}$. It is, however, not true that the Poisson process starts from scratch at *any* random timepoint. We shall return to this problem in Subsection 2.3.   □

Example 1.2. There are a number of (purely) mathematical relations for which there exist probabilistic proofs that require "no computation." For example, the formula

$$\int_t^\infty \frac{1}{\Gamma(k)} \lambda^k x^{k-1} e^{-\lambda x} \, dx = \sum_{j=0}^{k-1} e^{-\lambda t} \frac{(\lambda t)^j}{j!} \tag{1.21}$$

can be proved by partial integration (and induction). However, it is (also) an "immediate consequence" of (1.10). To see this we observe that the left-hand side equals $P(T_k > t)$ by Theorem 1.3(b) and the right-hand side equals $P(X(t) < k)$. Since these probabilities are the same, (1.21) follows. We shall point to further examples of this kind later on.   □

1.4 A Third Definition of the Poisson Process

So far we have given two equivalent definitions of the Poisson process and, in Theorem 1.3, determined some distributional properties of the (inter) occurrence times. Our final definition amounts to the fact that these properties, in fact, characterize the Poisson process.

Definition III. *Let* $\{X(t), t \geq 0\}$ *be a stochastic process with* $X(0) = 0$, *let* τ_1 *be the time of the first occurrence, and let* τ_k *be the time between the* $(k-1)th$ *and the* kth *occurrences for* $k \geq 2$. *If* $\{\tau_k, \, k \geq 1\}$ *are independent,* $\text{Exp}(\theta)$*-distributed random variables for some* $\theta > 0$ *and* $X(t) = $ *# occurrences in* $(0, t]$, *then* $\{X(t), t \geq 0\}$ *is a Poisson process with intensity* $\lambda = \theta^{-1}$.   □

Theorem 1.4. *Definitions I, II, and III are equivalent.*

Proof. In view of Theorems 1.1 and 1.3(a) we must show (for example) that a stochastic process $\{X(t), t \geq 0\}$, defined according to Definition III, has independent, stationary, Poisson-distributed increments.

We first show that

$$P(X(t) = k) = e^{-\lambda t} \frac{(\lambda t)^k}{k!} \quad \text{for} \quad k = 0, 1, 2, \ldots, \tag{1.22}$$

where $\lambda = \theta^{-1}$.

Thus, set $\lambda = \theta^{-1}$. For $k = 0$, it follows from (1.8) that

$$P(X(t) = 0) = P(\tau_1 > t) = \int_t^\infty \lambda e^{-\lambda x}\, dx = e^{-\lambda t},$$

which proves (1.22) for that case.

Now let $k \geq 1$ and set $T_k = \tau_1 + \tau_2 + \cdots + \tau_k$. Then $T_k \in \Gamma(k, 1/\lambda)$. This, together with (1.10) and (1.17), yields

$$\begin{aligned}
P(X(t) = k) &= P(X(t) \geq k) - P(X(t) \geq k+1) \\
&= P(T_k \leq t) - P(T_{k+1} \leq t) = P(T_{k+1} > t) - P(T_k > t) \\
&= \int_t^\infty \frac{1}{\Gamma(k+1)} \lambda^{k+1} x^k e^{-\lambda x}\, dx - \int_t^\infty \frac{1}{\Gamma(k)} \lambda^k x^{k-1} e^{-\lambda x}\, dx \\
&= \sum_{j=0}^{k} e^{-\lambda t} \frac{(\lambda t)^j}{j!} - \sum_{j=0}^{k-1} e^{-\lambda t} \frac{(\lambda t)^j}{j!} = e^{-\lambda t} \frac{(\lambda t)^k}{k!},
\end{aligned}$$

as desired.

The following, alternative derivation of (1.22), which is included here because we need an extension below, departs from (1.10), according to which

$$P(X(t) = k) = P(T_k \leq t < T_{k+1}) = \int_t^\infty \left(\int_0^t f_{T_k, T_{k+1}}(u, v)\, du \right) dv. \quad (1.23)$$

To determine $f_{T_k, T_{k+1}}(u, v)$, we use transformation. Since T_k and τ_{k+1} are independent with known distributions, we have

$$f_{T_k, \tau_{k+1}}(t, s) = \frac{1}{\Gamma(k)} \lambda^{k+1} t^{k-1} e^{-\lambda t} e^{-\lambda s}, \quad \text{for} \quad s, t \geq 0$$

(recall that $\lambda = \theta^{-1}$), so that an application of Theorem 1.2.1 yields

$$f_{T_k, T_{k+1}}(u, v) = \frac{1}{\Gamma(k)} \lambda^{k+1} u^{k-1} e^{-\lambda v}, \quad \text{for} \quad 0 \leq u \leq v. \quad (1.24)$$

By inserting this into (1.23) and integrating, we finally obtain

$$P(X(t) = k) = \frac{\lambda^k}{(k-1)!} \int_t^\infty \lambda e^{-\lambda v} \left(\int_0^t u^{k-1}\, du \right) dv = e^{-\lambda t} \frac{(\lambda t)^k}{k!}.$$

Next we consider the two time intervals $(0, s]$ and $(s, s+t]$ jointly. To begin with, let $i \geq 0$ and $j \geq 2$ be nonnegative integers. We have

$$\begin{aligned}
P(X(s) &= i,\ X(s+t) - X(s) = j) \\
&= P(X(s) = i,\ X(s+t) = i+j) \\
&= P(T_i \leq s < T_{i+1},\ T_{i+j} \leq s+t < T_{i+j+1}) \\
&= \int_{s+t}^\infty \int_s^{s+t} \int_s^{t_3} \int_0^s f_{T_i, T_{i+1}, T_{i+j}, T_{i+j+1}}(t_1, t_2, t_3, t_4)\, dt_1 dt_2 dt_3 dt_4.
\end{aligned}$$

In order to find the desired joint density, we extend the derivation of (1.24) as follows. Set $Y_1 = T_i$, $Y_2 = T_{i+1} - T_i$, $Y_3 = T_{i+j} - T_{i+1}$, and $Y_4 = T_{i+j+1} - T_{i+j}$. The joint density of Y_1, Y_2, Y_3, and Y_4 is easily found from the assumptions:

$$f_{Y_1, Y_2, Y_3, Y_4}(y_1, y_2, y_3, y_4)$$
$$= \frac{1}{\Gamma(i)} y_1^{i-1} \lambda^i e^{-\lambda y_1} \cdot \lambda e^{-\lambda y_2} \cdot \frac{1}{\Gamma(j-1)} y_3^{j-2} \lambda^{j-1} e^{-\lambda y_3} \cdot \lambda e^{-\lambda y_4},$$

for $y_1, y_2, y_3, y_4 > 0$. An application of Theorem 1.2.1 yields the desired density, which is inserted into the integral above. Integration (the details of which we omit) finally yields

$$P(X(s) = i, \; X(s+t) - X(s) = j) = e^{-\lambda s} \frac{(\lambda s)^i}{i!} \cdot e^{-\lambda t} \frac{(\lambda t)^j}{j!}. \qquad (1.25)$$

The obvious extension to an arbitrary finite number of time intervals concludes the proof for that case.

It remains to check the boundary cases $i = 0$, $j = 0$ and $i = 0$, $j = 1$ (actually, these cases are easier): Toward that end we modify the derivation of (1.25) as follows:

For $i = 0$ and $j = 0$, we have

$$P(X(s) = 0, \; X(s+t) - X(s) = 0) = P(X(s+t) = 0)$$
$$= P(T_1 > s+t) = e^{-\lambda(s+t)} = e^{-\lambda s} \cdot e^{-\lambda t},$$

which is (1.25) for that case.

For $i = 0$ and $j = 1$ we have

$$P(X(s) = 0, \; X(s+t) - X(s) = 1) = P(X(s) = 0, \; X(s+t) = 1)$$
$$= P(s < T_1 \le s+t < T_2) = \int_{s+t}^{\infty} \int_s^{s+t} f_{T_1, T_2}(t_1, t_2) \, dt_1 dt_2.$$

Inserting the expression for the density as given by (1.24) (with $k = 1$) and integration yields

$$P(X(s) = 0, \; X(s+t) - X(s) = 1) = e^{-\lambda s} \cdot \lambda t e^{-\lambda t},$$

which is (1.25) for that case.

The proof is complete. $\qquad\qquad\qquad\qquad\qquad\qquad\qquad\qquad\qquad\qquad\square$

2 Restarted Poisson Processes

We have now encountered three equivalent definitions of the Poisson process. In the following we shall use these definitions at our convenience in order to establish various properties of the process. At times we shall also give several proofs of some fact, thereby illustrating how the choice of definition affects the complexity of the proof.

2.1 Fixed Times and Occurrence Times

In the first result of this section we use the lack of memory property to assert that a Poisson process started at a fixed (later) time point is, again, a Poisson process. (Since the Poisson process always starts at 0, we have to subtract the value of the new starting point.)

Theorem 2.1. *If $\{X(t),\, t \geq 0\}$ is a Poisson process, then so is*

$$\{X(t+s) - X(s),\, t \geq 0\} \quad \text{for every fixed } s > 0.$$

Proof. Put $Y(t) = X(t+s) - X(s)$ for $t \geq 0$. By arguing as in the proof of Theorem 1.1, it follows that the Y-process has independent increments and that an occurrence in the Y-process during $(t, t+h]$ corresponds to an occurrence in the X-process during $(t+s, t+s+h]$. The properties of Definition II are thus satisfied, and the conclusion follows.   □

Next we prove the corresponding assertion for $\{X(T_k + t) - X(T_k),\, t \geq 0\}$, where T_k, as before, is the time of the kth occurrence in the original Poisson process. Observe that in this theorem we (re)start at the *random times* T_k, for $k \geq 1$.

Theorem 2.2. *If $\{X(t),\, t \geq 0\}$ is a Poisson process, then so is*

$$\{X(T_k + t) - X(T_k),\, t \geq 0\} \quad \text{for every fixed } k \geq 1.$$

First Proof. The first occurrence in the new process corresponds to the $(k+1)$th occurrence in the original process, the second to the $(k+2)$th occurrence, and so on; occurrence m in $\{X(T_k + t) - X(T_k),\, t \geq 0\}$ is the same as occurrence $k+m$ in the original process, for $m \geq 1$. Since the original durations are independent and $\text{Exp}(1/\lambda)$-distributed, it follows that the same is true for the durations of the new process. The conclusion follows in view of Definition III.

Second Proof. Put $Y(t) = X(T_k + t) - X(T_k)$ for $t \geq 0$. The following computation shows that the increments of the new process are Poisson-distributed. By the law of total probability, we have for $n = 0, 1, 2, \ldots$ and $t, s > 0$,

$$P(Y(t+s) - Y(s) = n) = P(X(T_k + t + s) - X(T_k + s) = n)$$

$$= \int_0^\infty P(X(T_k + t + s) - X(T_k + s) = n \mid T_k = u) \cdot f_{T_k}(u)\, du$$

$$= \int_0^\infty P(X(u + t + s) - X(u + s) = n \mid T_k = u) \cdot f_{T_k}(u)\, du$$

$$= \int_0^\infty P(X(u + t + s) - X(u + s) = n) \cdot f_{T_k}(u)\, du$$

$$= \int_0^\infty e^{-\lambda t} \frac{(\lambda t)^n}{n!} \cdot f_{T_k}(u)\, du = e^{-\lambda t} \frac{(\lambda t)^n}{n!}.$$

The crucial point is that the events $\{X(u+t+s)-X(u+s) = n\}$ and $\{T_k = u\}$ are independent (this is used for the fourth equality). This follows since the first event depends on $\{X(v),\ u+s < v \le u+t+s\}$, the second event depends on $\{X(v),\ 0 < v \le u\}$, and the intervals $(0, u]$ and $(u+s, u+t+s]$ are disjoint. An inspection of the integrands shows that, for the same reason, we further have

$$X(T_k + t + s) - X(T_k + s) \mid T_k = u \in \text{Po}(\lambda t). \qquad (2.1)$$

To prove that the process has independent increments, one considers finite collections of disjoint time intervals jointly. □

Exercise 2.1. Let $\{X(t),\ t \ge 0\}$ be a Poisson process with intensity 4.

(a) What is the expected time of the third occurrence?
(b) Suppose that the process has been observed during one time unit. What is the expected time of the third occurrence given that $X(1) = 8$?
(c) What is the distribution of the time between the 12th and the 15th occurrences? □

Example 2.1. Susan stands at a road crossing. She needs six seconds to cross. Cars pass by with a constant speed according to a Poisson process with an intensity of 15 cars a minute. Susan does not dare to cross the street before she has clear visibility, which means that there appears a gap of (at least) six seconds between two cars. Let N be the number of cars that pass before the necessary gap between two cars appears. Determine

(a) the distribution of N, and compute $E\,N$;
(b) the total waiting time T before Susan can cross the road.

Solution. (a) The car arrival intensity is $\lambda = 15$, which implies that the waiting time τ_1 until a car arrives is $\text{Exp}(1/15)$-distributed.

Now, with N as defined above, we have $N \in \text{Ge}(p)$, where

$$p = P(N = 0) = P(\tau_1 > \tfrac{1}{10}) = e^{-\frac{1}{10}(15)} = e^{-1.5}.$$

It follows that $E\,N = e^{1.5} - 1$.

(b) Let $\tau_1, \tau_2, \ldots$ be the times between cars. Then $\tau_1, \tau_2, \ldots$ are independent, $\text{Exp}(1/15)$-distributed random variables. The actual waiting times, however, are $\tau_k^* = \tau_k \mid \tau_k \le 0.1$, for $k \le 1$. Since there are N cars passing before she can cross, we obtain

$$T = \tau_1^* + \tau_2^* + \cdots + \tau_N^*,$$

which equals zero when N equals zero. It follows from Section 3.6 that

$$E\,T = E\,N \cdot E\,\tau_1^* = (e^{1.5} - 1) \cdot \left(\frac{1}{15} - \frac{0.1}{e^{1.5} - 1} \right) = \frac{e^{1.5} - 2.5}{15}. \qquad □$$

Exercise 2.2. (a) Next, suppose that Susan immediately wants to return. Determine the expected number of cars and the expected waiting time before she can return.
(b) Find the expected total time that has elapsed upon her return.

Exercise 2.3. This time, suppose that Susan went across the street to buy ice cream, which requires an Exp(2)-distributed amount of time, after which she returns. Determine the expected total time that has elapsed from her start until her return has been completed.

Exercise 2.4. Now suppose that Susan and Daisy went across the street, after which Daisy wanted to return immediately, whereas Susan wanted to buy ice cream. After having argued for 30 seconds about what to do, they decided that Susan would buy her ice cream (as above) and then return while Daisy would return immediately and wait for Susan. How long did Daisy wait for Susan? □

2.2 More General Random Times

In this subsection we generalize Theorem 2.2 in that we consider restarts at certain other random time points. The results will be used in Section 5 ahead.

Theorem 2.3. *Suppose that $\{X(t), t \geq 0\}$ is a Poisson process and that T is a nonnegative random variable that is independent of the Poisson process. Then $\{X(T+t) - X(T), t \geq 0\}$ is a Poisson process.*

First Proof. Set $Y(t) = X(T+t) - X(T)$ for $t \geq 0$. We show that Definition I applies. The independence of the increments is a simple consequence of the facts that they are independent in the original process and that T is independent of that process. Furthermore, computations analogous to those of the proof of Theorem 2.2 yield, for $0 \leq t_1 < t_2$ and $k = 0, 1, 2, \ldots$ (when T has a continuous distribution),

$$P(Y(t_2) - Y(t_1) = k) = P(X(T+t_2) - X(T+t_1) = k)$$

$$= \int_0^\infty P(X(T+t_2) - X(T+t_1) = k \mid T = u) \cdot f_T(u) \, du$$

$$= \int_0^\infty P(X(u+t_2) - X(u+t_1) = k \mid T = u) \cdot f_T(u) \, du$$

$$= \int_0^\infty P(X(u+t_2) - X(u+t_1) = k) \cdot f_T(u) \, du$$

$$= \int_0^\infty e^{-\lambda(u+t_2-(u+t_1))} \frac{\left(\lambda(u+t_2-(u+t_1))\right)^k}{k!} \cdot f_T(u) \, du$$

$$= e^{-\lambda(t_2-t_1)} \frac{(\lambda(t_2-t_1))^k}{k!} \int_0^\infty f_T(u) \, du$$

$$= e^{-\lambda(t_2-t_1)} \frac{(\lambda(t_2-t_1))^k}{k!}.$$

Once again, the proof for discrete T is analogous and is left to the reader.

In order to determine the distribution of an increment, we may, alternatively, use transforms, for example, generating functions. Let $0 \leq t_1 < t_2$. We first observe that

$$
\begin{aligned}
h(t) &= E(s^{Y(t_2)-Y(t_1)} \mid T = t) \\
&= E(s^{X(T+t_2)-X(T+t_1)} \mid T = t) = E(s^{X(t+t_2)-X(t+t_1)} \mid T = t) \\
&= E\, s^{X(t+t_2)-X(t+t_1)} = e^{\lambda(t_2-t_1)(s-1)},
\end{aligned}
$$

that is, $h(t)$ does not depend on t. An application of Theorem 2.2.1 yields

$$
\begin{aligned}
g_{Y(t_2)-Y(t_1)}(s) &= E\, s^{Y(t_2)-Y(t_1)} = E\big(E(s^{Y(t_2)-Y(t_1)} \mid T)\big) \\
&= E\, h(T) = e^{\lambda(t_2-t_1)(s-1)},
\end{aligned}
$$

which, in view of Theorem 3.2.1 (uniqueness of the generating function), shows that $Y(t_2) - Y(t_1) \in \mathrm{Po}(\lambda(t_2 - t_1))$, as required. $\qquad \square$

Remark 2.1. As for independence, the comments at the end of the second proof of Theorem 2.2 also apply here. $\qquad \square$

Second Proof. Let $Y(t)$, for $t \geq 0$, be defined as in the first proof. Independence of the increments follows as in that proof.

For T continuous, we further have

$P(\text{exactly one } Y\text{-occurrence during } (t, t+h])$

$$
\begin{aligned}
&= \int_0^\infty P(\text{exactly one } X\text{-occurrence during } (u+t, u+t+h]) \mid T = u) \\
&\qquad \times f_T(u)\, du \\
&= \int_0^\infty P(\text{exactly one } X\text{-occurrence during } (u+t, u+t+h]) \cdot f_T(u)\, du \\
&= \int_0^\infty (\lambda h + o(h)) \cdot f_T(u)\, du = \lambda h + o(h) \quad \text{as} \quad h \to 0,
\end{aligned}
$$

and, similarly,

$$
P(\text{at least two } Y\text{-occurrences during } (t, t+h]) = o(h) \quad \text{as} \quad h \to 0.
$$

Again, the computations for T discrete are analogous. The conditions of Definition II have thus been verified, and the result follows. $\qquad \square$

Remark 2.2. For the reader who is acquainted with Lebesgue integration, we remark that the proofs for T continuous and T discrete actually can be combined into one proof, which, in addition, is valid for T having an arbitrary distribution.

Remark 2.3. It is a lot harder to give a proof of Theorem 2.3 based on Definition III. This is because the kth occurrence in the new process corresponds to an occurrence with a *random* number in the original process. However, it is not so difficult to show that the time until the first occurrence in the Y-process, $T_1^{(y)} \in \text{Exp}(1/\lambda)$.

Explicitly, say, for T continuous,

$$
\begin{aligned}
P(T_1^{(y)} > t) &= P(Y(t) = 0) = P(X(T + t) - X(T) = 0) \\
&= \int_0^\infty P(X(T + t) - X(T) = 0 \mid T = u) \cdot f_T(u)\, du \\
&= \int_0^\infty P(X(u + t) - X(u) = 0 \mid T = u) \cdot f_T(u)\, du \\
&= \int_0^\infty P(X(u + t) - X(u) = 0) \cdot f_T(u)\, du \\
&= \int_0^\infty e^{-\lambda t} \cdot f_T(u)\, du = e^{-\lambda t}. \qquad \square
\end{aligned}
$$

In our second generalization of Theorem 2.2 we restart the Poisson process at $\min\{T_k, T\}$, where T_k is as before and T is independent of $\{X(t),\, t \geq 0\}$.

Theorem 2.4. *Let $\{X(t),\, t \geq 0\}$ be a Poisson process, let, for $k \geq 1$, T_k be the time of the kth occurrence, let T be a nonnegative random variable that is independent of the Poisson process, and set $T_k^* = \min\{T_k, T\}$. Then $\{X(T_k^* + t) - X(T_k^*),\, t \geq 0\}$ is a Poisson process.*

Proof. Put $Y(t) = X(T_k^* + t) - X(T_k^*)$, $t \geq 0$. We begin by determining the distribution of the increments of the new process. By arguing as in the second proof of Theorem 2.2, it follows that $X(u + t_2) - X(u + t_1)$ is independent of $\{T = u\}$ and that

$$
X(T + t_2) - X(T + t_1) \mid T = u \in \text{Po}(\lambda(t_2 - t_1)) \text{ for } 0 \leq t_1 < t_2 \qquad (2.2)
$$

(cf. also (2.1)). However, the same properties hold true with T replaced by T_k^*. To see this, we note that the event $\{T_k^* = u\}$ depends only on $\{X(t),\, 0 \leq t \leq u\}$ and T (which is independent of $\{X(t),\, t \geq 0\}$). As a consequence, the event $\{T_k^* = u\}$ is independent of everything occurring after time u, in particular of $X(u + t_2) - X(u + t_1)$. We thus have the same independence property as before. It follows that

$$
X(T_k^* + t_2) - X(T_k^* + t_1) \mid T_k^* = u \in \text{Po}(\lambda(t_2 - t_1)) \text{ for } 0 \leq t_1 < t_2. \qquad (2.3)
$$

The first proof of Theorem 2.3 thus applies (cf. also the second proof of Theorem 2.2), and we conclude that

$$
Y(t_2) - Y(t_1) \in \text{Po}(\lambda(t_2 - t_1)) \quad \text{for} \quad 0 \leq t_1 < t_2. \qquad (2.4)
$$

The independence of the increments follows from the above facts and from the fact that the original process has independent increments. $\qquad \square$

Remark 2.4. A minor variation of the proof is as follows (if T has a continuous distribution).

For $n = 0, 1, 2, \ldots$, we have

$$
\begin{aligned}
P(Y(t_2) - Y(t_1) = n) &= P(X(T_k^* + t_2) - X(T_k^* + t_1) = n) \\
&= P(X(T_k^* + t_2) - X(T_k^* + t_1) = n, \, T_k^* = T_k) \\
&\quad + P(X(T_k^* + t_2) - X(T_k^* + t_1) = n, \, T_k^* = T) \\
&= P(X(T_k^* + t_2) - X(T_k^* + t_1) = n, \, T_k < T) + \\
&\quad + P(X(T_k^* + t_2) - X(T_k^* + t_1) = n, \, T \leq T_k) \\
&= \int_0^\infty P(X(T_k^* + t_2) - X(T_k^* + t_1) = n \mid T_k = u < T) \, P(T > u) \\
&\qquad \times f_{T_k}(u) \, du \\
&\quad + \int_0^\infty P(X(T_k^* + t_2) - X(T_k^* + t_1) = n \mid T = u \leq T_k) \, P(T_k \geq u) \\
&\qquad \times f_T(u) \, du \\
&= \int_0^\infty P(X(u + t_2) - X(u + t_1) = n \mid T_k = u < T) \, P(T > u) \\
&\qquad \times f_{T_k}(u) \, du \\
&\quad + \int_0^\infty P(X(u + t_2) - X(u + t_1) = n \mid T = u \leq T_k) \, P(T_k \geq u) \\
&\qquad \times f_T(u) \, du \\
&= \int_0^\infty P(X(u + t_2) - X(u + t_1) = n) \, P(T > u) \cdot f_{T_k}(u) \, du \\
&\quad + \int_0^\infty P(X(u + t_2) - X(u + t_1) = n) \, P(T_k \geq u) \cdot f_T(u) \, du \\
&= e^{-\lambda(t_2 - t_1)} \frac{(\lambda(t_2 - t_1))^n}{n!} \\
&\quad \times \left(\int_0^\infty \int_u^\infty f_T(v) f_{T_k}(u) \, dv \, du + \int_0^\infty \int_u^\infty f_{T_k}(v) f_T(u) \, dv \, du \right) \\
&= e^{-\lambda(t_2 - t_1)} \frac{(\lambda(t_2 - t_1))^n}{n!},
\end{aligned}
$$

which shows that the increments are Poisson-distributed, as desired.

The removal of the conditioning is justified by the fact that the events $\{T_k = u < T\}$ and $\{T = u \leq T_k\}$ depend only on $\{X(t), \, 0 \leq t \leq u\}$ and T, which makes them independent of $X(u + t_2) - X(u + t_1)$.

By considering several disjoint time intervals jointly, one can prove independence of the increments. $\qquad\square$

2.3 Some Further Topics

Parts of the content of this subsection touch, or even cross, the boundary of the scope of this book. In spite of this, let us make some further remarks.

As for the difference between Theorem 2.1 and Theorems 2.2 to 2.4, we may make a comparison with Markov processes. Theorem 2.1, which is based on the "starting from scratch" property at *fixed* time points, is a consequence of what is called the *weak Markov property* (where one conditions on fixed times). Theorems 2.2 to 2.4, which establish the starting from scratch property for certain *random* time points, is a consequence of the so-called strong Markov property (which involves conditioning on (certain) random times).

A closer inspection of the proof of Theorem 2.4 shows that the hardest points were those required to prove relation (2.3) and the independence of the increments. For these conclusions we used the fact that "T_k^* does not depend on the future" in the sense that the event $\{T_k^* = u\}$ depends only on what happens to the original Poisson process up to time u, that is, on $\{X(t), t \le u\}$. Analogous arguments were made in the second proof of Theorem 2.2, the proof of Theorem 2.3, and Remark 2.4.

In view of this it is reasonable to guess that theorems of the preceding kind hold true for any T that is independent of the future in the same sense. Indeed, there exists a concept called *stopping time* based on this property. Moreover,

(a) the strong Markov property is satisfied if the usual (weak) Markov property holds with fixed times replaced by stopping times;

(b) Poisson processes start from scratch at stopping times, that is, Theorems 2.2 to 2.4 can be shown to hold true for T being an arbitrary stopping time; Theorems 2.2 to 2.4 are special cases of this more general result.

We conclude with an example, which shows that a restarted Poisson process is not always a Poisson process.

Example 2.2. Let $\{X(t), t \ge 0\}$ be a Poisson process, and set

$$T = \sup\{n : X(n) = 0\}.$$

This means that T is the last integral time point before the first occurrence. Further, let T' be the time of the first occurrence in the process $\{X(T + t) - X(T), t \ge 0\}$. Then, necessarily, $P(T' \le 1) = 1$, which, in particular, implies that T' does not follow an exponential distribution. The new process thus cannot be a Poisson process (in view of Definition III).   □

The important feature of the example is that the event $\{T = n\}$ depends on the future, that is, on $\{X(t), t > n\}$; T is not a stopping time.

3 Conditioning on the Number of Occurrences in an Interval

In this section we investigate how a given number of occurrences of a Poisson process during a fixed time interval are distributed within that time interval. For simplicity, we assume that the time interval is $(0, 1]$. As it turns out, all results are independent of the intensity of the Poisson process. The reason for this is that the intensity acts only as a scaling factor and that conditioning annihilates the scaling effects. Moreover, if $Y \in \text{Exp}(\theta)$, for $\theta > 0$, then $aY \in \text{Exp}(a\theta)$ for every $a > 0$. By exploiting these facts and the lack of memory property, it is easy (and a good exercise) to formulate and prove corresponding results for general intervals.

The simplest problem is to determine the distribution of T_1 given that $X(1) = 1$. A moment's thought reveals the following. In view of the lack of memory property, the process should not be able to remember *when* during the time interval $(0, 1]$ there was an occurrence. All time points should, in some sense, be equally likely. Our first result establishes that this is, indeed, the case.

Theorem 3.1. *The conditional distribution of T_1 given that $X(1) = 1$ is the $U(0, 1)$-distribution, that is,*

$$F_{T_1 \mid X(1)=1}(t) = P(T_1 \leq t \mid X(1) = 1) = \begin{cases} 0, & \text{for} \quad t < 0, \\ t, & \text{for} \quad 0 \leq t \leq 1, \\ 1, & \text{for} \quad t > 1, \end{cases} \tag{3.1}$$

or, equivalently,

$$f_{T_1 \mid X(1)=1}(t) = \begin{cases} 1, & \text{for} \quad 0 \leq t \leq 1, \\ 0, & \text{otherwise.} \end{cases} \tag{3.2}$$

First Proof. For $0 \leq t \leq 1$, we have

$$P(T_1 \leq t \mid X(1) = 1) = \frac{P(T_1 \leq t, X(1) = 1)}{P(X(1) = 1)}$$

$$= \frac{P(X(t) = 1, X(1) = 1)}{P(X(1) = 1)}$$

$$= \frac{P(X(t) = 1, X(1) - X(t) = 0)}{P(X(1) = 1)}$$

$$= \frac{P(X(t) = 1) \cdot P(X(1) - X(t) = 0)}{P(X(1) = 1)}$$

$$= \frac{\lambda t e^{-\lambda t} \cdot e^{-\lambda(1-t)}}{\lambda e^{-\lambda}} = t.$$

The cases $t < 0$ and $t > 1$ are, of course, trivial. This proves (3.1), from which (3.2) follows by differentiation.

Second Proof. This proof is similar to the second proof of Theorem 2.2. Let $0 \le t \le 1$.

$$
\begin{aligned}
P(T_1 \le t \mid X(1) = 1) &= \frac{P(T_1 \le t, X(1) = 1)}{P(X(1) = 1)} \\
&= \frac{\int_0^t P(X(1) = 1 \mid T_1 = s) \cdot f_{T_1}(s)\, ds}{P(X(1) = 1)} \\
&= \frac{\int_0^t P(X(1) - X(s) = 0 \mid T_1 = s) \cdot f_{T_1}(s)\, ds}{P(X(1) = 1)} \\
&= \frac{\int_0^t P(X(1) - X(s) = 0) \cdot f_{T_1}(s)\, ds}{P(X(1) = 1)} \\
&= \frac{\int_0^t e^{-\lambda(1-s)} \cdot \lambda e^{-\lambda s}\, ds}{\lambda e^{-\lambda}} = \int_0^t ds = t. \quad \square
\end{aligned}
$$

Now suppose that $X(1) = n$. Intuitively, we then have n points, each of which behaves according to Theorem 3.1. In view of the lack of memory property, it is reasonable to believe that they behave independently of each other. In the remainder of this section we shall verify these facts. We first show that the (marginal) distribution of T_k given that $X(1) = n$ is the same as that of the kth order variable in a sample of n independent, $U(0, 1)$-distributed random variables (cf. Theorem 4.1.1). Then we show that the joint conditional distribution of the occurrence times is the same as that of the order statistic of n independent, $U(0, 1)$-distributed random variables (cf. Theorem 4.3.1).

Theorem 3.2. *For $k = 1, 2, \ldots, n$,*

$$
T_k \mid X(1) = n \in \beta(k, n + 1 - k),
$$

that is,

$$
f_{T_k \mid X(1) = n}(t) = \begin{cases} \dfrac{\Gamma(n+1)}{\Gamma(k)\Gamma(n+1-k)} t^{k-1}(1-t)^{n-k}, & \text{for} \quad 0 \le t \le 1, \\ 0, & \text{otherwise.} \end{cases}
$$

Remark 3.1. For $n = k = 1$, we rediscover Theorem 3.1. $\quad\square$

Proof. We modify the second proof of Theorem 3.1. For $0 \le t \le 1$, we have

$$
\begin{aligned}
P(T_k \le t \mid X(1) = n) &= \frac{P(T_k \le t, X(1) = n)}{P(X(1) = n)} \\
&= \frac{\int_0^t P(X(1) = n \mid T_k = s) \cdot f_{T_k}(s)\, ds}{P(X(1) = n)} \\
&= \frac{\int_0^t P(X(1) - X(s) = n - k) \cdot f_{T_k}(s)\, ds}{P(X(1) = n)}
\end{aligned}
$$

$$= \frac{\int_0^t e^{-\lambda(1-s)} \frac{(\lambda(1-s))^{n-k}}{(n-k)!} \cdot \frac{1}{\Gamma(k)} \lambda^k s^{k-1} e^{-\lambda s} \, ds}{e^{-\lambda} \frac{\lambda^n}{n!}}$$

$$= \frac{n!}{\Gamma(k)\,(n-k)!} \int_0^t s^{k-1}(1-s)^{n-k} \, ds$$

$$= \frac{\Gamma(n+1)}{\Gamma(k)\Gamma(n+1-k)} \int_0^t s^{k-1}(1-s)^{n+1-k-1} \, ds.$$

The density is obtained via differentiation. $\qquad \square$

Theorem 3.3. *The joint conditional density of* $T_1, T_2, \ldots, T_n$ *given that* $X(1) = n$ *is*

$$f_{T_1,\ldots,T_n|X(1)=n}(t_1,\ldots,t_n) = \begin{cases} n!, & \text{for } 0 < t_1 < t_2 < \cdots < t_n < 1, \\ 0, & \text{otherwise.} \end{cases}$$

Proof. We first determine the distribution of $(T_1, T_2, \ldots, T_n)$. With τ_k, $1 \le k \le n$, as before, it follows from Theorem 1.3(a) that

$$f_{\tau_1,\ldots,\tau_n}(u_1,\ldots,u_n) = \prod_{k=1}^n \lambda e^{-\lambda u_k} = \lambda^n \exp\Big\{-\lambda \sum_{k=1}^n u_k\Big\}, \quad u_k > 0,$$

which, with the aid of Theorem 1.2.1, yields

$$f_{T_1,\ldots,T_n}(t_1,\ldots,t_n) = \lambda^n e^{-\lambda t_n} \quad \text{for } 0 < t_1 < t_2 < \cdots < t_n. \tag{3.3}$$

By proceeding as in the proof of Theorem 3.2, we next obtain, for $0 < t_1 < t_2 < \cdots < t_n < 1$,

$$P(T_1 \le t_1, T_2 \le t_2, \ldots, T_n \le t_n \mid X(1) = n)$$

$$= \frac{P(T_1 \le t_1, T_2 \le t_2, \ldots, T_n \le t_n, X(1) = n)}{P(X(1) = n)}$$

$$= \frac{\iint \cdots \int P(X(1) - X(s_n) = 0) \cdot f_{T_1,\ldots,T_n}(s_1,\ldots,s_n) \, ds_1 ds_2 \cdots ds_n}{P(X(1) = n)}$$

$$= \frac{\int_0^{t_1} \int_{s_1}^{t_2} \cdots \int_{s_{n-1}}^{t_n} e^{-\lambda(1-s_n)} \cdot \lambda^n e^{-\lambda s_n} \, ds_n ds_{n-1} \cdots ds_1}{e^{-\lambda} \frac{\lambda^n}{n!}}$$

$$= n! \int_0^{t_1} \int_{s_1}^{t_2} \cdots \int_{s_{n-1}}^{t_n} ds_n ds_{n-1} \cdots ds_1.$$

Differentiation yields the desired conclusion. $\qquad \square$

This establishes that the joint conditional distribution of the occurrence times is the same as that of the order statistic of a sample from the $U(0,1)$-distribution as claimed above. Another way to express this fact is as follows:

Theorem 3.4. *Let $U_1, U_2, \ldots, U_n$ be independent, $U(0,1)$-distributed random variables, and let $U_{(1)} \leq U_{(2)} \leq \cdots \leq U_{(n)}$ be the order variables. Then*

$$((T_1, T_2, \ldots, T_n) \mid X(1) = n) \overset{d}{=} (U_{(1)}, U_{(2)}, \ldots, U_{(n)}).$$ □

Remark 3.2. A problem related to these results is the computation of the conditional probability

$$P(X(s) = k \mid X(t) = n), \quad \text{for} \quad k = 0, 1, 2, \ldots, n \quad \text{and} \quad 0 \leq s \leq t.$$

One solution is to proceed as before (please do!). Another way to attack the problem is to use Theorem 3.3 as follows. Since the occurrences are uniformly distributed in $(0, t]$, it follows that the probability that a given occurrence precedes s equals s/t, for $0 \leq s \leq t$. In view of the independence we conclude that, for $0 \leq s \leq t$,

$$\# \text{ occurrences in } (0, s] \mid X(t) = n \in \text{Bin}(n, s/t),   \tag{3.4}$$

and hence, for $k = 0, 1, \ldots, n$ and $0 \leq s \leq t$, that

$$P(X(s) = k \mid X(t) = n) = \binom{n}{k} \left(\frac{s}{t}\right)^k \left(1 - \frac{s}{t}\right)^{n-k}.   \tag{3.5}$$

Remark 3.3. Just as (1.21) was obtained with the aid of (1.10), we may use a conditional version of (1.10) together with (3.4) to show that

$$\frac{\Gamma(n+1)}{\Gamma(k)\Gamma(n+1-k)} \int_0^t x^{k-1}(1-x)^{n-k}\, dx = \sum_{j=k}^{n} \binom{n}{j} t^j (1-t)^{n-j}.   \tag{3.6}$$

The appropriate conditional version of (1.10) is

$$P(T_k \leq t \mid X(1) = n) = P(X(t) \geq k \mid X(1) = n),   \tag{3.7}$$

for $k = 1, 2, \ldots, n$ and $0 \leq t \leq 1$. Since the left-hand sides of (3.6) and (3.7) are equal and since this is also true for the right-hand sides, (3.6) follows immediately from (3.7). Observe also that relation (3.6) was proved by induction (partial integration) during the proof of Theorem 4.1.1.

Remark 3.4. The result in Remark 3.2 can be generalized to several subintervals. Explicitly, by similar arguments one can, for example, show that the joint conditional distribution of

$$(X(t_1) - X(s_1), X(t_2) - X(s_2), \ldots, X(t_k) - X(s_k)) \mid X(1) = n$$

is multinomial with parameters $(n; p_1, \ldots, p_k)$, where $p_j = t_j - s_j$ for $j = 1, 2, \ldots, k$ and $0 \leq s_1 < t_1 \leq s_2 < t_2 \leq \cdots \leq s_k < t_k \leq 1$.   □

4 Conditioning on Occurrence Times

In the previous section we conditioned on the event $\{X(1) = n\}$, that is, on the event that there have been n occurrences at time 1. In this section we condition on T_n, that is, on the nth occurrence time. The conclusions are as follows:

Theorem 4.1. *For $k = 1, 2, \ldots, n$,*

(a) $T_k \mid T_n = 1 \in \beta(k, n - k)$;

(b) $T_k/T_n \in \beta(k, n - k)$;

(c) T_n *and* T_k/T_n *are independent.*

Proof. The conclusions are fairly straightforward consequences of Theorem 1.2.1.

(a) Let $Y_1 \in \Gamma(r, \theta)$ and $Y_2 \in \Gamma(s, \theta)$ be independent random variables, and set $V_1 = Y_1$ and $V_2 = Y_1 + Y_2$. By Theorem 1.2.1 we have

$$
\begin{aligned}
f_{V_1, V_2}(v_1, v_2) &= \frac{1}{\Gamma(r)} \frac{1}{\theta^r} v_1^{r-1} e^{-v_1/\theta} \cdot \frac{1}{\Gamma(s)} \frac{1}{\theta^s} (v_2 - v_1)^{s-1} e^{-(v_2 - v_1)/\theta} \cdot 1 \\
&= \frac{\Gamma(r + s)}{\Gamma(r)\Gamma(s)} \left(\frac{v_1}{v_2}\right)^{r-1} \left(1 - \frac{v_1}{v_2}\right)^{s-1} \frac{1}{v_2} \cdot \frac{1}{\Gamma(r + s)} \frac{1}{\theta^{r+s}} v_2^{r+s-1} e^{-v_2/\theta}, \quad (4.1)
\end{aligned}
$$

for $0 < v_1 < v_2$. Since $V_2 \in \Gamma(r + s, \theta)$, it follows that

$$
f_{V_1 \mid V_2 = 1}(v) = \frac{f_{V_1, V_2}(v, 1)}{f_{V_2}(1)} = \frac{\Gamma(r + s)}{\Gamma(r)\Gamma(s)} v^{r-1} (1 - v)^{s-1}, \text{ for } 0 < v < 1, \quad (4.2)
$$

that is, $V_1 \mid V_2 = 1 \in \beta(r, s)$. By observing that $T_n = T_k + (T_n - T_k)$ and by identifying T_k with Y_1 and $T_n - T_k$ with Y_2 (and hence k with r, $n - k$ with s, and $1/\lambda$ with θ), we conclude that (a) holds.

(b) and (c) It follows from Theorem 1.2.1 applied to (4.1) and the transformation $W_1 = V_1/V_2 (= Y_1/(Y_1 + Y_2))$ and $W_2 = V_2 (= Y_1 + Y_2)$ that

$$
\begin{aligned}
&f_{W_1, W_2}(w_1, w_2) \\
&= \frac{\Gamma(r + s)}{\Gamma(r)\Gamma(s)} w_1^{r-1} (1 - w_1)^{s-1} \cdot \frac{1}{\Gamma(r + s)} \frac{1}{\theta^{r+s}} w_2^{r+s-1} e^{-w_2/\theta}, \quad (4.3)
\end{aligned}
$$

for $0 < w_1 < 1$ and $w_2 > 0$ (cf. Example 1.2.5 and Problems 1.3.41 and 1.3.42). This proves the independence of W_1 and W_2 and that $W_1 \in \beta(r, s)$. The identification $T_k/T_n = W_1$ and $T_n = W_2$ and parameters as above concludes the proof of (b) and (c). $\qquad\square$

The results can also be generalized to joint distribution s. We provide only the statements here and leave the details to the reader.

By starting from the joint density of $(T_1, T_2, \ldots, T_n)$ in (3.3) and by making computations analogous to those that lead to (4.2), one can show that, for $0 < t_1 < t_2 < \cdots < t_{n-1} < 1$,

$$f_{T_1,\ldots,T_{n-1}|T_n=1}(t_1,\ldots,t_{n-1}) = (n-1)!. \tag{4.4}$$

This means that the conditional distribution of $(T_1, T_2, \ldots, T_{n-1})$ given that $T_n = 1$ is the same as that of the order statistic corresponding to a sample of size $n-1$ from a $U(0,1)$-distribution.

Furthermore, by applying a suitable transformation and Theorem 1.2.1 to the density in (3.3), we obtain, for $0 < y_1 < y_2 < \cdots < y_{n-1} < 1$ and $y_n > 0$,

$$f_{\frac{T_1}{T_n},\frac{T_2}{T_n},\ldots,\frac{T_{n-1}}{T_n},T_n}(y_1, y_2, \ldots, y_n) = \lambda^n y_n^{n-1} e^{-\lambda y_n}.$$

By viewing this as

$$f_{\frac{T_1}{T_n},\frac{T_2}{T_n},\ldots,\frac{T_{n-1}}{T_n},T_n}(y_1, y_2, \ldots, y_n) = (n-1)! \cdot \frac{1}{\Gamma(n)} \lambda^n y_n^{n-1} e^{-\lambda y_n}$$

(in the same domain), it follows that $(T_1/T_n, T_2/T_n, \ldots, T_{n-1}/T_n)$ is distributed as the order statistic corresponding to a sample of size $n-1$ from a $U(0,1)$-distribution and that the vector is independent of T_n ($\in \Gamma(n, 1/\lambda)$).

It is also possible to verify that the marginal densities of T_k and T_k/T_n are those given in Theorem 4.1.

The following result collects the above facts:

Theorem 4.2. *Let* $U_1, U_2, \ldots, U_n$ *be independent,* $U(0,1)$-*distributed random variables, and let* $U_{(1)} \leq U_{(2)} \leq \cdots \leq U_{(n)}$ *be the order variables. Then*

(a) $((T_1, T_2, \ldots, T_n) \mid T_{n+1} = 1) \overset{d}{=} (U_{(1)}, U_{(2)}, \ldots, U_{(n)})$;

(b) $(T_1/T_{n+1}, T_2/T_{n+1}, \ldots, T_n/T_{n+1}) \overset{d}{=} (U_{(1)}, U_{(2)}, \ldots, U_{(n)})$;

(c) $(T_1/T_{n+1}, T_2/T_{n+1}, \ldots, T_n/T_{n+1})$ *and* T_{n+1} *are independent.*   □

5 Several Independent Poisson Processes

Suppose that we are given m independent Poisson processes

$$\{X_1(t), t \geq 0\}, \{X_2(t), t \geq 0\}, \ldots, \{X_m(t), t \geq 0\}$$

with intensities $\lambda_1, \lambda_2, \ldots, \lambda_m$, respectively, and consider a new process $\{Y(t), t \geq 0\}$ defined as follows: The Y-occurrences are defined as the union of all X_k-occurrences, $k = 1, 2, \ldots, m$, that is, every Y-occurrence corresponds to an X_k-occurrence for some k, and vice versa.

As a typical example, we might consider a service station to which m kinds of customers arrive according to (m) independent Poisson processes. The Y-process then corresponds to arrivals (of any kind) to the service station.

Another example (discussed ahead) is that of a (radioactive) source emitting particles of different kinds according to independent Poisson processes. The Y-process then corresponds to "a particle is emitted."

A typical realization for $m = 5$ is given in the following figure:

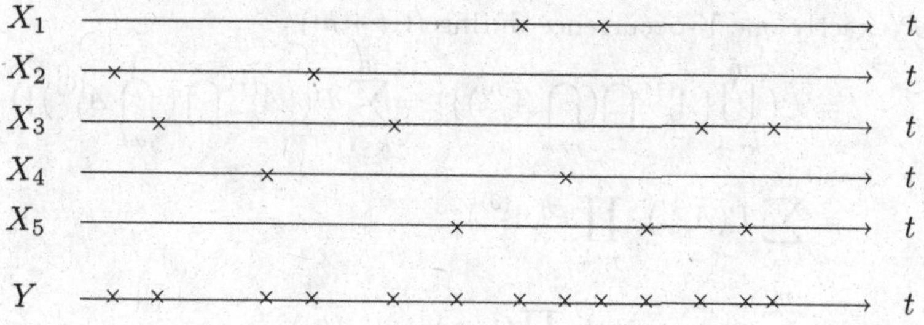

Figure 5.1

5.1 The Superpositioned Poisson Process

The Y-process we have just described is called a *superpositioned Poisson process*; the inclusion of "Poisson" in the name is motivated by the following result:

Theorem 5.1. $\{Y(t), t \geq 0\}$ *is a Poisson process with intensity* $\lambda = \lambda_1 + \lambda_2 + \cdots + \lambda_m$.

First Proof. We show that the conditions of Definition I are satisfied.

The Y-process has independent increments because all the X-processes do and also because the processes are independent. Further, since the sum of independent, Poisson-distributed random variables is Poisson-distributed with a parameter equal to the sum of the individual ones, it follows that

$$Y(t+s) - Y(s) = \sum_{k=1}^{m} \big(X_k(t+s) - X_k(s)\big) \in \text{Po}\big(\sum_{k=1}^{m} \lambda_k t\big), \qquad (5.1)$$

for all $s, t \geq 0$.

Second Proof. We show that Definition II is applicable.

The independence of the increments of the Y-process follows as before. Next we note that there is exactly one Y-occurrence during $(t, t+h]$ if (and only if) there is exactly one X-occurrence during $(t, t+h]$. Therefore, let

$$A_k^{(i)} = \{i\ X_k\text{-occurrences during } (t, t+h]\}, \qquad (5.2)$$

for $k = 1, 2, \ldots, m$ and $i = 0, 1, 2, \ldots$. Then

$$P(A_k^{(0)}) = 1 - \lambda_k h + o(h),$$
$$P(A_k^{(1)}) = \lambda_k h + o(h),$$
$$P\big(\bigcup_{i=2}^{\infty} A_k^{(i)}\big) = o(h), \qquad (5.3)$$

as $h \to 0$. Thus

$P(\text{exactly one } Y\text{-occurrence during } (t, t+h])$

$$= P\left(\bigcup_{k=1}^{m}\{A_k^{(1)} \bigcap (\bigcap_{j \neq k} A_j^{(0)})\}\right) = \sum_{k=1}^{m} P(\{A_k^{(1)} \bigcap (\bigcap_{j \neq k} A_j^{(0)})\})$$

$$= \sum_{k=1}^{m} P(A_k^{(1)}) \cdot \prod_{j \neq k} P(A_j^{(0)})$$

$$= \sum_{k=1}^{m} (\lambda_k h + o(h)) \cdot \prod_{j \neq k} (1 - \lambda_j h + o(h))$$

$$= \left(\sum_{k=1}^{m} \lambda_k\right) \cdot h + o(h) \quad \text{as} \quad h \to 0,$$

which shows that condition (b) in Definition II is satisfied with $\lambda = \sum_{k=1}^{m} \lambda_k$.

Finally, at least two Y-occurrences during $(t, t+h]$ means that we have either at least two X_k-occurrences for at least one k or exactly one X_k-occurrence for at least two different values of k. Thus

$P(\text{at least two } Y\text{-occurrences during } (t, t+h])$

$$= P\left(\left\{\bigcup_{k=1}^{m}\bigcup_{i=2}^{\infty} A_k^{(i)}\right\} \bigcup \left\{\bigcup_{j=2}^{m} \bigcup_{\substack{k_1,\ldots,k_j \\ k_i \text{ different}}} \bigcap_{i=1}^{j} A_{k_i}^{(1)}\right\}\right)$$

$$\leq P\left(\bigcup_{k=1}^{m}\bigcup_{i=2}^{\infty} A_k^{(i)}\right) + P\left(\bigcup_{j=2}^{m} \bigcup_{\substack{k_1,\ldots,k_j \\ k_i \text{ different}}} \bigcap_{i=1}^{j} A_{k_i}^{(1)}\right)$$

$$\leq \sum_{k=1}^{m} P\left(\bigcup_{i=2}^{\infty} A_k^{(i)}\right) + \sum_{j=2}^{m} \sum_{\substack{k_1,\ldots,k_j \\ k_i \text{ different}}} P\left(\bigcap_{i=1}^{j} A_{k_i}^{(1)}\right)$$

$$= m \cdot o(h) + \sum_{j=2}^{m} \sum_{\substack{k_1,\ldots,k_j \\ k_i \text{ different}}} \prod_{i=1}^{j} (\lambda_{k_i} h + o(h))$$

$$= o(h) \quad \text{as} \quad h \to 0,$$

since the dominating term in the product is $(\prod_{i=1}^{j} \lambda_{k_i}) \cdot h^j = o(h)$ as $h \to 0$, for all $j \geq 2$, and the number of terms in the double sum is finite.

This establishes that condition (c) in Definition II is satisfied, and the proof is, again, complete. $\square$

Just as for Theorem 2.3, it is cumbersome to give a complete proof based on Definition III. Let us show, however, that the durations in the Y-process are $\text{Exp}((\sum_{k=1}^{m} \lambda_k)^{-1})$-distributed; to prove independence requires more tools than we have at our disposal here.

We begin by determining the distribution of the time T_y until the first Y-occurrence. Let $T^{(k)}$ be the time until the first X_k-occurrence, $k = 1, 2, \ldots, m$. Then $T^{(1)}, T^{(2)}, \ldots, T^{(m)}$ are independent, $T^{(k)} \in \text{Exp}(1/\lambda_k)$, $k = 1, 2, \ldots, m$, and

$$T_y = \min_{1 \le k \le m} T^{(k)}. \tag{5.4}$$

It follows that

$$P(T_y > t) = P\Big(\bigcap_{k=1}^{m}\{T^{(k)} > t\}\Big) = \prod_{k=1}^{m} P(T^{(k)} > t)$$

$$= \prod_{k=1}^{m} e^{-\lambda_k t} = \exp\Big\{-\Big(\sum_{k=1}^{m} \lambda_k\Big)t\Big\}, \quad \text{for} \quad t \ge 0,$$

that is,

$$T_y \in \text{Exp}\Big(\Big(\sum_{k=1}^{m} \lambda_k\Big)^{-1}\Big). \tag{5.5}$$

Next, consider some fixed j, and set $\widetilde{T}^{(j)} = \min\{T^{(i)} : i \ne j\}$. Since

$$T_y = \min\{T^{(j)}, \widetilde{T}^{(j)}\} \tag{5.6}$$

and $\widetilde{T}^{(j)}$ is independent of the X_j-process, it follows from Theorem 2.4 (with $k = 1$) that $\{X_j(T_y + t) - X_j(T_y), t \ge 0\}$ is a Poisson process (with intensity λ_j).

Since j was arbitrary, the same conclusion holds for all j, which implies that the time between the first and second Y-occurrences is the same as the time until the first occurrence in the superpositioned process generated by the X-processes restarted at T_y (cf. the first proof of Theorem 2.2). By (5.5), however, we know that this waiting time has the desired exponential distribution. Finally, by induction, we conclude that the same is true for all durations. $\qquad\square$

Example 2.1 (continued). Recall Susan standing at a road crossing, needing 6 seconds to cross the road. Suppose that the following, more detailed description of the traffic situation is available. Cars pass from left to right with a constant speed according to a Poisson process with an intensity of 10 cars a minute, and from right to left with a constant speed according to a Poisson process with an intensity of 5 cars a minute. As before, let N be the number of cars that pass before the necessary gap between two cars appears. Determine

(a) the distribution of N, and compute $E\,N$;
(b) the total waiting time T before Susan can cross the road.

It follows from Theorem 5.1 that the process of cars passing by is a Poisson process with intensity $10 + 5 = 15$. The answers to (a) and (b) thus are the same as before.   □

Exercise 5.1. A radioactive substance emits particles of two kinds, A and B, according to two independent Poisson processes with intensities 10 and 15 particles per minute, respectively. The particles are registered in a counter, which is started at time $t = 0$. Let T be the time of the first registered particle. Compute ET.   □

5.2 Where Did the First Event Occur?

In connection with the results of the preceding subsection, the following is a natural question: What is the probability that the first Y-occurrence is caused by the X_k-process? Equivalently (in the notation of Subsection 5.1), what is the probability that $T^{(k)}$ is the smallest among $T^{(1)}$, $T^{(2)}$, ..., $T^{(m)}$? For the service station described in Example 1.1, this amounts to asking for the probability that the first customer to arrive is of some given kind. For the particles it means asking for the probability that a given type of particle is the first to be emitted.

In Figure 5.1 the X_2-process causes the first Y-occurrence.

Suppose first that $m = 2$ and that $\lambda_1 = \lambda_2$. By symmetry the probability that the first Y-occurrence is caused by the X_1-process equals

$$P(T^{(1)} < T^{(2)}) = \frac{1}{2}. \tag{5.7}$$

Similarly, if $\lambda_1 = \lambda_2 = \cdots = \lambda_m$ for some $m \geq 2$, the probability that the first Y-occurrence is caused by the X_k-process equals

$$P(\min\{T^{(1)}, T^{(2)}, \ldots, T^{(m)}\} = T^{(k)}) = \frac{1}{m}. \tag{5.8}$$

Now, let $\lambda_1, \lambda_2, \ldots, \lambda_m$ be arbitrary and $m \geq 2$. We wish to determine the probability in (5.8), that is,

$$P(T^{(k)} < \min_{j \neq k} T^{(j)}) = P(T^{(k)} < \widetilde{T}^{(k)}). \tag{5.9}$$

From the previous subsection we know that $\widetilde{T}^{(k)} \in \mathrm{Exp}((\sum_{j \neq k} \lambda_j)^{-1})$ and that $\widetilde{T}^{(k)}$ and $T^{(k)}$ are independent. The desired probability thus equals

$$\int_0^\infty \left(\int_x^\infty \lambda_k e^{-\lambda_k x} \cdot \widetilde{\lambda}_k e^{-\widetilde{\lambda}_k y} \, dy \right) dx = \frac{\lambda_k}{\lambda_k + \widetilde{\lambda}_k}, \tag{5.10}$$

where $\widetilde{\lambda}_k = \sum_{j \neq k} \lambda_j$. Thus, the answer to the question raised above is that, for $k = 1, 2, \ldots, m$,

$$P(\min\{T^{(1)}, T^{(2)}, \ldots, T^{(m)}\} = T^{(k)}) = \frac{\lambda_k}{\lambda_1 + \lambda_2 + \cdots + \lambda_m}. \qquad (5.11)$$

In particular, if all λ_k are equal, (5.11) reduces to (5.8) (and, for $m = 2$, to (5.7)).

Remark 5.1. Since the exponential distribution is continuous, there are no ties, that is, all probabilities such as $P(T^{(i)} = T^{(j)})$ with $i \neq j$ equal zero, in particular, $P(\text{all } T^{(j)} \text{ are different}) = 1$.   □

Example 2.1 (continued). What is the probability that the first car that passes runs from left to right?

Since the intensities from left to right and from right to left are 10 and 5, respectively, it follows that the answer is $10/(10 + 5) = 2/3$.

Example 5.1. A radioactive material emits α-, β-, γ-, and δ-particles according to four independent Poisson processes with intensities λ_α, λ_β, λ_γ, and λ_δ, respectively. A particle counter counts all emitted particles. Let $Y(t)$ be the number of emissions (registrations) during $(0, t]$, for $t \geq 0$.

(a) Show that $\{Y(t), t \geq 0\}$ is a Poisson process, and determine the intensity of the process.
(b) What is the expected duration T_y until a particle is registered?
(c) What is the probability that the first registered particle is a β-particle?
(d) What is the expected duration T_β until a β-particle is registered?

Solution. (a) It follows from Theorem 5.1 that $\{Y(t), t \geq 0\}$ is a Poisson process with intensity

$$\lambda = \lambda_\alpha + \lambda_\beta + \lambda_\gamma + \lambda_\delta.$$

(b) $T_y \in \text{Exp}(1/\lambda)$, that is, $E T_y = 1/\lambda$.
(c) Recalling formula (5.11), the answer is λ_β/λ.
(d) Since β-particles are emitted according to a Poisson process with intensity λ_β independently of the other Poisson processes, it follows that $T_\beta \in \text{Exp}(1/\lambda_\beta)$ and hence that $E T_\beta = 1/\lambda_\beta$.   □

Exercise 5.1. Compute the probability that the first registered particle is an α-particle.

Exercise 5.2. John and Betty are having a date tonight. They agree to meet at the opera house X_j and X_b hours after 7 p.m., where X_j and X_b are independent, $\text{Exp}(1)$-distributed random variables.

(a) Determine the expected arrival time of the first person.
(b) Determine his or her expected waiting time.
(c) Suppose that, in addition, they decide that they will wait at most 30 minutes for each other. What is the probability that they will actually meet?   □

We close this subsection by pointing out another computational method for finding the probability in (5.11) when $m = 2$. The idea is to view probabilities as expectations of indicators as follows:

$$P(T^{(1)} < T^{(2)}) = E\, I\{T^{(1)} < T^{(2)}\} = E\big(E(I\{T^{(1)} < T^{(2)}\} \mid T^{(1)})\big)$$

$$= E\, e^{-\lambda_2 T^{(1)}} = \psi_{T^{(1)}}(-\lambda_2) = \frac{1}{1 - \frac{1}{\lambda_1} \cdot (-\lambda_2)} = \frac{\lambda_1}{\lambda_1 + \lambda_2}.$$

For the third equality sign we used the fact that

$$E(I\{T^{(1)} < T^{(2)}\} \mid T^{(1)} = t) = E(I\{t < T^{(2)}\} \mid T^{(1)} = t)$$
$$= E\, I\{t < T^{(2)}\} = P(T^{(2)} > t) = e^{-\lambda_2 t}.$$

5.3 An Extension

An immediate generalization of the problem discussed in the previous subsection is given by the following question: What is the probability that there are n X_k-occurrences preceding occurrences of any other kind?

The following is an example for the case $m = 2$. A mathematically equivalent example formulated in terms of a game is given after the solution; it is instructive to reflect a moment on why the problems are indeed equivalent:

Example 5.2. A radioactive source emits a substance, which is a mixture of α-particles and β-particles. The particles are emitted as independent Poisson processes with intensities λ and μ particles per second, respectively. Let N be the number of emitted α-particles between two consecutive β-particles. Find the distribution of N.

First Solution. The "immediate" solution is based on conditional probabilities.

We first consider the number of α-particles preceding the *first* β-particle. Let T_β be the waiting time until the first β-particle is emitted. Then $T_\beta \in \mathrm{Exp}(1/\mu)$ and

$$P(N = n \mid T_\beta = t) = e^{-\lambda t}\frac{(\lambda t)^n}{n!} \quad \text{for} \quad n = 0, 1, 2, \ldots. \tag{5.12}$$

It follows that

$$P(N = n) = \int_0^\infty P(N = n \mid T_\beta = t) \cdot f_{T_\beta}(t)\, dt$$

$$= \int_0^\infty e^{-\lambda t}\frac{(\lambda t)^n}{n!} \cdot \mu e^{-\mu t}\, dt = \int_0^\infty \frac{\mu \lambda^n}{\Gamma(n+1)} t^n e^{-(\lambda+\mu)t}\, dt$$

$$= \frac{\mu \lambda^n}{(\lambda+\mu)^{n+1}} \int_0^\infty \frac{1}{\Gamma(n+1)}(\lambda+\mu)^{n+1} t^n e^{-(\lambda+\mu)t}\, dt$$

$$= \frac{\mu}{\lambda+\mu}\Big(\frac{\lambda}{\lambda+\mu}\Big)^n, \quad \text{for} \quad n = 0, 1, 2, \ldots,$$

that is, $N \in \mathrm{Ge}(\mu/(\lambda+\mu))$.

Observe that this, in particular, shows that the probability that the first emitted particle is a β-particle equals $\mu/(\lambda + \mu)$ (in agreement with (5.11)).

This answers the question of how many α-particles there are before the *first* β-particle is emitted. In order to answer the original, more general question, we observe that, by Theorem 2.4 (cf. also the third proof of Theorem 5.1), "everything begins from scratch" each time a particle is emitted. It follows that the number of α-particles *between* two β-particles follows the same geometric distribution.

Second Solution. We use (5.12) and transforms. Since

$$E(s^N \mid T_\beta = t) = e^{\lambda t(s-1)}, \tag{5.13}$$

we obtain, for $s < 1 + \mu/\lambda$,

$$g_N(s) = E s^N = E\big(E(s^N \mid T_\beta)\big) = E e^{\lambda T_\beta(s-1)} = \psi_{T_\beta}(\lambda(s-1))$$

$$= \frac{1}{1 - \frac{\lambda(s-1)}{\mu}} = \frac{\mu}{\mu + \lambda - \lambda s} = \frac{\frac{\mu}{\lambda+\mu}}{1 - \frac{\lambda}{\lambda+\mu}s},$$

which is the generating function of the $\mathrm{Ge}(\mu/(\lambda + \mu))$-distribution. By the uniqueness theorem (Theorem 3.2.1), we conclude that $N \in \mathrm{Ge}(\mu/(\lambda + \mu))$.

Third Solution. The probability that an α-particle comes first is equal to $\mu/(\lambda + \mu)$, by (5.11). Moreover, everything starts from scratch each time a particle is emitted. The event $\{N = n\}$ therefore occurs precisely when the first n particles are α-particles and the $(n + 1)$th particle is a β-particle. The probability of this occurring equals

$$\frac{\lambda}{\lambda + \mu} \cdot \frac{\lambda}{\lambda + \mu} \cdots \frac{\lambda}{\lambda + \mu} \cdot \frac{\mu}{\lambda + \mu},$$

with n factors $\mu/(\lambda + \mu)$. This shows (again) that

$$P(N = n) = \left(\frac{\lambda}{\lambda + \mu}\right)^n \frac{\mu}{\lambda + \mu} \quad \text{for} \quad n = 0, 1, 2, \ldots, \tag{5.14}$$

as desired.   □

Example 5.3. Patricia and Cindy are playing computer games on their computers. The duration of the games are $\mathrm{Exp}(\theta)$- and $\mathrm{Exp}(\mu)$-distributed, respectively, and all durations are independent. Find the distribution of the number of games Patricia wins between two consecutive wins by Cindy.   □

Now, let $m \geq 2$ be arbitrary. In Example 5.2 this corresponds to m different kinds of particles. The problem of finding the number of particles of type k preceding any other kind of particle is reduced to the case $m = 2$ by putting all other kinds of particles into one (big) category in a manner similar to that of Subsections 5.1 and 5.2. We thus create a superpositioned Y-process based on the X_j-processes with $j \neq k$. By Theorem 5.1, this yields a Poisson process with intensity $\widetilde{\lambda}_k = \sum_{j \neq k} \lambda_j$, which is independent of the X_k-process. The rest is immediate.

We collect our findings from Subsections 5.2 and 5.3 in the following result:

Theorem 5.2. *Let* $\{X_1(t),\, t \geq 0\}, \{X_2(t),\, t \geq 0\}, \ldots, \{X_m(t),\, t \geq 0\}$ *be independent Poisson processes with intensities* $\lambda_1, \lambda_2, \ldots, \lambda_m$, *respectively, and set* $p_k = \lambda_k/(\lambda_1 + \cdots + \lambda_m)$, *for* $k = 1, 2, \ldots, m$. *For every* k, $1 \leq k \leq m$, *we then have the following properties:*

(a) *The probability that the first occurrence is caused by the* X_k-*process equals* p_k.
(b) *The probability that the first* n *occurrences are caused by the* X_k-*process equals* p_k^n, $n \geq 1$.
(c) *The number of* X_k-*occurrences preceding an occurrence of any other kind is* $\text{Ge}(1 - p_k)$-*distributed.*
(d) *The number of occurrences preceding the first occurrence in the* X_k-*process is* $\text{Ge}(p_k)$-*distributed.*
(e) *The number of non-*X_k-*occurrences between two occurrences in the* X_k-*process is* $\text{Ge}(p_k)$-*distributed.* □

5.4 An Example

In Example 1.2 and Remark 3.3 two mathematical formulas were demonstrated with the aid of probabilistic arguments. Here is another example:

$$\int_0^\infty x n e^{-x}(1 - e^{-x})^{n-1}dx = 1 + \frac{1}{2} + \frac{1}{3} + \cdots + \frac{1}{n}. \tag{5.15}$$

One way to prove (5.15) is, of course, through induction and partial integration. Another method is to identify the left-hand side as $E\,Y_{(n)}$, where $Y_{(n)}$ is the largest of n independent, identically $\text{Exp}(1)$-distributed random variables, $Y_1, Y_2, \ldots, Y_n$. As for the right-hand side, we put $Z_1 = Y_{(1)}$ and $Z_k = Y_{(k)} - Y_{(k-1)}$, for $k \geq 2$, compute the joint distribution of these differences, and note that

$$Y_{(n)} = Z_1 + Z_2 + \cdots + Z_n. \tag{5.16}$$

This solution was suggested in Problem 4.4.21.

Here we prove (5.15) by exploiting properties of the Poisson process, whereby Theorems 2.4 and 5.1 will be useful.

Solution. Consider n independent Poisson processes with intensity 1, and let $Y_1, Y_2, \ldots, Y_n$ be the times until the first occurrences in the processes. Then Y_k, for $1 \leq k \leq n$, are independent, $\text{Exp}(1)$-distributed random variables. Further, $Y_{(n)} = \max\{Y_1, Y_2, \ldots, Y_n\}$ is the time that has elapsed when every process has had (at least) one occurrence.

We next introduce Z_1, Z_2, ..., Z_n as above as the differences of the order variables $Y_{(1)}, Y_{(2)}, \ldots, Y_{(n)}$. Then Z_1 is the time until the first overall occurrence, which, by Theorem 5.1, is $\text{Exp}(1/n)$-distributed. By Theorem 2.4, all processes start from scratch at time Z_1.

We now remove the process where something occurred. Then Z_2 is the time until an event occurs in one of the remaining $n-1$ processes. By arguing as above, it follows that $Z_2 \in \text{Exp}(1/(n-1))$, and we repeat as above. By (5.16) this finally yields

$$EY_{(n)} = \frac{1}{n} + \frac{1}{n-1} + \frac{1}{n-2} + \cdots + \frac{1}{2} + 1,$$

as desired (since $EY_{(n)}$ also equals the left-hand side of (5.15)).

6 Thinning of Poisson Processes

By a *thinned* stochastic process, we mean that not every occurrence is observed. The typical example is particles that are emitted from a source according to a Poisson process, but, due to the malfunctioning of the counter, only some of the particles are registered. Here we shall confine ourselves to studying the following, simplest case.

Let $X(t)$ be the number of *emitted* particles in $(0, t]$, and suppose that $\{X(t), t \geq 0\}$ is a Poisson process with intensity λ. Suppose, further, that the counter is defective as follows. Every particle is registered with probability p, where $0 < p < 1$ (and not registered with probability $q = 1-p$). Registrations of different particles are independent. Let $Y(t)$ be the number of *registered* particles in $(0, t]$. What can be said about the process $\{Y(t), t \geq 0\}$?

The intuitive guess is that $\{Y(t), t \geq 0\}$ is a Poisson process with intensity λp. The reason for this is that the registering process behaves like the emitting process, but with a smaller intensity; particles are registered in the same fashion as they are emitted, but more sparsely. The deficiency of the counter acts like a (homogeneous) filter or sieve. We now prove that this is, indeed, the case, and we begin by providing a proof adapted to fit Definition II.

Since $\{X(t), t \geq 0\}$ has independent increments, and particles are registered or not independently of each other, it follows that $\{Y(t), t \geq 0\}$ also has independent increments. Now, set

$$A_k = \{k \text{ } Y\text{-occurrences during } (t, t+h]\},$$
$$B_k = \{k \text{ } X\text{-occurrences during } (t, t+h]\},$$

for $k = 0, 1, 2, \ldots$. Then

$$P(A_1) = P\left(A_1 \bigcap \left(\bigcup_{k=1}^{\infty} B_k\right)\right) = P(B_1 \cap A_1) + P\left(\left(\bigcup_{k=2}^{\infty} B_k\right)\bigcap A_1\right)$$

$$= P(B_1) \cdot P(A_1 \mid B_1) + P\left(\left(\bigcup_{k=2}^{\infty} B_k\right)\bigcap A_1\right)$$

$$= (\lambda h + o(h)) \cdot p + o(h) \quad \text{as} \quad h \to 0,$$

since $P\left(\left(\bigcup_{k=2}^{\infty} B_k\right)\bigcap A_1\right) \leq P\left(\bigcup_{k=2}^{\infty} B_k\right) = o(h)$ as $h \to 0$. This proves that condition (b) of Definition II is satisfied. That condition (c) is satisfied follows from the fact that

$$P\left(\bigcup_{k=2}^{\infty} A_k\right) \leq P\left(\bigcup_{k=2}^{\infty} B_k\right) = o(h) \quad \text{as} \quad h \to 0.$$

We have thus shown that $\{Y(t), t \geq 0\}$ is a Poisson process with intensity λp.

Next we provide a proof based on Definition I. This can be done in two ways, either by conditioning or by using transforms. We shall do both, and we begin with the former (independence of the increments follows as before).

Let $0 \leq s < t$, and for $n = 0, 1, 2, \ldots$ and $k = 0, 1, 2, \ldots, n$, let $D_{n,k}$ be the event that "n particles are emitted during $(s,t]$ and k of them are registered." Then

$$P(D_{n,k}) = P(X(t) - X(s) = n)$$
$$\times P(k \text{ registrations during } (s,t] \mid X(t) - X(s) = n)$$
$$= e^{-\lambda(t-s)} \frac{(\lambda(t-s))^n}{n!} \cdot \binom{n}{k} p^k q^{n-k}. \tag{6.1}$$

Furthermore, for $k = 0, 1, 2, \ldots$, we have

$$P(Y(t) - Y(s) = k) = P\left(\bigcup_{n=k}^{\infty} D_{n,k}\right) = \sum_{n=k}^{\infty} P(D_{n,k})$$

$$= \sum_{n=k}^{\infty} e^{-\lambda(t-s)} \frac{(\lambda(t-s))^n}{n!} \cdot \binom{n}{k} p^k q^{n-k}$$

$$= e^{-\lambda(t-s)} \frac{(\lambda(t-s))^k p^k}{k!} \sum_{n=k}^{\infty} \frac{(\lambda(t-s))^{n-k} q^{n-k}}{(n-k)!}$$

$$= e^{-\lambda p(t-s)} \frac{(\lambda p(t-s))^k}{k!},$$

which shows that $Y(t) - Y(s) \in \text{Po}(\lambda p(t-s))$.

Alternatively, we may use indicator variables. Namely, let

$$Z_k = \begin{cases} 1, & \text{if particle } k \text{ is registered,} \\ 0, & \text{otherwise.} \end{cases} \tag{6.2}$$

Then $\{Z_k, \, k \geq 1\}$ are independent, $\text{Be}(p)$-distributed random variables and

$$Y(t) = Z_1 + Z_2 + \cdots + Z_{X(t)}. \tag{6.3}$$

Thus,

$$P(Y(t) = k) = \sum_{n=k}^{\infty} P(Y(t) = k \mid X(t) = n) \cdot P(X(t) = n)$$

$$= \sum_{n=k}^{\infty} \binom{n}{k} p^k q^{n-k} e^{-\lambda t} \frac{(\lambda t)^n}{n!},$$

which leads to the same computations as before (except that here we have assumed that $s = 0$ for simplicity).

The last approach, generating functions and Theorem 3.6.1 together yield

$$g_{Y(t)}(u) = g_{X(t)}\big(g_Z(u)\big) = e^{\lambda t(q+pu-1)} = e^{\lambda pt(u-1)} = g_{\text{Po}(\lambda pt)}(u), \tag{6.4}$$

and the desired conclusion follows.

Just as for Theorem 5.1, it is harder to give a complete proof based on Definition III. It is, however, fairly easy to prove that the time T_y until the first registration is $\text{Exp}(1/\lambda p)$-distributed:

$$P(T_y > t) = P\Big(\bigcup_{k=0}^{\infty} \{\{X(t) = k\} \cap \{\text{no registration}\}\}\Big)$$

$$= \sum_{k=0}^{\infty} e^{-\lambda t} \frac{(\lambda t)^k}{k!} q^k = e^{-\lambda t} \sum_{k=0}^{\infty} \frac{(\lambda q t)^k}{k!} = e^{-\lambda pt}.$$

Remark 6.1. We have (twice) used the fact that, for $k = 1, 2, \ldots, n$,

$$P(k \text{ registrations during } (s,t] \mid X(t) - X(s) = n) = \binom{n}{k} p^k q^{n-k} \tag{6.5}$$

without proof. We ask the reader to check this formula. We also refer to Problem 9.8, where further properties are given.   □

Exercise 6.1. Cars pass by a gas station. The periods between arrivals are independent, $\text{Exp}(1/\lambda)$-distributed random variables. The probability that a passing car needs gas is p, and the needs are independent.

(a) What kind of process can be used to describe the phenomenon "cars coming to the station"?

Now suppose that a car that stops at the station needs gas with probability p_{g} and oil with probability p_{o} and that these needs are independent. What kind of process can be used to describe the phenomena:

(b) "cars come to the station for gas"?

(c) "cars come to the station for oil"?
(d) "cars come to the station for gas and oil"?
(e) "cars come to the station for gas or oil"?

Exercise 6.2. Suppose that the particle counter in Example 5.1 is unreliable in the sense that particles are registered with probabilities p_α, p_β, p_γ, and p_δ, respectively, and that all registrations occur (or not) independently of everything else.

(a) Show that the registration process is a Poisson process and determine the intensity.
(b) What is the expected duration until a particle is emitted?
(c) What is the expected duration until a particle is registered?
(d) What is the probability that the first registered particle is a γ-particle?
(e) What is the expected duration until a γ-particle is emitted?
(f) What is the expected duration until a γ-particle is registered?   □

We conclude this section with a classical problem called the *coupon collector's problem*. Each element in a finite population has a "bonus" attached to it. Elements are drawn from the population by simple random sampling with replacement and with equal probabilities. Each time a new element is obtained, one receives the corresponding bonus. One object of interest is the bonus sum after all elements have been obtained. Another quantity of interest is the *random* sample size, that is, the total number of draws required in order for all elements to have appeared.

Here we shall focus on the latter quantity, but we first give a concrete example. There exist n different pictures (of movie stars, baseball players, statisticians, etc.). Each time one buys a bar of soap, one of the pictures (which is hidden inside the package) is obtained. The problem is to determine how many bars of soap one needs to buy in order to obtain a complete collection of pictures.

We now use a Poisson process technique to determine the expected sample size or the expected number of soaps one has to buy. To this end, we assume that the bars of soap are bought according to a Poisson process with intensity 1. Each buy corresponds to an event in this process. Furthermore, we introduce n independent Poisson processes (one for each picture) such that if a soap with picture k is bought, we obtain an event in the kth process. When (at least) one event has occurred in all of these n processes, one has a complete collection of pictures.

Now, let T be the time that has elapsed at that moment, let N be the number of soaps one has bought at time T, and let $Y_1, Y_2, \ldots$ be the periods between the buys. Then

$$T = Y_1 + Y_2 + \cdots + Y_N. \tag{6.6}$$

Next we consider the process "the kth picture is obtained," where $1 \leq k \leq n$. This process may be viewed as having been obtained by observing the original Poisson process with intensity 1, "registering" only the observation corresponding to the kth picture. Therefore, these n processes are Poisson processes with intensity $1/n$, which, furthermore, are independent of each other.

The next step is to observe that

$$T \overset{d}{=} \max\{X_1, X_2, \ldots, X_n\}, \tag{6.7}$$

where $X_1, X_2, \ldots, X_n$ are independent, $\mathrm{Exp}(n)$-distributed random variables, from which it follows that

$$ET = n\left(1 + \frac{1}{2} + \frac{1}{3} + \cdots + \frac{1}{n}\right). \tag{6.8}$$

Here we have used the scaling property of the exponential distribution (if $Z \in \mathrm{Exp}(1)$ and $V \in \mathrm{Exp}(a)$, then $aZ \overset{d}{=} V$) and Problem 4.4.21 or formula (5.15) of Subsection 5.4.

Finally, since N and $Y_1, Y_2, \ldots$ are independent, it follows from the results of Section 3.6 that

$$ET = EN \cdot EY_1 = EN \cdot 1 = n\left(1 + \frac{1}{2} + \frac{1}{3} + \cdots + \frac{1}{n}\right). \tag{6.9}$$

If $n = 100$, for example, then $EN = ET \approx 518.74$. Note also that the expected number of soaps one has to buy in order to obtain the *last* picture equals n, that is, 100 in the numerical example.

Remark 6.2. For large values of n, $EN = ET \approx n(\log n + \gamma)$, where $\gamma = 0.57721566\ldots$ is Euler's constant. For $n = 100$, this approximation yields $EN \approx 518.24$. □

Let us point out that this is not the simplest solution to this problem. A simpler one is obtained by considering the number of soaps bought in order to obtain "the next new picture." This decomposes the total number N of soaps into a sum of Fs-distributed random variables (with parameters $1, 1/2, 1/3, \ldots, 1/n$, respectively) from which the conclusion follows; the reader is asked to fill in the details. The Poisson process approach, however, is very convenient for generalizations.

Exercise 6.3. Jesper has a CD-player with a "random" selections function. This means that the different selections on a CD are played in a random order. Suppose that the CD he got for his birthday contains 5 Mozart piano sonatas consisting of 3 movements each, and suppose that all movements are exactly 4 minutes long. Find the expected time until he has listened to everything using the random function.

Exercise 6.4. Margaret and Elisabeth both collect baseball pictures. Each time their father buys a candy bar he gives them the picture. Find the expected number of bars he has to buy in order for both of them to have a complete picture collection (that is, they share all pictures and we seek the number of candy bars needed for two complete sets of pictures). □

7 The Compound Poisson Process

Definition 7.1. *Let $\{Y_k,\ k \geq 1\}$ be i.i.d. random variables, let $\{N(t),\ t \geq 0\}$ be a Poisson process with intensity λ, which is independent of $\{Y_k,\ k \geq 1\}$, and set*

$$X(t) = Y_1 + Y_2 + \cdots + Y_{N(t)}.$$

Then $\{X(t),\ t \geq 0\}$ is a compound Poisson process. $\qquad\square$

If the Y-variables are $\text{Be}(p)$-distributed, then $\{X(t),\ t \geq 0\}$ is a Poisson process. For the general case we know from Theorem 3.6.4 that

$$\varphi_{X(t)}(u) = g_{N(t)}\big(\varphi_Y(u)\big) = e^{\lambda t(\varphi_Y(u)-1)}. \tag{7.1}$$

The probability function of $X(t)$ can be expressed as follows. Let $S_n = \sum_{k=1}^{n} Y_k,\ n \geq 1$. Then

$$P(X(t) = k) = \sum_{n=0}^{\infty} P(S_n = k) \cdot e^{-\lambda t}\frac{(\lambda t)^n}{n!}, \quad \text{for } k = 0, 1, 2, \ldots. \tag{7.2}$$

Example 6.1 (thinning) was of this kind (cf. (6.3) and (6.4)).

Exercise 7.1. Verify formula (7.2). $\qquad\square$

Example 7.1 (The randomized random walk). Consider the following generalization of the simple, symmetric random walk. The jumps, $\{Y_k,\ k \geq 1\}$, are still independent and equal ± 1 with probability $1/2$ each, but the times of the jumps are generated by a Poisson process, that is, the times between the jumps are not 1, but rather are independent, equidistributed (nonnegative) random variables. In this model $S_n = \sum_{k=1}^{n} Y_k$ is the position of the random walk after n *steps* and $X(t)$ is the position at *time t*.

Example 7.2. (Risk theory). An insurance company is subject to claims from its policyholders. Suppose that claims are made at time points generated by a Poisson process and that the sizes of the claims are i.i.d. random variables. If $\{Y_k,\ k \geq 1\}$ are the amounts claimed and $N(t)$ is the number of claims made up to time t, then $X(t)$ equals the total amount claimed at time t. If, in addition, the initial capital of the company is u and the gross premium rate is β, then the quantity

$$u + \beta t - X(t) \tag{7.3}$$

equals the capital of the company at time t. In particular, if this quantity is negative, then financial ruin has occurred.

In order to avoid negative values in examples like the one above, we may use the quantity $\max\{0, u + \beta t - X(t)\}$ instead of (7.3). $\qquad\square$

Example 7.3. (Storage theory). In this model the stock—for example, the water in a dam or the stock of merchandise in a store—is refilled at a constant rate, β. The starting level is $X(0) = u$. The stock decreases according to a Poisson process, and the sizes of the decreases are i.i.d. random variables. The quantity (7.3) then describes the content of water in the dam or the available stock in the store, respectively. A negative quantity implies that the dam is empty or that the store has zero stock. □

In general, $N(t)$ denotes the *number of occurrences* in $(0, t]$ for $t > 0$, and the sequence $\{Y_k, k \geq 1\}$ corresponds to the *values* (prices, rewards) associated with the occurrences. We therefore call $X(t)$ the *value of the Poisson process* at time t, for $t > 0$.

Remark 7.1. The compound Poisson process is also an important process in its own right, for example, in the characterization of classes of limit distributions. □

8 Some Further Generalizations and Remarks

There are many generalizations and extensions of the Poisson process. In this section we briefly describe some of them.

8.1 The Poisson Process at Random Time Points

As we have noted, the Poisson process $\{X(t),\, t \geq 0\}$ has the property that the increments are Poisson-distributed, in particular, $X(t) \in \mathrm{Po}(\lambda t)$, for $t > 0$. We first remark that this need not be true at random time points.

Example 8.1. Let $T = \min\{t : X(t) = k\}$. Then $P(X(T) = k) = 1$, that is, $X(T)$ is degenerate.

Example 8.2. A less trivial example is obtained by letting $T \in \mathrm{Exp}(\theta)$, where T is independent of $\{X(t),\, t \geq 0\}$. Then

$$P(X(T) = n) = \int_0^\infty P(X(T) = n \mid T = t) \cdot f_T(t)\, dt$$
$$= \frac{1}{1 + \lambda\theta}\left(\frac{\lambda\theta}{1 + \lambda\theta}\right)^n \quad \text{for} \quad n = 0, 1, 2, \ldots, \tag{8.1}$$

that is, $X(T) \in \mathrm{Ge}(1/(1 + \lambda\theta))$ (cf. also Section 2.3 and Subsection 5.3).

Alternatively, by proceeding as in Section 3.5 or Subsection 5.3, we obtain, for $s < 1 + 1/\lambda\theta$,

$$g_{X(T)}(s) = E\big(E(s^{X(T)} \mid T)\big) = \psi_T\big(\lambda(s - 1)\big) = \frac{\frac{1}{1+\lambda\theta}}{1 - \frac{\lambda\theta}{1+\lambda\theta}s},$$

which is the generating function of the $\mathrm{Ge}(1/(1 + \lambda\theta))$-distribution. □

The fact that the same computations were performed in Subsection 5.3 raises the question if there is any connection between Examples 5.2 and 8.2. The answer is, of course, yes, because in Example 5.2 we were actually interested in determining the distribution of the number of α-particles at time $T_\beta \in \mathrm{Exp}(1/\mu)$, which is precisely what Example 8.2 is all about.

8.2 Poisson Processes with Random Intensities

Computations similar to those of the previous subsection also occur when the intensity is random.

Example 8.3. Suppose we are given a Poisson process with an exponential intensity, that is, let $\{X(t),\, t \geq 0\}$ be a Poisson process with intensity $\Lambda \in \mathrm{Exp}(\theta)$. Determine the distribution of $X(t),\, t \geq 0$.

What is meant here is that *conditional on* $\Lambda = \lambda$, $\{X(t), t \geq 0\}$ is a Poisson process with intensity λ (recall Sections 2.3 and 3.5). For $s, t \geq 0$, this means that

$$X(t+s) - X(s) \mid \Lambda = \lambda \in \mathrm{Po}(\lambda t) \quad \text{with} \quad \Lambda \in \mathrm{Exp}(\theta). \qquad (8.2)$$

By arguing as in the cited sections, it follows (please check!), for $s, t \geq 0$ and $n = 0, 1, 2, \ldots$, that

$$P(X(t+s) - X(s) = n) = \frac{1}{1 + \theta t}\left(\frac{\theta t}{1 + \theta t}\right)^n.$$

Using generating functions as before yields

$$g_{X(t+s)-X(t)}(u) = \frac{\frac{1}{1+\theta t}}{1 - \frac{\theta t}{1+\theta t}u} \quad \text{for} \quad u < 1 + \frac{1}{\theta t}.$$

In either case, the conclusion is that $X(t+s) - X(s) \in \mathrm{Ge}(1/(1 + \theta t))$. $\qquad \square$

Remark 8.1. The process $\{X(t),\, t \geq 0\}$ thus is not a Poisson process in general. The expression "Poisson process with random intensity" is to be interpreted as in (8.2) and the sentence preceding that formula. $\qquad \square$

Now that we know that the process of Example 8.3 is not a Poisson process, it might be of interest to see at what point(s) the conditions of Definition II break down.

Let us begin by computing the probabilities of one and at least two events, respectively, during $(t, t+h]$. We have, as $h \to 0$,

$$P(X(t+h) - X(t) = 1) = \frac{1}{1 + \theta h} \cdot \frac{\theta h}{1 + \theta h} = \theta h + o(h) \qquad (8.3)$$

and

$$P(X(t+h) - X(t) \geq 2) = \sum_{n=2}^{\infty} \frac{1}{1+\theta h}\left(\frac{\theta h}{1+\theta h}\right)^n = \left(\frac{\theta h}{1+\theta h}\right)^2 = o(h), \quad (8.4)$$

that is, conditions (b) and (c) in Definition II are satisfied. The only remaining thing to check, therefore, is the independence of the increments (a check that must necessarily end in a negative conclusion).

Let $0 \leq s_1 < s_1 + t_1 \leq s_2 < s_2 + t_2$. For $m, n = 0, 1, 2, \ldots$, we have

$$P(X(s_1 + t_1) - X(s_1) = m, X(s_2 + t_2) - X(s_2) = n)$$

$$= \int_0^{\infty} P(X(s_1 + t_1) - X(s_1) = m, X(s_2 + t_2) - X(s_2) = n \mid \Lambda = \lambda)$$

$$\times f_\Lambda(\lambda)\, d\lambda$$

$$= \int_0^{\infty} e^{-\lambda t_1} \frac{(\lambda t_1)^m}{m!} e^{-\lambda t_2} \frac{(\lambda t_2)^n}{n!} \frac{1}{\theta} e^{-\frac{\lambda}{\theta}} d\lambda$$

$$= \binom{m+n}{m} \frac{t_1^m t_2^n}{(t_1 + t_2 + \frac{1}{\theta})^{m+n+1}} \cdot \frac{1}{\theta}. \qquad (8.5)$$

By dividing with the marginal distribution of the first increment, we obtain the conditional distribution of $X(t_2 + s_2) - X(s_2)$ given that $X(t_1 + s_1) - X(s_1) = m$. Namely, for $n, m = 0, 1, 2, \ldots$, we have

$$P(X(s_2 + t_2) - X(s_2) = n \mid X(s_1 + t_1) - X(s_1) = m)$$

$$= \binom{n+m}{m}\left(\frac{t_2}{t_1 + t_2 + \frac{1}{\theta}}\right)^n \left(\frac{t_1 + \frac{1}{\theta}}{t_1 + t_2 + \frac{1}{\theta}}\right)^{m+1}. \qquad (8.6)$$

This shows that the increments are not independent. Moreover, since

$$\binom{n+m}{m} = \binom{n + (m+1) - 1}{(m+1) - 1},$$

we may identify the conditional distribution in (8.6) as a negative binomial distribution.

One explanation of the fact that the increments are not independent is that if the number of occurrences in the first time interval is known, then we have obtained some information on the intensity, which in turn provides information on the number of occurrences in later time intervals. Note, however, that *conditional on* $\Lambda = \lambda$, the increments are indeed independent; this was, in fact, tacitly exploited in the derivation of (8.5).

The following example illustrates how a Poisson process with a random intensity may occur (cf. also Example 2.3.1):

Example 8.4. Suppose that radioactive particles are emitted from a source according to a Poisson process such that the intensity of the process depends on the kind of particle the source is emitting. That is, given the kind of particles, they are emitted according to a Poisson process.

For example, suppose we have m boxes of radioactive particles, which (due to a past error) have not been labeled; that is, we do not know which kind of particles the source emits.   □

Remark 8.2. Note the difference between this situation and that of Example 5.1.   □

8.3 The Nonhomogeneous Poisson Process

A *nonhomogeneous Poisson process* is a Poisson process with a *time-dependent intensity*. With this process one can model time-dependent phenomena, for example, phenomena that depend on the day of the week or on the season. In the example "telephone calls arriving at a switchboard," it is possible to incorporate into the model the assumption that the intensity varies during the day.

The strict definition of the nonhomogeneous Poisson process is Definition II with condition (b) replaced by

(b′) P(exactly one occurrence during $(t, t+h]) = \lambda(t)h + o(h)$ as $h \to 0$.

The case $\lambda(t) \equiv \lambda$ corresponds, of course, to the ordinary Poisson process.

In Example 7.2, risk theory, one can imagine seasonal variations; for car insurances one can, for example, imagine different intensities for summers and winters. In queueing theory one might include rush hours in the model, that is, the intensity may depend on the time of the day.

By modifying the computations that led to (1.5) and (1.6) (as always, check!), we obtain

$$X(t_2) - X(t_1) \in \text{Po}\Big(\int_{t_1}^{t_2} \lambda(u)\,du \Big) \quad \text{for} \quad 0 \le t_1 < t_2. \tag{8.7}$$

If, for example, $\lambda(t) = t^2$, for $t \ge 0$, then $X(2) - X(1)$ is $\text{Po}(\int_1^2 u^2 du)$-distributed, that is, $\text{Po}(7/3)$-distributed.

8.4 The Birth Process

The *(pure) birth process* has a *state-dependent intensity*. The definition is Definition II with (b) replaced by

(b″) $P(X(t+h) = k+1 \mid X(t) = k) = \lambda_k h + o(h)$ as $h \to 0$
    for $k = 0, 1, 2, \ldots$.

As the name suggests, a jump from k to $k+1$ is called a *birth*.

Example 8.5. A typical birthrate is $\lambda_k = k \cdot \lambda$, $k \ge 1$. This corresponds to the situation where the event $\{X(t) = k\}$ can be interpreted as k "individuals" exist and each individual gives birth according to a Poisson process with intensity λ; $\lambda_k = k\lambda$ is the cumulative intensity when $X(t) = k$. Note also the connection with the superpositioned Poisson process described in Section 5.

Example 8.6. Imagine a waiting line in a store to which customers arrive according to a Poisson process with intensity λ. However, if there are k persons in the line, they join the line with probability $1/(k+1)$ and leave (for another store) with probability $k/(k+1)$. Then $\lambda_k = k/(k+1)$. $\qquad\square$

One may, analogously, introduce deaths corresponding to downward transitions (from k to $k-1$ for $k \geq 1$). For the formal definition we need the obvious assumption of type (b) for the deaths. A process thus obtained is called a *(pure) death process*. If both births and deaths may occur, we have a *birth and death process*.

In the telephone switchboard example, births might correspond to arriving calls and deaths to ending conversations. If one studies the number of customers in a store, births might correspond to arrivals and deaths to departures.

Remark 8.3. For these processes the initial condition $X(0) = 0$ is not always the natural one. Consider, for example, the following situation. One individual in a population of size N has been infected with some dangerous virus. One wishes to study how the infection spreads. If $\{X(t) = k\}$ denotes the event that there are k infected individuals, the obvious initial condition is $X(0) = 1$. If instead $\{X(t) = k\}$ denotes the event that there are k noninfected individuals, the natural initial condition is $\{X(0) = N - 1\}$. $\qquad\square$

8.5 The Doubly Stochastic Poisson Process

This process is defined as a nonhomogeneous Poisson process with an intensity function that is a stochastic process. It is also called a *Cox process*. In particular, the intensity (process) may itself be a Poisson process. In the time homogeneous case, the process reduces to that of Subsection 8.2.

An example is the pure birth process. More precisely, let $\{X(t), t \geq 0\}$ be the pure birth process with intensity $\lambda_k = k\lambda$ of Example 8.5. A reinterpretation of the discussion there shows that the intensity actually is the stochastic process $\Lambda(t) = \lambda X(t)$.

8.6 The Renewal Process

By modifying Definition III of a Poisson process in such a way that the durations $\{\tau_k, k \geq 1\}$ are just i.i.d. nonnegative random variables, we obtain a *renewal process*. Conversely, a renewal process with exponentially distributed durations is a Poisson process.

More precisely, a *random walk* $\{S_n, n \geq 0\}$ is a sequence of random variables, starting at $S_0 = 0$, with i.i.d. increments $X_1, X_2, \ldots$. A renewal process is a random walk with nonnegative increments. The canonical application is a lightbulb that whenever it fails is instantly replaced by a new, identical one, which, upon failure is replaced by another one, which, in turn, ... The central object of interest is the *(renewal) counting process*, which counts the number of replacements during a given time.

Technically, we let $X_1, X_2, \ldots$ be the individual lifetimes, more generally, the *durations* of the individual objects, and set $S_n = \sum_{k=1}^{n} X_k$, $n \geq 1$. The number of replacements during the time interval $(0, t]$, that is, the counting process, then becomes

$$N(t) = \max\{n : S_n \leq t\}, \quad t \geq 0.$$

If, in particular, the lifetimes have an exponential distribution, the counting process reduces to a Poisson process.

A discrete example is the *binomial process*, in which the durations are independent, $\mathrm{Be}(p)$-distributed random variables. This means that with probability p there is a new occurrence after one time unit and with probability $1-p$ after zero time (an instant occurrence). The number of occurrences $X(t)$ up to time t follows a (translated) *negative binomial* distribution.

Formally, if $Z_0, Z_1, Z_2, \ldots$ are the number of occurrences at the respective time points $0, 1, 2, \ldots$, then $Z_0, Z_1, Z_2, \ldots$ are independent, $\mathrm{Fs}(p)$-distributed random variables and $X(t) = X(n) = Z_0 + Z_1 + Z_2 + \cdots + Z_n$, where n is the largest integer that does not exceed t ($n = [t]$). It follows that $X(n) - n \in$ $\mathrm{NBin}(n, p)$ (since the negative binomial distribution is a sum of independent, geometric distributions).

Although there are important differences between renewal counting processes and the Poisson process, such as the lack of memory property, which does *not* hold for general renewal processes, their asymptotic behavior is in many respects similar.

For example, for a Poisson $\{X(t), t \geq 0\}$ one has

$$E\,X(t) = \lambda t \quad \text{and} \quad \frac{X(t)}{t} \xrightarrow{p} \lambda \quad \text{as} \quad t \to \infty, \tag{8.8}$$

and one can show that if $E\,X_1 = \mu < \infty$, then, for a renewal counting process $\{N(t), t \geq 0\}$, one has

$$\frac{E\,N(t)}{t} \to \frac{1}{\mu} \quad \text{and} \quad \frac{N(t)}{t} \xrightarrow{p} \frac{1}{\mu} \quad \text{as} \quad t \to \infty, \tag{8.9}$$

where, in order to compare the results, we observe that the intensity λ of the Poisson process corresponds to $1/\mu$ in the renewal case.

Remark 8.4. The first result in (8.9) is called the *elementary renewal theorem*.

Remark 8.5. One can, in fact prove that convergence in probability may be sharpened to almost sure convergence in both cases. □

A more general model, one that allows for repair times, is the *alternating renewal process*. In this model $X_1, X_2, \ldots$, the lifetimes, can be considered as the time periods during which some device functions, and an additional sequence $Y_1, Y_2, \ldots$ may be interpreted as the successive, intertwined, repair times. In, for example, queueing theory, lifetimes might correspond to busy times and repair times to idle times.

In this model one may, for example, derive expressions for the relative amount of time the device functions or the relative amount of time the queueing system is busy.

8.7 The Life Length Process

In connection with the nonhomogeneous Poisson process, it is natural to mention the *life length process*. This process has two states, 0 and 1, corresponding to life and death, respectively. The connection with the Poisson process is that we may interpret the life length process as a truncated nonhomogeneous Poisson process, in that the states 1, 2, 3, ... are lumped together into state 1.

Definition 8.1. *A life length process $\{X(t), t \geq 0\}$ is a stochastic process with states 0 and 1, such that $X(0) = 0$ and*

$$P(X(t+h) = 1 \mid X(t) = 0) = \lambda(t)h + o(h) \quad as \quad h \to 0.$$

The function λ is called the intensity function. □

To see the connection with the nonhomogeneous Poisson process, let $\{X(t), t \geq 0\}$ be such a process and let $\{X^*(t), t \geq 0\}$ be defined as follows:

$$X^*(t) = \begin{cases} 0, & \text{when} \quad X(t) = 0, \\ 1, & \text{when} \quad X(t) \geq 1. \end{cases}$$

The process $\{X^*(t), t \geq 0\}$ thus defined is a life length process. The following figure illustrates the connection.

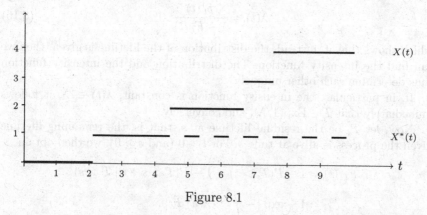

Figure 8.1

We now derive some properties of life length processes. With the notations $P_0(t) = P(X(t) = 0)$ and $P_1(t) = P(X(t) = 1)$, we have

$$P_0(0) = 1 \quad \text{and} \quad P_0(t) + P_1(t) = 1. \tag{8.10}$$

By arguing as in the proof of Theorem 1.1 (check the details!), we obtain

$$P_0(t + h) = P_0(t)(1 - \lambda(t)h) + o(h) \quad \text{as} \quad h \to 0,$$

from which it follows that

$$P_0'(t) = -\lambda(t)P_0(t), \tag{8.11}$$

and hence that

$$P_0(t) = \exp\left\{-\int_0^t \lambda(s)\,ds\right\} \quad \text{and} \quad P_1(t) = 1 - \exp\left\{-\int_0^t \lambda(s)\,ds\right\}. \tag{8.12}$$

Now, let T be the lifetime (life length) of the process. Since

$$\{T > t\} = \{X(t) = 0\} \tag{8.13}$$

(cf. (1.8)), the distribution function of T is

$$F_T(t) = 1 - P(T > t) = P_1(t) = 1 - \exp\left\{-\int_0^t \lambda(s)\,ds\right\}. \tag{8.14}$$

Differentiation yields the density:

$$f_T(t) = \lambda(t)\exp\left\{-\int_0^t \lambda(s)\,ds\right\}. \tag{8.15}$$

The computations above show how the distribution of the lifetime can be obtained if the intensity function is given. On the other hand, (8.14) and (8.15) together yield

$$\lambda(t) = \frac{f_T(t)}{1 - F_T(t)}, \tag{8.16}$$

which shows that if, instead, the distribution of the lifetime is given, then we can find the intensity function. The distribution and the intensity function thus determine each other uniquely.

If, in particular, the intensity function is constant, $\lambda(t) \equiv \lambda$, it follows immediately that $T \in \text{Exp}(1/\lambda)$, and conversely.

Now, let T_s be the residual lifetime at s, that is, the remaining lifetime given the process is alive at time s. For $t > 0$ (and $s \geq 0$), we then obtain

$$F_{T_s}(t) = 1 - P(T_s > t) = 1 - P(T > s + t \mid T > s)$$
$$= 1 - \exp\left\{-\int_s^{s+t} \lambda(u)\,du\right\}.$$

Remark 8.6. Note that the life length process does not have independent increments. Why is this "obvious"?

Remark 8.7. The function $R_T(t) = 1 - F_T(t)$ provides the probability that the life length exceeds t; it is called the *survival function*. Using this function, we may rewrite (8.16) as

$$\lambda(t) = \frac{f_T(t)}{R_T(t)}. \qquad \Box$$

If, for example, the intensity function is constant, λ, then $R_T(t) = e^{-\lambda t}$, for $t > 0$. For $\lambda(t) = t^2$, we obtain

$$R_T(t) = \exp\left\{ - \int_0^t s^2 \, ds \right\} = \exp\{-t^3/3\}, \quad \text{for} \quad t > 0.$$

We conclude with a heuristic explanation of the nature of (8.14). The left-hand side in the definition equals the probability that the process dies during $(t, t+h]$ given that it is still alive at time t. According to the right-hand side, this probability equals $\lambda(t)h + o(h) \approx \lambda(t)h$ for h small.

Another way to describe this probability is

$$P(t < T \le t + h \mid T > t) = \frac{P(t < T \le t + h)}{P(T > t)} = \frac{F_T(t+h) - F_T(t)}{1 - F_T(t)}.$$

By the mean value theorem, it follows, for $0 \le \theta \le 1$ and f "nice", that

$$P(t < T \le t + h \mid T > t) = \frac{h \cdot f_T(t + \theta h)}{1 - F_T(t)} \approx \frac{h f_T(t)}{1 - F_T(t)} = \frac{h f_T(t)}{R_T(t)}.$$

A comparison with the definition finally "shows" that

$$\lambda(t)h \approx \frac{h f_T(t)}{1 - F_T(t)}, \quad \text{for} \quad h \quad \text{small}, \tag{8.17}$$

which "justifies" (8.16).

9 Problems

1. Let $\{X_1(t), t \ge 0\}$ and $\{X_2(t), t \ge 0\}$ be independent Poisson processes with common intensity λ. Suppose that $X_1(3) = 9$ and $X_2(3) = 5$. What is the probability that the X_1-process reaches level 10 before the X_2-process does?
2. Solve the same problem under the assumption that the processes have intensities λ_1 and λ_2, respectively.
3. Consider two independent Poisson processes $\{X_1(t), t \ge 0\}$ and $\{X_2(t), t \ge 0\}$ with common intensity. What is the probability that the two-dimensional process $\{(X_1(t), X_2(t)), t \ge 0\}$ passes through the point
   (a) $(1, 1)$?
   (b) $(1, 2)$?
   (c) (i, j)?

4. Susan likes pancakes very much, but Tom does not. The time to eat a pancake can be assumed to be exponential. Their mother has studied them over the years and estimates the parameters ($= 1$/expected time) to be 7 and 2, respectively. Compute the probability that Susan finishes 10 pancakes before Tom has finished his first one.

5. Suppose that customers arrive at a counter or server according to a Poisson process with intensity λ and that the service times are independent, $\text{Exp}(1/\mu)$-distributed random variables. Suppose also that a customer arrives at time zero and finds the counter free.
   (a) Determine the distribution of the number of customers that arrive while the first customer is served.
   (b) Compute the probability that the server will be busy forever.

   Remark. We may interpret the situation as follows: We are given a branching process where the lifetimes of the individuals are independent, $\text{Exp}(1/\mu)$-distributed random variables and the reproduction is such that individuals give birth at a constant rate λ throughout their lives under the usual independence assumptions. Furthermore, the initial population consists of one individual (i.e., $X(0) = 1$). In (a) we wish to find the distribution of the number of children obtained by an individual, and in (b) we ask for the probability of nonextinction (i.e., $1 - \eta$).

6. Fredrik and Ulrich both received soap bubble machines for Christmas. The machines emit bubbles according to independent Poisson processes with intensities 3 and 2 (bubbles per minute), respectively. Suppose they turn them on at the same time.
   (a) Find the probability that Fredrik's machine produces the first bubble.
   (b) Find the probability that Ulrich's machine produces 3 bubbles before Fredrik's first bubble.

7. At the center of espionage in Kznatropsk one is thinking of a new method for sending Morse telegrams. Instead of using the traditional method, that is, to send letters in groups of 5 according to a Poisson process with intensity 1, one might send them one by one according to a Poisson process with intensity 5. Before deciding which method to use one would like to know the following: What is the probability that it takes less time to send one group of 5 letters the traditional way than to send 5 letters the new way (the actual transmission time can be neglected).

8. Consider a Poisson process with intensity λ. We start observing at time $t = 0$. Let T be the time that has elapsed at the first occurrence. Continue to observe the process T further units of time. Let $N(T)$ be the number of occurrences during the latter period (i.e., during $(T, 2T]$). Determine the distribution of $N(T)$.

9. A particle source A emits one particle at a time, according to a Poisson process with an intensity of two particles a minute. Another particle source B emits *two* particles at a time, according to a Poisson process with an intensity of one pair of particles a minute. The sources are independent of each other. We begin to observe the sources at time zero. Compute the

probability that source A has emitted two particles before source B has done so.

10. A specific component in a cryptometer has an $\text{Exp}(\mu)$-distributed lifetime, $\mu > 0$. If replacement is made as soon as a component fails, and if $X(t) = \#$ failures during $(0,t] = \#$ replacements during $(0,t]$, then $\{X(t),\, t \geq 0\}$ is, of course, a Poisson process. Let $\{V_n,\, n \geq 1\}$ be these usual interreplacement times, and suppose, instead, that the nth component is replaced;

   (a) After time $\min\{V_n, a\}$, that is, as soon as the component fails or reaches age a, whichever comes first. Show that the replacement process is not a Poisson process.

   (b) After time $\min\{V_n, W_n\}$, where $\{W_n,\, n \geq 1\}$ is a sequence of independent, $\text{Exp}(\theta)$-distributed random variables, $\theta > 0$, which is independent of $\{V_n,\, n \geq 1\}$. Show that the replacement process is a Poisson process and determine the intensity.

11. Karin arrives at the post office, which opens at 9:00 a.m., at 9:05 a.m. She finds two cashiers at work, both serving one customer each. The customers started being served at 9:00 and 9:01, respectively. The service times are independent and $\text{Exp}(8)$-distributed. Let T_k be the time from 9:05 until service has been completed for k of the two customers, $k = 1, 2$. Find $E\,T_k$ for $k = 1$ and 2.

12. Måns waits for the bus. The waiting time, T, until a bus comes is $U(0, a)$-distributed. While he waits he tries to get a ride from cars that pass by according to a Poisson process with intensity λ. The probability of a passing car picking him up is p. Determine the probability that Måns is picked up by some car before the bus arrives.

   Remark. All necessary independence assumptions are permitted.

13. Consider a sender that transmits signals according to a Poisson process with intensity λ. The signals are received by a receiver, however, in such a way that every signal is registered with probability p, $0 < p < 1$, and "missed" with probability $q = 1 - p$. Registrations are independent. Let $X(t)$ be the number of transmitted signals during $(0,t]$, let $Y(t)$ be the number of registered signals, and let $Z(t)$ be the number of nonregistered signals during this period, where $t \geq 0$.

   (a) Show that $Y(t)$ and $Z(t)$ are independent, and determine their distributions.

   (b) Determine the distribution of the number of signals that have been transmitted when the first signal is registered.

   (c) Determine the distribution of the number of signals that have been transmitted when the kth signal is registered.

   (d) Determine the conditional distribution of the number of registered signals given the number of transmitted signals, that is, compute $P(Y(t) = k \mid X(t) = n)$ for suitable choices of k and n.

(e) Determine the conditional distribution of the number of transmitted signals given the number of registered signals, that is, compute $P(X(t) = n \mid Y(t) = k)$ for suitable choices of k and n.

Remark. It thus follows from (a) that the number of registered signals during a given time period provides *no information* about the actual number of nonregistered signals.

14. We have seen that a thinned Poisson process, is, again, a Poisson-process. Prove the following analog for a "geometric process." More precisely:

   (a) Show that, if N and $X, X_1, X_2, \ldots$ are independent random variables, $N \in \text{Ge}(\alpha)$, and $X \in \text{Be}(\beta)$, then $Y = X_1 + X_2 + \cdots + X_N$ has a geometric distribution, and determine the parameter.

   (b) Safety check by computing mean and variance with the "usual" formulas for mean and variance of sums of a random number of independent random variables.

15. A radio amateur wishes to transmit a message. The frequency on which she sends the Morse signals is subject to random disturbances according to a Poisson process with intensity λ per second. In order to succeed with the transmission, she needs a time period of a seconds without disturbances. She stops as soon as she is done. Let T be the total time required to finish. Determine $E\,T$.

16. Peter wishes to take a picture of his girlfriend Sheila. Since they are in a rather dark room, he needs a rather long exposure time, during which Sheila must not move. The following model can be used to describe the situation. The success of a photo is called an "*A*-event." Each time Sheila moves, she causes a disturbance called a "*D*-event." *A*-events and *D*-events occur according to independent Poisson processes with intensities λ_A and λ_D, respectively. The experiment is started at time $t = 0$. Let T be the time of the first *A*-occurrence. The experiment is deemed successful if $T \geq 1$ and if no *D*-event occurs during the time interval $(0, T+2]$. What is the probability of a successful photo?

17. People arrive at an automatic teller machine (ATM) according to a Poisson process with intensity λ. The service time required at the ATM is constant, a seconds. Unfortunately, this machine does not allow for any waiting customers (i.e., no queue is allowed), which means that persons who arrive while the ATM is busy have to leave. When the a seconds of a customer have elapsed, the ATM is free to serve again, and so on. Suppose that the ATM is free at time zero, and let T_n be the time of the arrival of the nth customer. Find the distribution of T_n, and compute $E\,T_n$ and $\text{Var}\,T_n$.

   Remark. Customers arriving (and leaving) while the ATM is busy thus do not affect the service time.

18. Suppose that we are at time zero. Passengers arrive at a train station according to a Poisson process with intensity λ. Compute the expected value of the total waiting time of all passengers who have come to the station in order to catch a train that leaves at time t.

19. Suppose that electrical pulses having i.i.d. random amplitudes $A_1, A_2, \ldots$ arrive at a counter in accordance with a Poisson process with intensity λ. The amplitude of a pulse is assumed to decrease exponentially, that is, if a pulse has amplitude A upon its arrival, then its amplitude at time t is $Ae^{-\alpha t}$, where α is some positive parameter. We finally assume that the initial amplitudes of the pulses are independent of the Poisson process. Compute the expected value of the total amplitude at time t.

20. Customers arrive at a computer center at time points generated by a Poisson process with intensity λ. The number of jobs brought to the center by the customers are independent random variables whose common generating function is $g(u)$. Compute the generating function of the number of jobs brought to the computer center during the time interval $(s, t]$.

21. We have seen that if we superposition a fixed number of Poisson process we obtain a new Poisson process. This need however not be true if we superposition a *random number* of such processes. More precisely, let us superposition $N \in \mathrm{Fs}(p)$ independent Poisson processes, each with the same intensity λ, where N is independent of the Poisson processes.
    (a) Show that the new process is not a Poisson process, e.g., by computing its generating function, or by computing the mean and the variance (which are equal for the Poisson distribution).
    (b) Find (nevertheless) the probability that the first occurrence occurs in process number 1.

22. Let $X_1, X_2, \ldots$ be the i.i.d. lifetimes of some component in some large machine. The simplest replacement policy is to change a component as soon as it fails. In this case it may be necessary to call a repairman at night, which might be costly. Another policy, called *replacement based on age*, is to replace at failure or at some given age, a, say, whichever comes first, in which case the interreplacement times are

$$W_k = \min\{X_k, a\}, \quad k \geq 1.$$

Suppose that c_1 is the cost for replacements due to failure and that c_2 is the cost for replacements due to age. In addition, let Y_k be the cost attached to replacement k, $k \geq 1$, and let $N(t)$ be the number of replacements made in the time interval $(0, t]$, where $\{N(t), t \geq 0\}$ is a Poisson process, which is independent of $X_1, X_2, \ldots$. This means that

$$Z(t) = \sum_{k=1}^{N(t)} Y_k$$

is the total cost caused by the replacements in the time interval $(0, t]$ (with $Z(t) = 0$ whenever $N(t) = 0$).
    (a) Compute $E\, Y_1$ and $\mathrm{Var}\, Y_1$.
    (b) Compute $E\, Z(t)$ and $\mathrm{Var}\, Z(t)$.

23. Let $\{X(t), t \geq 0\}$ be a Poisson process with random intensity $\Lambda \in \Gamma(m, \theta)$.

(a) Determine the distribution of $X(t)$.

(b) Why is the conclusion in (a) reasonable?

Hint. Recall Example 8.3.

24. Let $\{X(t), t \geq 0\}$ be a Poisson process with intensity λ that is run N time units, where $N \in \mathrm{Fs}(p)$.

    (a) Compute $E\,X(N)$ and $\mathrm{Var}X(N)$.

    (b) Find the limit distribution of $X(N)$ as $\lambda \to 0$ and $p \to 0$ in such a way that $\lambda/p \to 1$.

25. A Poisson process is observed during n days. The intensity is, however, not constant, but varies randomly day by day, so that we may consider the intensities during the n days as n independent, $\mathrm{Exp}(1/\alpha)$-distributed random variables. Determine the distribution of the total number of occurrences during the n days.

26. Let $\{X(t), t \geq 0\}$ be a Poisson process and let $\{T_k, k \geq 1\}$ be the occurrence times. Suppose that we know that $T_3 = 1$ and that $T_1 = x$, where $0 < x < 1$. Our intuition then tells us that the conditional distribution of T_2 should be $U(x, 1)$-distributed. Prove that this is indeed the case, i.e., show that

$$T_2 \mid T_1 = x, T_3 = 1 \in U(x, 1) \quad \text{for} \quad 0 < x < 1.$$

27. Suppose that X_1, X_2, and X_3 are independent, $\mathrm{Exp}(1)$-distributed random variables, and let $X_{(1)}$, $X_{(2)}$, $X_{(3)}$ be the order variables. Determine $E(X_{(3)} \mid X_{(1)} = x)$. (Recall Example 4.2.3.)

28. Consider a queueing system where customers arrive according to a Poisson process with intensity λ customers per minute. Let $X(t)$ be the total number of customers that arrive during $(0, t]$. Compute the correlation coefficient of $X(t)$ and $X(t + s)$.

29. A particle is subject to hits at time points generated by a Poisson process with intensity λ. Every hit moves the particle a horizontal, $N(0, \sigma^2)$-distributed distance. The displacements are independent random variables, which, in addition, are independent of the Poisson process. Let S_t be the location of the particle at time t (we begin at time zero).

    (a) Compute $E\,S_t$.

    (b) Compute $\mathrm{Var}(S_t)$.

    (c) Show that

    $$\frac{S_t - E\,S_t}{\sqrt{\mathrm{Var}(S_t)}} \xrightarrow{d} N(0, a^2) \quad \text{as} \quad t \to \infty,$$

    and determine the value of the constant a.

30. Consider a Poisson process with intensity λ, and let T be the time of the *first* occurrence in the time interval $(0, t]$. If there is no occurrence during $(0, t]$, we set $T = t$. Compute $E\,T$.

31. In the previous example, let, instead, T be the time of the *last* occurrence in the time interval $(0, t]$. If there is no occurrence during $(0, t]$, we set $T = 0$. Compute $E\,T$.

32. A further (and final) definition of the Poisson process runs as follows: A nondecreasing stochastic process $\{X(t), t \geq 0\}$ is a Poisson process iff
    (a) it is nonnegative, integer-valued, and $X(0) = 0$;
    (b) it has independent, stationary increments;
    (c) it increases by jumps of unit magnitude only.
    Show that a process satisfying these conditions is a Poisson process.
    Remark. Note that if $\{X(t), t \geq 0\}$ is a Poisson process, then conditions (a)–(c) are obviously satisfied. We thus have a fourth, equivalent, definition of a Poisson process.

A

Suggestions for Further Reading

A natural first step for the reader who wishes to further penetrate the world of probability theory, stochastic processes, and statistics is to become acquainted with statistical theory at some moderate level in order to learn the fundamentals of estimation theory, hypothesis testing, analysis of variance, regression, and so forth.

In order to go deeper into probability theory one has to study the topic from a measure-theoretic point of view. A selection of books dealing with this viewpoint includes Billingsley (1986), Breiman (1968), Chow and Teicher (1988), Chung (1974), Dudley (1989), Durrett (1991), Gnedenko (1968), and Gut (2007) at various stages of modernity. Kolmogorov's treatise *Grundbegriffe der Wahrscheinlichkeitsrechnung* is, of course, the fundamental, seminal reference.

Loève (1977), Petrov (1975, 1995), Stout (1974), and Gut (2007) are mainly devoted to classical probability theory, including general versions of the law of large numbers, the central limit theorem, and the so-called law of the iterated logarithm.

For more on martingale theory, we recommend Neveu (1975), Hall and Heyde (1980), Williams (1991) and Gut (2007), Chapter 10. Doob (1953) contains the first systematic treatment of the topic.

Billingsley (1968, 1999), Grenander (1963), Parthasarathy (1967), and Pollard (1984) are devoted to convergence in more general settings. For example, if in the central limit theorem one considers the joint distribution of all partial sums $(S_1, S_2, \ldots, S_n)$, suitably normalized and linearly interpolated, one can show that in the limit this polygonal path behaves like the so-called Wiener process or Brownian motion.

Feller's two books (1968, 1971) contain a wealth of information and are pleasant reading, but they are not very suitable as textbooks.

An important application or part of probability theory is the theory of stochastic processes. Some books dealing with the general theory of stochastic processes are Gikhman and Skorokhod (1969, 1974, 1975, 1979), Grimmett and Stirzaker (1992), Resnick (1992), Skorokhod (1982), and, to some extent,

Doob (1953). The focus in Karatzas and Shreve (1991) is on Brownian motion. Protter (2005) provides an excellent introduction to the theory of stochastic integration and differential equations. So does Steele (2000), where the focus is mainly on financial mathematics, and Øksendal (2003).

Some references on applied probability theory, such as queueing theory, renewal theory, regenerative processes, Markov chains, and processes with independent increments, are Asmussen (2000, 2003), Çinlar (1975), Gut (2009), Prabhu (1965), Resnick (1992), Ross (1996), and Wolff (1989); Doob (1953) and Feller (1968, 1971) also contain material in this area. Leadbetter et al. (1983) and Resnick (2008) are mainly devoted to extremal processes.

The first real book on statistical theory is Cramér (1946), which contains a lot of information and is still most readable. Casella and Berger (1990) and, at a somewhat higher level, Rao (1973) are also adequate reading. The books by Lehmann and coauthors (1998, 2005) require a deeper prerequisite from probability theory and should be studied at a later stage. Liese and Miescke (2008) as well as Le Cam and Yang (2005) are more modern and advanced books in the area and also contain a somewhat different approach. And beyond those, there are of course, many more ....

In addition to those mentioned above the following reference list contains a selection of further literature.

References

1. Asmussen, S. (2000), *Ruin probabilities*, World Scientific Publishing, Singapore.
2. Asmussen, S. (2003), *Applied probability and queues*, 2nd ed, Springer-Verlag.
3. Barbour, A.D., Holst, L., and Janson, S. (1992). *Poisson approximation*, Oxford Science Publications, Clarendon Press, Oxford.
4. Billingsley, P. (1968), *Convergence of probability measures*, 2nd ed., Wiley, New York, 2nd ed.(1999).
5. Billingsley, P. (1986), *Probability and measure*, 2nd ed., Wiley, New York.
6. Bingham, N.H., Goldie, C.M., and Teugels, J.L. (1987), *Regular variation*. Cambridge University Press, Cambridge.
7. Breiman, L. (1968), *Probability*, Addison-Wesley, Reading, MA.
8. Casella, G., and Berger, R.L. (1990), *Statistical inference*, Wadsworth & Brooks/Cole, Belmont, CA.
9. Chow, Y.S., and Teicher, H. (1988), *Probability theory*, 2nd ed., Springer-Verlag, New York.
10. Chung, K.L. (1974), *A course in probability theory*, 2nd ed., Academic Press, Cambridge, MA.
11. Çinlar, E. (1975), *Introduction to stochastic processes*, Prentice-Hall, Englewood Cliffs, NJ.
12. Cramér, H. (1946), *Mathematical methods of statistics*, Princeton University Press, Princeton, NJ.
13. Doob, J.L. (1953), *Stochastic processes*, Wiley, New York.
14. Dudley, R. (1989), *Real analysis and probability*, Wadsworth & Brooks/Cole, Belmont, CA.

15. Durrett, R. (1991), *Probability: Theory and examples*, Wadsworth & Brooks/ Cole, Belmont, CA.
16. Embrechts, P., Klüppelberg, C., and Mikosch, T. (2008), *Modelling extremal events for insurance and finance*, Corr. 4th printing, Springer-Verlag, Berlin.
17. Feller, W. (1968), *An introduction to probability theory and its applications, Vol 1.*, 3rd ed., Wiley, New York.
18. Feller, W. (1971), *An introduction to probability theory and its applications, Vol 2.*, 2nd ed., Wiley, New York.
19. Gikhman, I.I., and Skorokhod, A.V. (1969), *Introduction to the theory of random processes*, Saunders, Philadelphia, PA.
20. Gikhman, I.I., and Skorokhod, A.V. (1974), *The theory of stochastic processes I*, Springer-Verlag, New York.
21. Gikhman, I.I., and Skorokhod, A.V. (1975), *The theory of stochastic processes II*, Springer-Verlag, New York.
22. Gikhman, I.I., and Skorokhod, A.V. (1979), *The theory of stochastic processes III*, Springer-Verlag, New York.
23. Gnedenko, B.V. (1967), *Theory of probability*, 4th ed. Chelsea, New York.
24. Gnedenko, B.V., and Kolmogorov, A.N. (1968), *Limit distributions for sums of independent random variables*, 2nd ed., Addison-Wesley, Cambridge, MA.
25. Grenander, U. (1963), *Probabilities on algebraic structures*, Wiley, New York.
26. Grimmett, G.R., and Stirzaker, D.R. (1992), *Probability theory and random processes*, 2nd ed., Oxford University Press, Oxford.
27. Gut, A. (2007), *Probability: A graduate course*, corr. 2nd printing. Springer-Verlag, New York.
28. Gut, A. (2009), *Stopped random walks*, 2nd ed., Springer-Verlag, New York.
29. Hall, P., and Heyde, C.C. (1980), *Martingale limit theory and its applications*, Academic Press, Cambridge, MA.
30. Karatzas, I., and Shreve, S.E. (1991), *Brownian motion and stochastic calculus*, Springer-Verlag, New York.
31. Kolmogorov, A.N. (1933), *Grundbegriffe der Wahrscheinlichkeitsrechnung.* English transl: *Foundations of the theory of probability*, Chelsea, New York (1956).
32. Leadbetter, M.R., Lindgren, G., and Rootzén, H. (1983), *Extremes and related properties of random sequences and processes*, Springer-Verlag, New York.
33. Le Cam, L., and Yang, G.L. (2000), *Asymptotics in statistics*, 2nd ed., Springer-Verlag, New York.
34. Lehmann, E.L., and Casella, G. (1998), *Theory of point estimation*, 2nd ed., Springer-Verlag, New York.
35. Lehmann, E.L., and Romano, J.P. (2005), *Testing statistical hypothesis*, 3rd ed., Springer-Verlag, New York.
36. Lévy, P. (1925), *Calcul des probabilités*, Gauthier-Villars, Paris.
37. Lévy, P. (1954), *Théorie de l'addition des variables aléatoires*, 2nd ed., Gauthier-Villars, Paris.
38. Liese, F., and Miescke, K.-J. (2008), *Statistical decision theory*, 3rd ed., Springer-Verlag, New York.
39. Loève, M. (1977), *Probability theory*, 4th ed., Springer-Verlag, New York.
40. Meyn, S.P., and Tweedie, R.L. (1993), *Markov chains and stochastic stability*, Springer-Verlag, London.
41. Neveu, J. (1975), *Discrete-parameter martingales*, North-Holland, Amsterdam.
42. Øksendal, B. (2003), *Stochastic differential equations*, 6th ed., Springer-Verlag, Berlin.

43. Parthasarathy, K.R. (1967), *Probability measures on metric spaces*, Academic Press, Cambridge, MA.

44. Petrov, V.V. (1975), *Sums of independent random variables*, Springer-Verlag, New York.

45. Petrov, V.V. (1995), *Limit theorems of probability theory*, Oxford University Press, Oxford.

46. Pollard, D. (1984), *Convergence of stochastic processes*, Springer-Verlag, New York.

47. Prabhu, N.U. (1965), *Stochastic processes*, Macmillan, New York.

48. Protter, P. (2005), *Stochastic integration and differential equations*, 2nd ed., Version 2.1, Springer-Verlag, Heidelberg.

49. Rao, C.R. (1973), *Linear statistical inference and its applications*, 2nd ed., Wiley, New York.

50. Resnick, S.I. (1992), *Adventures in stochastic processes*, Birkhäuser, Boston, MA.

51. Resnick, S.I. (1999), *A probability path*. Birkhäuser, Boston, MA.

52. Ross, S.M. (1996), *Stochastic processes*, 2nd ed., Wiley, New York.

53. Resnick, S.I. (2008), *Extreme values, regular variation, and point processes*, 2nd printing. Springer-Verlag.

54. Samorodnitsky, G., and Taqqu, M.S. (1994), *Stable non-Gaussian random processes*. Chapman & Hall, New York.

55. Skorokhod, A.V. (1982), *Studies in the theory of random processes*, Dover Publications, New York.

56. Spitzer, F. (1976), *Principles of random walk*, 2nd ed., Springer-Verlag, New York.

57. Steele, J.M. (2000), *Stochastic calculus and financial applications*. Springer-Verlag, New York.

58. Stout, W.F. (1974), *Almost sure convergence*, Academic Press, Cambridge, MA.

59. Williams, D. (1991), *Probability with martingales*, Cambridge University Press, Cambridge.

60. Wolff, R.W. (1989), *Stochastic modeling and the theory of queues*, Prentice-Hall, Englewood Cliffs, NJ.

B

Some Distributions and Their Characteristics

Discrete Distributions

Following is a list of discrete distributions, abbreviations, their probability functions, means, variances, and characteristic functions. An asterisk (*) indicates that the expression is too complicated to present here; in some cases a closed formula does not even exist.

Distribution, notation	Probability function	EX	$\operatorname{Var} X$	$\varphi_X(t)$
One point $\delta(a)$	$p(a)=1$	a	0	e^{ita}
Symmetric Bernoulli	$p(-1)=p(1)=\frac{1}{2}$	0	1	$\cos t$
Bernoulli $\mathrm{Be}(p)$, $0\le p\le 1$	$p(0)=q,\ p(1)=p;\ q=1-p$	p	pq	$q+pe^{it}$
Binomial $\mathrm{Bin}(n,p)$, $n=1,2,\dots,\ 0\le p\le 1$	$p(k)=\binom{n}{k}p^k q^{n-k},\ k=0,1,\dots,n;\ q=1-p$	np	npq	$(q+pe^{it})^n$
Geometric $\mathrm{Ge}(p)$, $0\le p\le 1$	$p(k)=pq^k,\ k=0,1,2,\dots;\ q=1-p$	$\dfrac{q}{p}$	$\dfrac{q}{p^2}$	$\dfrac{p}{1-qe^{it}}$
First success $\mathrm{Fs}(p)$, $0\le p\le 1$	$p(k)=pq^{k-1},\ k=1,2,\dots;\ q=1-p$	$\dfrac{1}{p}$	$\dfrac{q}{p^2}$	$\dfrac{pe^{it}}{1-qe^{it}}$
Negative binomial $\mathrm{NBin}(n,p)$, $n=1,2,3,\dots,$ $0\le p\le 1$	$p(k)=\binom{n+k-1}{k}p^n q^k,\ k=0,1,2,\dots;$ $q=1-p$	$n\dfrac{q}{p}$	$n\dfrac{q}{p^2}$	$\left(\dfrac{p}{1-qe^{it}}\right)^n$
Poisson $\mathrm{Po}(m)$, $m>0$	$p(k)=e^{-m}\dfrac{m^k}{k!},\ k=0,1,2,\dots$	m	m	$e^{m(e^{it}-1)}$
Hypergeometric $H(N,n,p)$, $n=0,1,\dots,N,$ $N=1,2,\dots,$ $p=0,\frac{1}{N},\frac{2}{N},\dots,1$	$p(k)=\dfrac{\dbinom{Np}{k}\dbinom{Nq}{n-k}}{\dbinom{N}{n}},\ k=0,1,\dots,Np;$ $q=1-p;$ $n-k=0,\dots,Nq$	np	$npq\,\dfrac{N-n}{N-1}$	$*$

Continuous Distributions

Following is a list of some continuous distributions, abbreviations, their densities, means, variances, and characteristic functions. An asterisk (*) indicates that the expression is too complicated to present here; in some cases a closed formula does not even exist.

Distribution, notation	Density	EX	$\operatorname{Var}X$	$\varphi_X(t)$
Uniform/Rectangular				
$U(a,b)$	$f(x)=\dfrac{1}{b-a},\ a<x<b$	$\frac{1}{2}(a+b)$	$\frac{1}{12}(b-a)^2$	$\dfrac{e^{itb}-e^{ita}}{it(b-a)}$
$U(0,1)$	$f(x)=1,\ 0<x<1$	$\frac{1}{2}$	$\frac{1}{12}$	$\dfrac{e^{it}-1}{it}$
$U(-1,1)$	$f(x)=\frac{1}{2},\ |x|<1$	0	$\frac{1}{3}$	$\dfrac{\sin t}{t}$
Triangular				
$\mathrm{Tri}(a,b)$	$f(x)=\dfrac{2}{b-a}\left(1-\dfrac{2}{b-a}\left|x-\dfrac{a+b}{2}\right|\right)$ $a<x<b$	$\frac{1}{2}(a+b)$	$\frac{1}{24}(b-a)^2$	$\left(\dfrac{e^{itb/2}-e^{ita/2}}{\frac{1}{2}it(b-a)}\right)^2$
$\mathrm{Tri}(-1,1)$	$f(x)=1-|x|,\ |x|<1$	0	$\frac{1}{6}$	$\left(\dfrac{\sin\frac{t}{2}}{\frac{t}{2}}\right)^2$
Exponential				
$\mathrm{Exp}(a),\ a>0$	$f(x)=\dfrac{1}{a}e^{-x/a},\ x>0$	a	a^2	$\dfrac{1}{1-ait}$
Gamma				
$\Gamma(p,a),\ a>0,p>0$	$f(x)=\dfrac{1}{\Gamma(p)}x^{p-1}\dfrac{1}{a^p}e^{-x/a},\ x>0$	pa	pa^2	$\dfrac{1}{(1-ait)^p}$
Chi-square				
$\chi^2(n),\ n=1,2,3,\ldots$	$f(x)=\dfrac{1}{\Gamma(\frac{n}{2})}x^{\frac{1}{2}n-1}\left(\frac{1}{2}\right)^{n/2}e^{-x/2},\ x>0$	n	$2n$	$\dfrac{1}{(1-2it)^{n/2}}$
Laplace				
$L(a),\ a>0$	$f(x)=\dfrac{1}{2a}e^{-|x|/a},\ -\infty<x<\infty$	0	$2a^2$	$\dfrac{1}{1+a^2t^2}$
Beta				
$\beta(r,s),\ r,s>0$	$f(x)=\dfrac{\Gamma(r+s)}{\Gamma(r)\Gamma(s)}x^{r-1}(1-x)^{s-1},$ $0<x<1$	$\dfrac{r}{r+s}$	$\dfrac{rs}{(r+s)^2(r+s+1)}$	$*$

Continuous Distributions (continued)

Distribution, notation	Density	EX	Var X	$\varphi_X(t)$
Weibull $W(\alpha,\beta)$, $\alpha,\beta > 0$	$f(x) = \frac{1}{\alpha\beta} x^{(1/\beta)-1} e^{-x^{1/\beta}/\alpha}$, $x > 0$	$\alpha^\beta \Gamma(\beta+1)$	$\alpha^{2\beta}(\Gamma(2\beta+1) - \Gamma(\beta+1)^2)$	*
Rayleigh $Ra(\alpha)$, $\alpha > 0$	$f(x) = \frac{2}{\alpha} x e^{-x^2/\alpha}$, $x > 0$	$\frac{1}{2}\sqrt{\pi\alpha}$	$\alpha(1 - \frac{1}{4}\pi)$	*
Normal $N(\mu,\sigma^2)$, $-\infty < \mu < \infty$, $\sigma > 0$	$f(x) = \frac{1}{\sigma\sqrt{2\pi}} e^{-\frac{1}{2}(x-\mu)^2/\sigma^2}$, $-\infty < x < \infty$	μ	σ^2	$e^{i\mu t - \frac{1}{2}t^2\sigma^2}$
$N(0,1)$	$f(x) = \frac{1}{\sqrt{2\pi}} e^{-x^2/2}$, $-\infty < x < \infty$	0	1	$e^{-t^2/2}$
Log-normal $LN(\mu,\sigma^2)$, $-\infty < \mu < \infty$, $\sigma > 0$	$f(x) = \frac{1}{\sigma x\sqrt{2\pi}} e^{-\frac{1}{2}(\log x - \mu)^2/\sigma^2}$, $x > 0$	$e^{\mu+\frac{1}{2}\sigma^2}$	$e^{2\mu}(e^{2\sigma^2} - e^{\sigma^2})$	*
(Student's) t $t(n)$, $n = 1,2,\ldots$	$f(x) = \frac{\Gamma(\frac{n+1}{2})}{\sqrt{\pi n}\Gamma(\frac{n}{2})} \cdot \frac{1}{(1+\frac{x^2}{n})^{(n+1)/2}}$, $-\infty < x < \infty$	0	$\frac{n}{n-2}$, $n > 2$	*
(Fisher's) F $F(m,n)$, $m,n = 1,2,\ldots$	$f(x) = \frac{\Gamma(\frac{m+n}{2})(\frac{m}{n})^{m/2}}{\Gamma(\frac{m}{2})\Gamma(\frac{n}{2})} \cdot \frac{x^{m/2-1}}{(1+\frac{mx}{n})^{(m+n)/2}}$, $x > 0$	$\frac{n}{n-2}$, $n > 2$	$\frac{n^2(m+2)}{m(n-2)(n-4)} - \left(\frac{n}{n-2}\right)^2$, $n > 4$	*

Continuous Distributions (continued)

Distribution, notation	Density	EX	$\operatorname{Var} X$	$\varphi_X(t)$
Cauchy				
$C(m,a)$	$f(x)=\frac{1}{\pi}\cdot\frac{a}{a^2+(x-m)^2}$, $-\infty<x<\infty$	$\nexists$	$\nexists$	$e^{imt-a\lvert t\rvert}$
$C(0,1)$	$f(x)=\frac{1}{\pi}\cdot\frac{1}{1+x^2}$, $-\infty<x<\infty$	$\nexists$	$\nexists$	$e^{-\lvert t\rvert}$
Pareto				
$Pa(k,\alpha)$, $k>0$, $\alpha>0$	$f(x)=\frac{\alpha k^\alpha}{x^{\alpha+1}}$, $x>k$	$\frac{\alpha k}{\alpha-1}$, $\alpha>1$	$\frac{\alpha k^2}{(\alpha-2)(\alpha-1)^2}$, $\alpha>2$	*

C

Answers to Problems

Chapter 1

2. $f_{1/X}(x) = \frac{1}{\pi} \cdot \frac{a}{a^2 x^2 + (mx-1)^2}$, $\quad -\infty < x < \infty$

13. $f_X(x) = x$ for $0 < x < 1$, $2 - x$ for $1 < x < 2$; $\quad X \in \mathrm{Tri}(0,2)$
    $f_Y(y) = 1$ for $0 < y < 1$; $\quad Y \in U(0,1)$
    $F(x,y) = 1$ for $x > 2$, $y > 1$; y for $x - 1 > y$, $0 < y < 1$; $xy - \frac{y^2}{2} - \frac{(x-1)^2}{2}$
    for $x - 1 < y < 1$, $1 < x < 2$; $xy - \frac{y^2}{2}$ for $0 < y < x$, $0 < x < 1$; $\frac{x^2}{2}$ for
    $x < y$, $0 < x < 1$; $1 - \frac{(2-x)^2}{2}$ for $y > 1$, $1 < x < 2$; 0, otherwise
    $F_X(x) = \frac{x^2}{2}$ for $0 < x < 1$, $1 - \frac{(2-x)^2}{2}$ for $1 < x < 2$, 1 for $x \geq 2$,
    0 otherwise
    $F_Y(y) = y$ for $0 < y < 1$, 1 for $y \geq 1$, 0 for $y \leq 0$

14. $Y \in \mathrm{Ge}(1 - e^{-1/a})$, $\quad f_Z(z) = \frac{\frac{1}{a} e^{-1/a}}{1 - e^{-1/a}}$, $0 < z < 1$

15. $\frac{1}{24}$

16. $EY = \log EX - \frac{1}{2}\log(1 + \frac{\mathrm{Var}X}{(EX)^2})$, $\quad \mathrm{Var}Y = \log(1 + \frac{\mathrm{Var}X}{(EX)^2})$

17. $f_Z(z) = \frac{z+1+a}{a(z+1)^2} e^{-z/a}$, $\quad z > 0$

18. $f(u) = 1 - e^{-u}$ for $0 < u < 1$, $e^{-u}(e - 1)$ for $u > 1$

19. $f_{X_1 \cdot X_2 \cdot X_3}(y) = \frac{1}{2e - 5}(1 - y)^2 e^y$, $\quad 0 < y < 1$

20. (a) $f_{Y_1, Y_2}(y_1, y_2) = \frac{64}{3}(y_1 y_2)^{5/3}$, $\quad 0 < y_1^2 < y_2 < \sqrt{y_1} < 1$ $\qquad$ (b) No

21. $f_X(x) = f_{X \cdot Y}(x) = \frac{1}{(1+x)^2}$, $\quad x > 0$, $\qquad (F(2,2))$

22. $f(u) = 4(1 - u)^3$, $\quad 0 < u < 1$

23. $\mathrm{Exp}(1)$

24. $Y \in \Gamma(2, \frac{1}{\lambda})$, $\quad \frac{X}{Y-X} \in F(2,2)$

25. $f(u) = \frac{20}{3}(u^{1/3} - u^{2/3})$, $\quad 0 < u < 1$

26. $E X = 1$, $\operatorname{Var} X = 3$

27. $f(u) = 5(u^{1/4} - u^{2/3})$, $0 < u < 1$

28. $f(u) = \frac{5}{3} u^{2/3}$, $0 < u < 1$

29. (a) $f_{X+Y}(u) = \frac{2u}{(1+u)^3}$, $u > 0$

   (b) $f_{X-Y}(u) = \frac{1}{2(1+|u|)^2}$, $-\infty < u < \infty$,   (symmetric $F(2,2)$)

30. $f(u) = \frac{u^2}{15}$ for $0 < u < 2$, $\frac{2u}{15}$ for $2 < u < 3$, $\frac{1}{15}u(5-u)$ for $3 < u < 5$

31. $\Gamma(2,1)$

32. $f(u) = \frac{u}{(2\log 2 - 1)(1-u)^3}$ for $0 < u < \frac{1}{2}$, $\frac{1}{(2\log 2 - 1)u^2}$ for $\frac{1}{2} < u < 1$

33. $f(u) = \frac{3u^2}{(1+u)^4}$, $u > 0$

34. $N(0, \frac{1}{2})$

35. $f(u) = \frac{1}{1-\log 2} \cdot \frac{1-u}{(1+u)^2}$, $0 < u < 1$

36. $U(0,1)$

37. $f_{XY}(u) = \frac{\log u}{u^2}$ for $u > 1$, $f_{X/Y}(v) = \frac{1}{2}$ for $0 < v < 1$, $\frac{1}{2v^2}$ for $v > 1$

38. $f_Z(z) = \frac{1}{3}e^{-z}(2z+1)$, $z > 0$

39. $f_{X_1/X_2}(u) = \frac{\Gamma(a_1+a_2)}{\Gamma(a_1)\Gamma(a_2)} \left(\frac{u}{1+u}\right)^{a_1-1} \left(\frac{1}{1+u}\right)^{a_2+1}$, $u > 0$,

   $X_1 + X_2 \in \Gamma(a_1 + a_2, b)$

40. (c) Mean $= \frac{r}{r+s}$,   variance $= \frac{rs}{(r+s)^2(r+s+1)}$

41. $f_{\mathbf{Y}}(\mathbf{y}) = \frac{1}{\Gamma(r_1)\Gamma(r_2)\Gamma(r_3)} y_1^{r_1-1} y_2^{r_1+r_2-1} y_3^{r_1+r_2+r_3-1} (1-y_1)^{r_2-1}$

   $\times (1-y_2)^{r_3-1} e^{-y_3}$, $0 < y_1 < 1$, $0 < y_2 < 1$, $y_3 > 0$;

   $Y_1 \in \beta(r_1, r_2)$, $Y_2 \in \beta(r_1 + r_2, r_3)$, $Y_3 \in \Gamma(r_1 + r_2 + r_3, 1)$, independent

42. (a) $\chi^2(2)$ (b) Yes   (c) $C(0,1)$

43. (a) $U(-1,1)$ (b) $C(0,1)$

44. $N(0,1)$,   independent

Chapter 2

1. $U(0,c)$

2. $f_{X|X+Y=2}(x) = \frac{3x}{2}(1 - \frac{x}{2})$, $0 < x < 2$

3. $P_n = \left(\frac{n}{n+1}\right)^n \to \frac{1}{e}$ as $n \to \infty$

4. $\exp\{-12(1 - e^{-1/10})\}$

5. Y_1 and $Y_2 \in \operatorname{Po}(\frac{\lambda}{2})$.

7. (a) $\frac{p}{2-p}$, $\frac{4(1-p)}{(2-p)^2}$ (b) $P(X = 1) = \frac{1}{2-p}$, $P(X = -1) = \frac{1-p}{2-p}$

8. (a) $U(0,1)$ (b) $\Gamma(3,y)$ (c) $E X = \frac{3}{2}$, $\operatorname{Var} X = 1.75$

9. (a) $c = 6$ (b) $E(Y \mid X = x) = \frac{1-x}{2}$, $E(X \mid Y = y) = \frac{2}{3}(1 - y)$

10. $c = 12$; $f_X(x) = 12x^2(1 - x)$, $0 < x < 1$, $f_Y(y) = 4y^3$, $0 < y < 1$;
    $EX = \frac{3}{5}$, $EY = \frac{4}{5}$; $E(Y \mid X = x) = \frac{1+x}{2}$, $E(X \mid Y = y) = \frac{3}{4}y$

11. $c = 10$; $f_X(x) = 5x^4$, $0 < x < 1$, $f_Y(y) = \frac{10}{3}y(1 - y^3)$, $0 < y < 1$;
    $EX = \frac{5}{6}$, $EY = \frac{5}{9}$; $E(Y \mid X = x) = \frac{2}{3}x$, $E(X \mid Y = y) = \frac{3}{4} \cdot \frac{1-y^4}{1-y^3}$

12. $E(Y \mid X = x) = \frac{2x}{3}$ and $E(X \mid Y = y) = \frac{2}{3}\frac{y^2+y+1}{y+1}$

13. $E(Y \mid X = x) = \frac{2x}{3}$ and $E(X \mid Y = y) = \frac{y+2}{2}$

14. $E(Y \mid X = x) = \frac{4+3x-7x^2}{6(1+x-2x^2)}$ and $E(X \mid Y = y) = \frac{8y}{15}$

15. $E(Y \mid X = x) = \frac{2x+2}{4x+3}$ and $E(X \mid Y = y) = \frac{4+9y}{6(1+3y)}$

16. $E(Y \mid X = x) = \frac{x+3}{2x+3}$ and $E(X \mid Y = y) = \frac{2+3y}{1+3y}$

17. $E(Y \mid X = x) = \frac{1}{x}$ and $E(X \mid Y = y) = \frac{2}{1+y}$

18. $c = 2$; $f_X(x) = 1 + 2x - 3x^2$, $0 < x < 1$, $f_Y(y) = 3y^2$, $0 < y < 1$;
    $EX = \frac{5}{12}$, $EY = \frac{3}{4}$; $E(Y \mid X = x) = \frac{2+5x+5x^2}{3(3x+1)}$, $E(X \mid Y = y) = \frac{5y}{9}$

19. $c = 6$; $f_X(x) = 6x(1-x)$, $0 < x < 1$, $f_Y(y) = 6y^{\frac{1}{2}}(1-y^{\frac{1}{2}})$, $0 < y < 1$;
    $E(Y \mid X = x) = \frac{1}{2}(x + x^2)$, $E(X \mid Y = y) = \frac{1}{2}(y + \sqrt{y})$

20. $E(Y \mid X = x) = \frac{x^3+x^{1/3}}{2}$ and $E(X \mid Y = y) = \frac{2}{3}\frac{y^{1/3}-y^{25/3}}{1-y^{14/3}}$

21. $E(Y \mid X = x) = \frac{2}{3}\frac{x^{.25}-x^{11.5}}{1-x^{7.5}}$ and $E(X \mid Y = y) = \frac{y^4+y^{1/4}}{2}$

22. $c = 24$; $f_X(x) = 12x^3(1 - x^2)$, $0 < x < 1$,
    $f_Y(y) = 6(1 - y^2)^2 y$, $0 < y < 1$;
    $E(Y \mid X = x) = \frac{2}{3}\sqrt{1-x^2}$, $E(X \mid Y = y) = \frac{4}{5}\sqrt{1-y^2}$

23. $c = 32$; $f_X(x) = 16x(1 - 4x^2)$, $0 < x < \frac{1}{2}$,
    $f_Y(y) = 4y(1 - y^2)$, $0 < y < 1$;
    $E(Y \mid X = x) = \frac{2}{3}\sqrt{1-4x^2}$, $E(X \mid Y = y) = \frac{1}{3}\sqrt{1-y^2}$

24. $E(Y \mid X = x) = \frac{x-1}{\log x}$ and $E(X \mid Y = y) = 2y$

25. $E(Y \mid X = x) = \frac{x-1}{\log x}$ and $E(X \mid Y = y) = \frac{3y}{2}$

26. $E(Y \mid X = x) = (1 + x)\left(1 - \frac{\log(1+x)}{x}\right)$,
    $E(X \mid Y = y) = \frac{2-y}{1-y}\log(2 - y) - 1 + y$

27. $E(Y \mid X = x) = \frac{x}{2}$ and $E(X \mid Y = y) = \frac{\frac{\pi}{2}-y\sin y-\cos y}{1-\sin y}$

28. $E(Y \mid X = x) = \frac{2x\log x-x}{4(\log x-1)}$ and $E(X \mid Y = y) = \frac{y+1}{2}$

29. $\frac{m}{2}$

30. $f(u) = \dfrac{\left(\frac{1}{\theta}\right)^p}{\left(x+\frac{1}{\theta}\right)^{p+1}}$, $u > 0$

31. $f(u) = \dfrac{\left(\frac{1}{a}\right)^p}{\left(x+\frac{1}{a}\right)^{p+1}}, \quad u > 0$

32. (a) $f_Y(y) = \frac{2}{(1+y)^3}, \quad y > 0$ (b) 1

33. $f_X(x) = -\frac{1}{2}\log|x|, \quad -1 < x < 1; \quad EX = 0, \quad \operatorname{Var}X = \frac{1}{9}$

34. (a) $\frac{n}{2}, \quad \frac{n^2}{20} + \frac{n}{5}$ (b) $P(X_n = k) = \frac{6(k+1)(n-k+1)}{(n+3)(n+2)(n+1)}, \quad k = 0,1,2,\dots,n$

35. $EY = \frac{n}{2}, \quad \operatorname{Var}Y = \frac{n^2}{12} + \frac{n}{6}, \quad \operatorname{Cov}(X,Y) = \frac{n}{12}$

36. $EY = \frac{3}{2}, \quad \operatorname{Var}Y = \frac{9}{4}, \quad \operatorname{Cov}(X,Y) = -\frac{1}{8},$
    $P(Y = n) = \frac{18}{(n+3)(n+2)(n+1)n}, \quad n \geq 1$

37. (a) $P(X = k) = \frac{1}{k(k+1)}, \quad k = 1,2,\dots$
    (b) EX does not exist $(= +\infty)$
    (c) $\beta(2,n)$

38. $P(Y = 0) = \frac{1-p}{2-p}, \quad P(Y = k) = \frac{(1-p)^{k-1}}{(2-p)^{k+1}}, \quad k = 1,2,\dots$

39. $f_{X,Y}(x,y) = n^2 \frac{y^{n-1}}{x^n}, \quad 0 < y < x < 1,$
    $E(Y \mid X = x) = \frac{nx}{n+1}, \quad E(X \mid Y = y) = -\frac{1-y}{\log y}$

Chapter 3

1. $P(X = k) = \frac{(1-e^{-1})^k}{k}, \quad k = 1,2,\dots, \quad EX = \operatorname{Var}X = e - 1$

2. $X \in \operatorname{Be}(c)$ for $0 \leq c \leq 1$. No solution for $c \notin [0,1]$

3. $U(0,2)$

4. $P(X = 0) = P(X = 1) = \frac{1}{4}, \ P(X = 2) = \frac{1}{2}$

5. (a) $\frac{(n+m-1)(n+m-2)\cdots(n+1)n}{(n+m-t-1)(n+m-t-2)\cdots(n-t+1)(n-t)} \quad \left(= \prod_{k=0}^{m-1}\left(1 - \frac{t}{n+k}\right)^{-1}\right), \quad t < n$

8. $\psi_{(X,\log X)}(t,u) = \frac{\Gamma(p+u)}{\Gamma(p)} \cdot \frac{a^u}{(1-at)^{p+u}}$

9. $np + \binom{n}{2}14p^2 + \binom{n}{3}36p^3 + \binom{n}{4}24p^4$

15. (b) $k_1 = EX, \quad k_2 = EX^2 - (EX)^2 = \operatorname{Var}X,$
    $k_3 = EX^3 - 3EXEX^2 + 2(EX)^3 = E(X - EX)^3$

17. (a) $g_X(s) = g_{X,Y}(s,1), \quad g_Y(t) = g_{X,Y}(1,t)$
    (b) $g_{X+Y}(t) = g_{X,Y}(t,t)$. No

18. $\operatorname{Exp}(\frac{1}{pa})$

19. (a) $\beta = \frac{\alpha p}{1-\alpha(1-p)}$ (b) $EY = \frac{1-\beta}{\beta}, \quad \operatorname{Var}Y = \frac{1-\beta}{\beta^2}$

20. (a) $P(Z = k) = \frac{(1-p)^2}{(2-p)^{k+1}}$ for $k \geq 1$, $P(Z = 0) = \frac{1}{2-p}$ (b) $\frac{1-p}{2-p}$

22. $EZ = 2, \operatorname{Var}Z = 6; \quad P(Z = 0) = \exp\{e^{-2} - 1\}$

23. $EY \cdot \operatorname{Var}N$

24. $\frac{m}{m-1}\left(e^{\lambda(m-1)} - 1\right)$

25. (a) Po(nm) (b) $EY = \text{Var}\,Y = nm$

26. $g_Y(t) = \exp\{b(e^{p(t-1)} - 1)\}$, $EY = bp$, $\text{Var}\,Y = bp(1+p)$

27. $g_T(t) = \frac{t}{3 - t^3 - t^5}$, $ET = 9$, $\text{Var}\,T = 98$

28. Ge$\left(\frac{1}{1+p}\right)$

29. $\frac{7}{4p}$ and $\frac{35}{24p} + \frac{49}{16p^2}$

30. (a) 3   and   9   (b) 3   and   9   (c) 6   and   30

31. $P(Y = 0) = \frac{1-p}{2-p}$, $P(Y = k) = \frac{(1-p)^{k-1}}{(2-p)^{k+1}}$, $k = 1, 2, \ldots$

33. 21000   and   69600

35. (a) $\eta = 1$ for $p \geq 1/2$, $\frac{p}{q}$ for $p < 1/2$ (b) $P(X(2) = 0) = \frac{p}{1-pq}$

36. $E\,X(2) = (np)^2$, $\text{Var}\,X(2) = (np)^2 q(1 + np)$,

$\quad P(X(2) = 0) = (q + pq^n)^n$, $P(X(2) = 1) = (np)^2(q + pq^n)^{n-1}q^{n-1}$

37. (a) $\frac{1}{3}$ (b) $P(X(2) = 0) = \frac{17}{72}$ (c) $E(X(1) \mid X(2) = 0) = \frac{6}{17}$

38. $\frac{3}{8}$, $\frac{1}{2}$, $\frac{1}{4}$,   respectively

39. (a) $\frac{p}{q}$ for $p < q$,   1 for $p \geq q$ (g) $E\,N$ does not exist $(= +\infty)$

40. (a) 7   (b) 0.0206

41. (a) $P(Y = 0) = \frac{1}{3}$, $P(Y = 2) = \frac{2}{3}$ (b) $\frac{1}{2}$

42. (a) $g_{X(1)}(t) = 1 - \alpha + \alpha g(t)$, $g_{X(2)}(t) = 1 - \alpha + \alpha g(1 - \alpha + \alpha g(t))$

$\quad$ (b) $\eta = 1$ for $\frac{\alpha q}{p} \leq 1$, $\frac{1 - \alpha q}{q}$ for $\frac{\alpha q}{p} > 1$

43. (a) $2m$ (b) $g_{X(1)}(t) = (g(t))^2$, $g_{X(2)}(t) = \left(g(g(t))\right)^2 g(t)$

$\quad$ (c) $(g(p_0))^2 \cdot p_0$

44. (a) 1   (b) $\frac{1}{1-m}$ (c) 1   and $\frac{k}{1-m}$

45. (a) $g_Z(t) = \exp\{\lambda(t \cdot e^{\mu(t-1)} - 1)\}$

$\quad$ (b) $E\,Z = \lambda(1 + \mu)$, $\text{Var}\,Z = \lambda(1 + 3\mu + \mu^2)$

47. (a) $e^{m(t-1)}$ (b) $\exp\{m(e^{m(t-1)} - 1)\}$

$\quad$ (c) $\exp\{m(te^{m(t-1)} - 1)\}$ (d) m^2

48. $P(X = k) = \frac{1}{k(k+1)}$, $k = 1, 2, \ldots$

Chapter 4

1. $\frac{1}{6}$ and   0

2. $\frac{\pi}{2}$

3. $f_Y(y) = 2 - 4y$ for $0 < y < \frac{1}{2}$, $4y - 2$ for $\frac{1}{2} < y < 1$

4. $P(N = n) = \frac{1}{n(n+1)}$, $n = 1, 2, \ldots$; $E\,N$ does not exist $(= +\infty)$

5. $\frac{n-1}{n+1}$

6. (a) $\frac{1}{8}$ (b) $\frac{1}{2}$

7. $\frac{1}{2}$

8. (a) $\frac{1}{2}$ (b) $\frac{2}{3}$

9. (a) $f(u) = 12u(1-u)^2, 0 < u < 1$ (b) $f(u) = 12u(1-u)^2, 0 < u < 1$

10. $\frac{1}{2}$

11. (a) $\frac{3}{4}$ (b) $a = 2 + \sqrt{2}$

12. $\frac{a+b}{2}$

13. $\frac{65}{6}$

14. $f_{X_{(1)}, X_{(3)})|X_{(2)}=x}(y,z) = \frac{1}{x(1-x)}, 0 < y < x < z < 1$

15. $\frac{2}{7}$

16. (a) $X_{(1)} \in \text{Exp}(\frac{a}{2}),\quad X_{(2)} - X_{(1)} \in \text{Exp}(a)$ (b) $y+a,\quad a - \frac{x}{e^{x/a}-1}$

17. $\frac{3-4x}{4(1-x)^2}$ for $0 < x \leq \frac{1}{2}$,   1 for $x > \frac{1}{2}$

19. $\text{Exp}(1)$

20. (a) $Y_k \in \text{Exp}(\frac{a}{n+1-k})$

    (b) $E X_{(n)} = a \cdot \sum_{k=1}^{n} \frac{1}{k},\quad \text{Var} X_{(n)} = a^2 \cdot \sum_{k=1}^{n} \frac{1}{k^2}$

21. (a) $Y_k \in \text{Exp}(\frac{1}{n+1-k})$

22. $E Z_n = \sum_{k=1}^{n} \frac{k+1}{2} = \frac{n(n+3)}{4},\quad \text{Var} Z_n = \sum_{k=1}^{n} \frac{(k+1)^2}{4} = \frac{n(2n^2+9n+13)}{24}$

24. $\Gamma(n, a)$

25. $\frac{1}{n} \sum_{k=1}^{n} X_k$

26. $f(x) = \frac{ape^{-ax}}{(1-(1-p)e^{-ax})^2},\quad x > 0$

27. $P(V=0) = e^{-\lambda},\quad f_V(x) = \lambda e^{-\lambda(1-x)}, 0 < x < 1,\quad E V = 1 - \frac{1}{\lambda} + \frac{e^{-\lambda}}{\lambda}$

28. $F_Y(y) = \exp\{-\lambda e^{-y/\theta}\}$ for $y \geq 0$ and 0 for $y < 0$

Chapter 5

2. $f_\Theta(\theta) = \frac{\sqrt{1-\rho^2}}{2\pi(1-\rho\sin 2\theta)},\quad 0 < \theta < 2\pi$

5. $E XY = \rho\sigma_y\sigma_x,\quad \text{Var} XY = \sigma_x^2\sigma_y^2(1+\rho^2)$

6. (a) $\frac{1}{\sqrt{\pi}}$ (b) 0

8. $N(1, \frac{1}{1+x^2})$ and $N(1, \frac{1}{1+y^2})$

9. (c) $C(0,1)$

10. $\frac{4}{3}$

12. (a) $N\left(\begin{pmatrix} 2 \\ -1 \end{pmatrix}, \begin{pmatrix} 10 & 5 \\ 5 & 5 \end{pmatrix}\right)$ (b) $N(y+3, 5)$

13. $N(6.5, \frac{25}{2})$

14. $N(4.2, 13.8)$

15. $N(-\frac{6x}{5}, \frac{9}{5})$

16. $N(0, \frac{4}{3})$

17. $N(\frac{18}{41}, \frac{25}{41})$

18. $N(-1, 6)$

19. $N(\frac{1}{5}, \frac{14}{5})$

20. (a) $N(0, 4)$ (b) $N(-2, 4)$

21. $N(2, 1)$

22. $N(2, \frac{2}{3})$

23. (a) $N(3, \frac{23}{4})$ (b) $N(3, \frac{7}{4})$

24. $N(\frac{2}{3}, \frac{8}{3})$

25. $N(1, 3)$

26. $N(0, \frac{11}{13})$

27. $N(0, 1)$

28. $N\left(0, \begin{pmatrix} \frac{3}{7} & -\frac{3}{14} \\ -\frac{3}{14} & \frac{31}{28} \end{pmatrix}\right)$

29. $N(\frac{8}{5}, \frac{28}{5})$

30. $N(\frac{4}{5}, \frac{18}{5})$

31. $N(-\frac{13}{79}, \frac{12}{79})$

32. $N(\frac{1}{2}, \frac{1}{4})$

33. (a) μ, $\sigma^2 \tau^2$, σ^2 (b) $N\left(\begin{pmatrix} \mu \\ \mu \end{pmatrix}, \begin{pmatrix} \sigma^2 & \sigma^2 \\ \sigma^2 & \sigma^2 + \tau^2 \end{pmatrix}\right)$

(c) $N\left(\mu + \frac{(y-\mu)\sigma^2}{\sigma^2+\tau^2}, \frac{\sigma^2\tau^2}{\sigma^2+\tau^2}\right)$

34. (a) $\frac{1}{\sqrt{1-t^2}}$ (b) $\frac{1}{\sqrt{1-2\rho t - t^2(1-\rho^2)}} = \frac{1}{\sqrt{(1-\rho t)^2 - t^2}}$

35. $\frac{1}{(1+t)\sqrt{1-2t}}$

36. $\frac{1}{\sqrt{1-4t+4(1-\rho^2)t^2}} = \frac{1}{\sqrt{(1-2t)^2 - 4\rho^2 t^2}}$

37. $\psi(t) = \frac{1}{1-2t}$, $t < 2$ $(\chi^2(2))$

38. $n - 1$

39. $a = b = 3/7$

Chapter 6

5. $Y_n \xrightarrow{d} Y$ as $n \to \infty$, where $F_Y(y) = \exp\{-\frac{1}{\pi y}\}$, $y > 0$

6. $\text{Exp}(\frac{1}{2})$

7. (a) $\text{Fs}(p_t)$

8. $\text{Exp}(\frac{1}{\lambda})$

9. $U(0,1)$

13. $\varphi_{S_n/n^2}(t) \to e^{-\sqrt{2|t|}}$ (symmetric stable $(1/2)$)

14. 2

15. (a) $\varphi_Y(t) = \exp\{\lambda(e^{-t^2/2} - 1)\}$

16. $N(0,2)$

17. $\text{Po}(\gamma)$

18. $\text{Ge}(\frac{\alpha}{\alpha+1})$

19. $\text{Exp}(\mu)$

20. σ^2

21. (a) $F_{V_m}(x) = e^{m(x-1)}, \quad 0 \le x \le 1$ (b) $\psi_{V_m}(t) = \frac{m}{m+t}e^t + \frac{t}{m+t}e^{-m}$
    (d) $m(1 - V_m) \xrightarrow{d} \text{Exp}(1)$ as $m \to \infty$

22. $\text{Po}(\lambda)$

24. (a) 0   (b) $\sqrt{n}Y_n \xrightarrow{d} N(0,3)$ as $n \to \infty$

26. $\frac{\sigma^2}{\mu}$

27. (a) 0   (b) $\frac{\sigma^2}{4\mu^2}$

31. $\frac{1}{a}\sqrt{\frac{3}{n}} \cdot \frac{S_n Z_n}{V_n} \xrightarrow{d} N(0,1)$ as $n \to \infty$

32. $\text{Exp}(1)$

34. $b^2 = \frac{\sigma^2}{4}$

35. $\sigma^2(g'(\mu))^2$

39. (a) $U(-1,1)$ (b) false, false.

40. (a) $U(-a,a)$ (b) $E X_n \to \frac{1}{2}$ for $\alpha > 1$, $= 1$ for $\alpha = 1$, $\to \infty$ for $\alpha < 1$,
    $\text{Var}(X_n) \to \frac{1}{12}$ for $\alpha > 2$, $\to \frac{5}{12}$ for $\alpha = 2$, $\to \infty$ for $\alpha < 2$

41. (b) $b_n P(X_1 > a_n) \to 0$ as $n \to \infty$

42. $f_n(x) = 1 - \cos(2n\pi x), 0 < x < 1$;   the density oscillates.

48. $N(\frac{1}{2}, \frac{1}{2})$

Chapter 7

2. The outcome of A/B is completely decisive for the rest.

4. $\mu = 0, \quad \sigma^2 = \frac{1}{9}$

5. (a) $\frac{1}{3}$ (b) $\frac{1}{3}$ (c) 0   (d) $\frac{2}{15}$ (e) $\frac{1}{9}$

13. (b) $\exp\{tS_n - \frac{nt^2}{2}\}$

Chapter 8

1. $\frac{31}{32}$

2. $1 - \left(\frac{\lambda_2}{\lambda_1 + \lambda_2}\right)^5$

3. (a) $\frac{1}{2}$ (b) $\frac{3}{8}$ (c) $\binom{i+j}{i}\left(\frac{1}{2}\right)^{i+j}$

4. $\left(\frac{7}{9}\right)^{10}$

5. (a) $\text{Ge}\left(\frac{\mu}{\lambda + \mu}\right)$ (b) $1 - \frac{\mu}{\lambda}$ for $\mu < \lambda$, 0 for $\mu \geq \lambda$

6. (a) $\frac{3}{5}$ (b) $\left(\frac{2}{5}\right)^3$

7. $1 - \left(\frac{5}{6}\right)^5 = \frac{4651}{7776}$

8. $P(N(T) = n) = \left(\frac{1}{2}\right)^{n+1}$, $n = 0, 1, 2, \ldots$, $(\text{Ge}(\frac{1}{2}))$

9. $\frac{4}{9}$

10. (a) Bounded interreplacement times (b) $\frac{1}{\mu} + \frac{1}{\theta}$

11. $ET_1 = 4$, $ET_2 = 12$

12. $1 - (1 - e^{-\lambda pa})/\lambda pa$

13. (a) $Y(t) \in \text{Po}(\lambda pt)$, $Z(t) \in \text{Po}(\lambda qt)$

    (b) $P(N = n) = p(1 - p)^{n-1}$, $n = 1, 2, \ldots$, $(\text{Fs}(p))$

    (c) $P(N = n) = \binom{n-1}{k-1}p^k(1-p)^{n-k}$, $n = k, k+1, \ldots$, $(k + \text{NBin}(k, p))$

    (d) $P(Y(t) = k \mid X(t) = n) = \binom{n}{k}p^k(1-p)^{n-k}$, $k = 0, 1, \ldots, n$,
    $n = 0, 1, 2, \ldots$, $(\text{Bin}(n, p))$

    (e) $P(X(t) = n \mid Y(t) = k) = e^{-\lambda(1-p)t}\frac{(\lambda(1-p)t)^{n-k}}{(n-k)!}$, $n = k, k+1, \ldots$,
    $(k + \#$ nonregistered particles$)$

14. (a) $\gamma = \frac{\alpha}{\alpha + \beta(1-\alpha)}$ (b) $EY = \frac{1-\gamma}{\gamma}$, $\text{Var}\, Y = \frac{1-\gamma}{\gamma^2}$

15. $(e^{a\lambda} - 1)/\lambda$

16. $\frac{\lambda_A}{\lambda_A + \lambda_D} \cdot e^{-(\lambda_A + 3\lambda_D)}$

17. $T_n = (n - 1)a + V_n$, where $V_n \in \Gamma(n, \frac{1}{\lambda})$;
    $ET_n = (n - 1)a + \frac{n}{\lambda}$, $\text{Var}T_n = \frac{n}{\lambda^2}$

18. $\lambda t^2/2$

19. $\frac{\lambda E A}{\alpha}(1 - e^{-\alpha t})$

20. $e^{\lambda(t-s)(g(u)-1)}$

21. (b) $\frac{-p \log p}{1-p}$

22. (a) $EY_1 = c_1 P(X_1 < a) + c_2 P(X_1 \geq a)$,
    $\text{Var}\, Y_1 = (c_1 - c_2)^2 P(X_1 < a) \cdot P(X_1 \geq a)$

    (b) $E Z_1(t) = \lambda t\big(c_1 P(X_1 < a) + c_2 P(X_1 \geq a)\big)$,
    $\text{Var}\,(Z_1(t) = \lambda t\big(c_1^2 P(X_1 < a) + c_2^2 P(X_1 \geq a)\big)$

23. (a) $P(X(t) = k) = \binom{m+k-1}{k}(\frac{1}{1+\theta t})^m(\frac{\theta t}{1+\theta t})^k$, $k = 0, 1, 2, \ldots,$

(NBin$(m, \frac{1}{1+\theta t})$)

24. (a) $\frac{\lambda}{p}$, $\frac{\lambda}{p} + \frac{\lambda^2 q}{p^2}$ (b) Ge$(\frac{1}{2})$

25. $P(N_n = k) = \binom{n+k-1}{k}(\frac{\alpha}{1+\alpha})^n(\frac{1}{1+\alpha})^k$, $k = 0, 1, 2, \ldots,$ (NBin$(n, \frac{\alpha}{1+\alpha})$)

27. $x + 1.5$

28. $\sqrt{\frac{t}{t+s}}$

29. (a) 0   (b) $\lambda t\sigma^2$ (c) 1.

30. $\frac{1}{\lambda}(1 - e^{-\lambda t})$

31. $t - (1 - e^{-\lambda t})/\lambda$

Index